D1198961

PREVENTION AND CONTROL OF SEWER SYSTEM OVERFLOWS

Prepared by the **Prevention and Control of Sewer System Overflows Task Force** of the **Water Environment Federation®**

Eric M. Harold, P.E., *Chair*
Sean W. FitzGerald, P.E., *Vice-Chair*

Gary S. Beck, P.E.
Jeffrey L. Bertacchi
Patrick J. Bradley
Chein-Chi Chang, Ph.D., P.E.
Laurie Chase, P.E.
Alan Cherubin
Gregory R. Chol, P.E.
Bruce J. Corwin, P.E.
Curtis D. Courter, P.E.
Thomas Curl, P.E.
Robert G. Decker, P.E.
Shawn A. Dent, P.E.
Jeffery Frey, P.E.
Gary Fujita, P.E.
William S. Gonwa, Ph.D., P.E.
Gunilla Goulding, P.E.
Geoffrey M. Grant, P.E.
Adel H. Hagekhalil, P.E., BCEE
Richard Helm, Ph.D., P.E.
Roy A. Herwig, P.E.
Suresh Hettiarachchi, P.E. (MN)
James Andrew Hewitt
Steven Hilderhoff, P.E., BCEE
Alan J. Hollenbeck, P.E.
Ross Homeniuk, MBA, P.Eng.
Phillip L. Hubbard, P.E.
Carol L. Hufnagel, P.E.
Lawrence P. Jaworski, P.E.
Peter Keefe
Michael D. Kyser, P.E.
Paul Louis
Mary Maciejowski
Waldo A. Margheim, P.E.
Daniel V. Markowitz, Ph.D.
Keith D. McCormack, P.E.
Charles McDowell
William McElhaney
Jane McLamarrah, Ph.D., P.E.
Drew S. Mihocko, P.E.
S. Wayne Miles, P.E., BCEE
Bill Moeller
Richard E. Nelson, P.E.
Anthony C. Offerman, P.E., R.L.S.
Gregory J. Osthues, P.E.
Sanath Palipana
Suzanne P. Pickett, P.E.
William C. Pisano, Ph.D., P.E.
Robert J. Rafferty, P.E.
Steven M. Ravel, P.E., BCEE
Reggie Rowe
Jeff Scarano, P.E.
Ross E. Schlobohm, P.E.
Nancy Schultz, P.E., D.WRE
Matthew E. Schultze, P.E.
Uzair (Sam) Shamsi, Ph.D., P.E., D.WRE
James Shelton
Marsha Slaughter, P.E.
Terry G. Soster, P.E.
Jimmy Stewart
Scott D. Struck, Ph.D.
Stephen Tilson
Martin M. Umberg, P.E.
K.C. "Kumar" Upendrakumar, P.E., BCEE
George Vania
Marc P. Walch, P.E.
Thomas M. Walski, Ph.D., P.E., D.WRE
Michel Wanna
Tina Wolff, P.E.

Under the Direction of the **Automation and Information Technology Subcommittee** of the **Technical Practice Committee**

2011

Water Environment Federation®
601 Wythe Street
Alexandria, VA 22314-1994 USA
http://www.wef.org

PREVENTION AND CONTROL OF SEWER SYSTEM OVERFLOWS

WEF Manual of Practice No. FD-17
Third Edition

Prepared by the Prevention and Control of Sewer System Overflows Task Force of the Water Environment Federation®

WEF Press

Water Environment Federation   Alexandria, Virginia

New York Chicago San Francisco Lisbon London Madrid
Mexico City Milan New Delhi San Juan Seoul
Singapore Sydney Toronto

The McGraw-Hill Companies

Prevention and Control of Sewer System Overflows, Third Edition

1 2 3 4 5 6 7 8 9 0 DOC/DOC 1 7 6 5 4 3 2 1

ISBN 978-0-07-173860-6
MHID 0-07-173860-6

Printed and bound by RR Donnelley.

This book is printed on acid-free paper.

About WEF

Formed in 1928, the Water Environment Federation® (WEF®) is a not-for-profit technical and educational organization with members from varied disciplines who work toward WEF's vision to preserve and enhance the global water environment.

For information on membership, publications, and conferences, contact:

Water Environment Federation®
601 Wythe Street
Alexandria, VA 22314-1994 USA
(703) 684-2400
http://www.wef.org

Manuals of Practice of the Water Environment Federation®

The WEF Technical Practice Committee (formerly the Committee on Sewage and Industrial Wastes Practice of the Federation of Sewage and Industrial Wastes Associations) was created by the Federation Board of Control on October 11, 1941. The primary function of the Committee is to originate and produce, through appropriate subcommittees, special publications dealing with the technical aspects of the broad interests of the Federation. These publications are intended to provide background information through a review of technical practices and detailed procedures that research and experience have shown to be functional and practical.

Water Environment Federation Technical Practice Committee
Control Group

Contents

Chapter 3 Regulatory Overview

Chapter 4 Information Management

Chapter 5 System Characterization

Chapter 6 System Maintenance and Management

Chapter 7 Overflow Mitigation Technologies

Chapter 8 Overflow Mitigation Plan Development and Implementation

List of Figures

Figure **Page**

List of Tables

Table **Page**

Preface

The purpose of this Manual of Practice is to provide the information necessary to help managers and engineers understand and analyze an overflow problem (dry weather, CSO, or SSO) and provide guidance in finding the best strategies to reduce or eliminate it. It is also intended to serve as a planning guide in developing long-term control plans for CSOs and SSOs.

This Manual of Practice was produced under the direction of Eric M. Harold, P.E., *Chair*, and Sean W. FitzGerald, P.E., *Vice-Chair*.

The principal authors of this Manual of Practice are as follows:

Chapter 1	Eric M. Harold, P.E.
	Sean W. FitzGerald, P.E.
Chapter 2	Gary S. Beck, P.E.
	Waldo A. Margheim, P.E.
Chapter 3	Martin M. Umberg, P.E.
	Adel H. Hagekhalil, P.E., BCEE
	Nancy Schultz, P.E., D.WRE
Chapter 4	Uzair (Sam) Shamsi, Ph.D., P.E., D.WRE
	Gregory J. Osthues, P.E.
Chapter 5	Carol L. Hufnagel, P.E.
	Daniel V. Markowitz, Ph.D.
	Keith D. McCormack, P.E.
Chapter 6	Ross E. Schlobohm, P.E.
	Gary Fujita, P.E.
	Phillip L. Hubbard, P.E.
	Marsha Slaughter, P.E.
Chapter 7	Carol L. Hufnagel, P.E.
	William S. Gonwa, Ph.D., P.E.
	Scott D. Struck, Ph.D
Chapter 8	Shawn A. Dent, P.E.

Authors' and reviewers' efforts were supported by the following organizations:

AECOM Water, Austin, Texas; Cincinnati, Ohio; and Winnipeg, Manitoba, Canada
Aurora Water, Aurora, Colorado
Bentley Systems, Inc., Exton, Pennsylvania
Black & Veatch Water, Kansas City, Missouri, and Gaithersburg, Maryland
Blue Heron Engineering Services, Ltd., Dublin, Ohio
Burns & McDonnell, Kansas City, Missouri
Camp Dresser & McKee, Cambridge, Massachusetts, and Rancho Murieta, California
CDM, Raleigh, North Carolina
Central Contra Costa Sanitary District, Martinez, California
CH2M HILL, Cincinnati, Ohio; Kansas City, Missouri; and Tuscaloosa, Alabama
City of Akron, Ohio
City of Columbus Division of Sewerage and Drainage, Ohio
City of Fayetteville, Arkansas
City of Goodyear, Arizona
City of Los Angeles Bureau of Sanitation, California
City of Oklahoma Utilities, Oklahoma City, Oklahoma
Compliance EnviroSystems, LLC, Baton Rouge, Louisiana
D&B/Guarino Engineers, LLC, Trevose, Pennsylvania
DC Water & Sewer Authority, Washington, D.C.
Farnsworth Group, Inc., Peoria, Illinois
George Butler Associates, Inc., Lenexa, Kansas
Hampton Roads Sanitation District, Virginia Beach, Virginia
Hazen and Sawyer, P.C., Chesapeake, Virginia, and Cincinnati, Ohio
HDR Engineering, Inc., Minneapolis, Minnesota
Hubbell, Roth & Clark, Bloomfield Hills, Michigan
KLH Engineers, Inc., Pittsburgh, Pennsylvania
Limno-Tech, Washington, D.C.
Macon Water Authority, Macon, Georgia
Malcolm Pirnie, Inc., Fort Wayne, Indiana; Akron, Cincinnati, and Cleveland, Ohio; and Arlington, Virginia
Metro Council Environmental Services, Eagan, Minnesota
Michael Baker Jr., Inc., Moon Township, Pennsylvania

MSD of Greater Cincinnati, Cincinnati, Ohio
MWH, Atlanta, Georgia, and Boston, Massachusetts
NEORSD, Cleveland, Ohio
Optimatics, Chicago, Illinois
PBS&J, Orlando, Florida
Professional Engineering Consultants, Tulsa, Oklahoma
Richard P. Arber Associates, Lakewood, Colorado
RJN Group, Dallas, Texas, and Wheaton, Illinois
San Jacinto River Authority, Texas
Stearns and Wheeler, PLLC, Raleigh, North Carolina
Symbiont, West Allis, Wisconsin
Tetra Tech, Golden, Colorado; Ann Arbor, Michigan; and Pittsburgh, Pennsylvania
Tilson & Associates, LLC, Torrington, Connecticut
Unitywater, Queensland, Australia
Veolia Water North America, Indianapolis, Indiana
Vermont ANR, DEC, Waterbury, Vermont
Village of Ruidoso, New Mexico
Wade Trim, Detroit, Michigan
Woodard & Curran, Andover, Massachusetts
Woolpert, Atlanta, Georgia

Chapter 1

Introduction

1.0 INTRODUCTION

This manual summarizes practices to help managers and engineers understand and analyze overflow problems (either dry weather or wet weather related) from a collection systems perspective and the manual provides guidance on finding the best strategies to reduce or eliminate them. Topics covered in this manual target the crux of managing collection systems, that is, preventing and controlling overflows. This chapter discusses the general purpose of the manual, outlines the intended audience and end users, provides a broad background for sewer overflow issues and concerns, and outlines the overall organization of the manual.

2.0 GENERAL PURPOSE OF THE MANUAL

This manual is intended to serve as a planning guide in developing long-term control plans (LTCPs) for control of combined sewer overflows (CSOs) and control or elimination of sanitary sewer overflows (SSOs). The focus of this manual is controlling overflows from collection systems and, as such, will not substantively address discharges from municipal separate storm sewer systems (MS4s). It is not intended to be a stand-alone detailed design manual; rather, it covers a broad spectrum of topics, each of which could easily be extrapolated into a separate manual.

The practice of sewer system management and overflow control has evolved significantly during the last decade. Recently, new requirements in federal guidance for wastewater collection have been developed, including guidance requiring capture of wet weather flows in sanitary sewers. Through programs such as capacity, management, operation, and maintenance, growing attention is being focused on daily operations and management of collection systems to prevent and control not only wet weather overflows, but also dry weather overflows. This manual provides detailed, practical guidance on wet weather related CSO and SSO control and on more traditional dry weather overflow prevention, which reflects newly evolved practices and newly developed regulatory requirements.

Throughout this manual, there are discussions or descriptions of particular products such as mathematical models and specific treatment technologies. The authors have presented this information in an attempt to be as descriptive and specific as possible. Their objective is to give the reader an indication of where and/or when such a technology has been used and to relay comments on its reported success.

This manual augments previously published documents on sewer overflow and capacity planning and regulatory guidance, including the following:

- *Existing Sewer Evaluation and Rehabilitation* (WEF et al., 2009),
- *Guide to Managing Peak Wet Weather Flows in Municipal Wastewater Collection and Treatment Systems* (WEF, 2006),
- *Combined Sewer Overflows: Guidance for Monitoring and Modeling* (U.S. EPA, 1999), and
- *Combined Sewer Overflows: Guidance for Long-Term Control Plans* (U.S. EPA, 1995).

The field of sewer management incorporates a distinctive lexicon and set of acronyms. Table 1.1 provides the reader with a list of acronyms and definitions used throughout this manual. Table 1.2 provides a glossary of terms that take on specific meanings in the context of overflow management and control.

TABLE 1.1 Acronyms.

Acronym	Definition
AM/FM	Automated mapping/facilities management
AMSA	Association of Metropolitan Sewerage Agencies (now NACWA)
APWA	American Public Works Association
ASCE	American Society of Civil Engineers
BAT	Best available technology economically achievable
BCT	Best conventional control technology currently available
BWF	Base wastewater flow
CCTV	Closed-circuit television
CIP	Capital improvement program
CIPP	Cured-in-place pipe
CMMS	Computerized maintenance management system
CMOM	Capacity, management, operations, and maintenance
CSO	Combined sewer overflow
CSS	Combined sewer system. As defined in the 1994 CSO Control Policy (EPA 1994), a CSS is a wastewater collection system owned by a state of municipality (as defined by Section 502(4) of the Clean Water Act) that conveys domestic, commercial, and industrial wastewaters and stormwater runoff through a single pipe system to a publicly owned treatment works.
CWA	Clean Water Act. The act passed by the U.S. Congress to control water pollution. Formally, the Federal Water Pollution Control Act (or Amendments) of 1972 and all subsequent amendments [(P.L. 92-500), 33 U.S.C. 1251 et. seq., as amended by: P.L. 96-483; P.L. 97-117; P.L. 95-217, 97-117, 97-440, and 100-04].
DIP	Ductile iron pipe
FACA	Federal Advisory Committee Act

(continued)

Table 1.1 (Continued)

Acronym	Definition
FIPP	Formed-in-place pipe
FOG	Fats, oils, and grease
FROG	Fats, roots, oils, and grease
FRP	Fiber-glass-reinforced plastic
GIS	Geographic information system
GPS	Global positioning system
GWI	Groundwater infiltration
HDD	Horizontal directional drilling
HDPE	High-density polyethylene
HGL	Hydraulic grade line
I/I	Inflow and infiltration
LID	Low-impact development
LTCP	Long-term control plan
MACP	Manhole Assessment and Certification Program (NASSCO)
MMS	Maintenance management system
MS4	Municipal separate storm sewer system
MTBM	Micro-tunneling boring machine
NAAPI	North American Association of Pipeline Inspectors
NACWA	National Association of Clean Water Agencies (formerly AMSA)
NASSCO	National Association of Sewer Service Companies
NMC	Nine minimum control
NPDES	National Pollutant Discharge Elimination System
PACP	Pipeline Analysis Certification Program (NASSCO)
POTW	Publicly owned treatment works
PVC	Polyvinyl chloride
QA/QC	Quality assurance/quality control
RCP	Reinforced concrete pipe
RDII	Rainfall-derived infiltration and inflow

Table 1.1 (Continued)

Acronym	Definition
RPM	Reinforced plastic mortar
RTC	Real-time control
SCADA	Supervisory control and data acquisition
SECAP	System evaluation and capacity assurance plan
SSES	Sewer system evaluation survey
SSET	Sewer scanning and evaluation technology
SSMP	Sewer system management plan
SSO	Sanitary sewer overflow
SSS	Sanitary sewer system. As defined in the 2004 U.S. EPA report to Congress, an SSS is "… a municipal wastewater collection system that conveys domestic, commercial, and industrial wastewater, and limited amounts of infiltrated groundwater and storm water, to a POTW."
TTC	Trenchless Technology Center
U.S. EPA	U.S. Environmental Protection Agency
VCP	Vitrified clay pipe
WEF	Water Environment Federation
WERF	Water Environment Research Foundation
WWTP	Wastewater treatment plant

Table 1.2 Glossary.

Term	Definition in this context
Adaptive management	A structured, iterative process of optimal decision making in the face of uncertainty, with an aim to reducing uncertainty over time via system monitoring.
Avoidable	Legal term of art meaning that a consequence could have been prevented with the exercise of reasonable engineering judgment in facilities planning/implementation and/or adequate management, operations, and maintenance practices.

(continued)

TABLE 1.2 (Continued)

Term	Definition in this context
Bacteria	Microscopic, unicellular organisms, some of which are pathogenic and can cause infection and disease in animals and humans. Most often, non-pathogenic bacteria, such as fecal coliform and enterococci, are used to indicate the likely presence of disease-causing, fecal-borne microbial pathogens (U.S. EPA, 2004).
Biochemical oxygen demand (BOD)	A measurement of the amount of oxygen used by the decomposition of organic material, over a specified time period (typically 5 days) in a wastewater sample; it is used as a measurement of the readily decomposable organic content of wastewater (U.S. EPA, 1996).
Biological treatment system	That portion of a treatment system that is based on biotic activity.
Blending	The practice of diverting a part of peak wet weather flows at WWTPs around biological treatment units and combining effluent from all processes before discharge from a permitted outfall. Although POTWs sometimes divert around tertiary facilities, the "blending" discussion has focused specifically on secondary treatment, because of the CWA technology requirement.
Bypass	The intentional diversion of waste streams from any portion of a treatment facility [40 CFR 122.42(m)].
Capacity	The design maximum flow or loading that a wastewater system and its components can handle in a specified period of time with predictable and consistent performance. Also, the peak flow is equal to the maximum flow only when the time periods are the same. Maximum flow may be greater than or equal to the peak flow.
Chemical oxygen demand (COD)	A measure of the oxygen-consuming capacity of inorganic and organic matter present in wastewater. The COD is expressed as the amount of oxygen consumed in milligrams per liter. Results do not necessarily correlate to the BOD because the chemical oxidant may react with substances that bacteria do not stabilize (U.S. EPA, 1996).
Chemical treatment	Any water or wastewater treatment process involving the addition of chemicals to obtain a desired result, such as precipitation, coagulation, flocculation, sludge conditioning, disinfection, or odor control.
Clean Water Act (CWA)	Act passed by the U.S. Congress to control water pollution. Formally, the Federal Water Pollution Control Act (or Amendments) of 1972 and all subsequent amendments [(P.L. 92-500), 33 U.S.C. 1251 et. seq., as amended by: P.L. 96-483; P.L. 97-117; P.L. 95-217, 97-117, 97-440, and 100-04].

Table 1.2 (Continued)

Term	Definition in this context
Coliform bacteria	Rod-shaped bacteria living in the intestines of humans and other warm-blooded animals.
Collection system	Conveyance system including intercepting sewers, sewers, pipes, pumping stations, and other structures that convey liquid waste for treatment. Where sewers also convey stormwater, some agencies include combined sewer outfalls in the definition of *collection system*.
Combined sewer overflow (CSO)	A discharge of untreated wastewater from a combined sewer system at a point before the headworks of a POTW (U.S. EPA, 2004).
Combined sewer system	A wastewater collection system owned by a municipality (as defined by Section 502(4) of the CWA) that conveys domestic, commercial, and industrial wastewater and stormwater runoff through a single pipe system to a POTW (U.S. EPA, 2004).
Controls	Processes and/or activities that contribute to removal of pollutants from wastewater or to containing and conveying wastewater for treatment and discharge.
Conventional pollutants	As defined by the CWA, conventional pollutants include BOD, TSS, fecal coliform, pH, and oil and grease (U.S. EPA, 2004).
Cryptosporidium	A protozoan parasite that can live in the intestines of humans and animals. *Cryptosporidium parva* is a species of *Cryptosporidium* known to be infective to humans.
Designated uses	Those uses specified in state or tribal water quality standards regulations for a particular segment, whether or not they are being attained (40 CFR 131.3.(g)). Uses so designated in water quality standards are not meant to specify those activities or processes that the water body is currently able to fully support. Rather, they are the uses/processes that the state or tribe wishes the water body to be clean enough to support, whether or not the water body can, in its current conditions, fully support them (U.S. EPA, 2005).
Disinfection	The selective destruction of disease-causing microbes through the application of chemicals or energy.
Enterococci	*Streptococcus* bacteria of fecal origin used as an indicator organism in the determination of wastewater pollution.
Escherichia coli (*E. coli*)	Coliform bacteria of fecal origin used as an indicator organism in the determination of wastewater pollution.

(*continued*)

Table 1.2 (Continued)

Term	Definition in this context
Feasible alternatives	The legal term of art used in the "Bypass" regulation to identify alternative controls that are both technically achievable and affordable [40 CFR 122.42(m)].
Fecal coliform	Coliforms present in the feces of warm-blooded animals.
Flow equalization	Transient storage of wastewater for release to a sewer system or treatment process at a controlled rate to provide a reasonably uniform flow.
Giardia lamblia	A protozoan parasite that is responsible for giardiasis. Giardiasis is a gastrointestinal disease caused by ingestion of waterborne *Giardia lamblia*, often resulting from the activities of beavers, muskrats, or other warm-blooded animals in surface water used as a potable water source.
Green infrastructure	Term used in wet weather management to describe a number of technologies that are viewed as more natural to the environment than traditional concrete and steel structures. Green infrastructure can include site specific stormwater best management technology such as green roofs, rain gardens, and pervious pavement, and also larger, more regional facilities in scope such as bio-retention facilities.
High-rate clarification (chemically enhanced clarification)	A physical–chemical treatment process used to treat wet weather flows. Coagulants and/or flocculants are added to the wastewater before entering a gravity settling process. The process creates conditions under which dense flocs with a high settling velocity are formed, allowing them to be removed efficiently at high surface overflow rates with corresponding high TSS and BOD removal.
Impaired waters	A water body that does not meet criteria to protect the designated use(s) as specified in the water quality standards.
Inch-diameter-mile	A unit of measure computed by summing the product of sewer diameter (in inches) times the length of sewer (in miles), for successive reaches of connected sewer pipe.
Indicator bacteria (indicator organisms)	Bacteria that is common in human waste. Indicator bacteria are not harmful in themselves, but their presence is used to indicate the likely presence of disease-causing, fecal-borne microbial pathogens that are more difficult to detect (U.S. EPA, 2004). Coliform, fecal coliform, *E. coli*, and *enterococcus* are all used as indicator bacteria in water quality standards. While bacteria have generally been used as indicators of human fecal contamination, other organisms may be used.

TABLE 1.2 (Continued)

Term	Definition in this context
Infiltration	The water entering a sewer system, including building sewers, from the ground, through such means as defective pipes, pipe joints, connections, or manhole walls. Infiltration does not include and is distinguished from inflow.
Inflow	The water discharged to a sewer system, including service connections, from such sources as roof leaders; cellar, yard, and area drains; foundation drains; cooling water discharges; drains from springs and swampy areas; manhole covers; cross-connections from storm sewers, combined sewers, catch basins; stormwaters; surface runoff; street wash waters; or drainage. Inflow does not include and is distinguished from infiltration.
Optimum management of wet weather flows	Site- and process-specific. Guidance to clarify based on industry experience.
Pathogen	Formal definition: "Any disease-producing organism" (Random House, 1968). As used in the industry, organisms capable of causing disease, including disease-causing bacteria, protozoa, and viruses.
Performance	The manner or efficiency an activity or process functions, operates, or is accomplished to fulfill a purpose. Performance has goals or objectives against which measures are established to make corrections to confirm that what is about to happen is what was intended and that it conforms to plan, for example, reducing pollutants from wastewater to a targeted amount.
Physical treatment	A water or wastewater treatment process that uses only physical methods, such as filtration or sedimentation.
Physical–chemical treatment	Nonbiological treatment processes that use a combination of physical and chemical treatment methods.
Practices	Assembly or grouping of protocols.
Preliminary treatment	Treatment steps including comminution, screening, grit removal, pre-aeration, and/or flow equalization that prepare wastewater influent for further treatment.
Primary treatment	First steps in wastewater treatment, wherein screens and sedimentation tanks are used to remove most materials that float or will settle (U.S. EPA, 2004).

(*continued*)

TABLE 1.2 (Continued)

Term	Definition in this context
Principles	Criteria, engineering rules, and other underlying facts and factors used to evaluate and select controls to achieve the municipality's goals and objectives.
Processes/ activities	Facilities, treatment systems, management activities, and other activities/actions to contain and convey wastewater and treat that wastewater to remove pollutants.
Protocol	A sequenced set of controls (activities and processes) prescribed for a particular purpose that implements the principles.
Publicly owned treatment works (POTW)	A treatment works, as defined by section 212 of the CWA, that is owned by a state or municipality. This definition includes any devices and systems used in the storage, treatment, recycling, and reclamation of municipal wastewater or industrial wastes of a liquid nature. It also includes sewers, pipes, and other conveyances only if they convey wastewater to a POTW treatment plant [40 CFR §403.3], (U.S. EPA, 2004). A POTW also refers to the municipality or agency, as defined in CWA §502(4), which has jurisdiction over the treatment works and its operation [40 CFR 403.3(o), CWA § 502(4)]. This manual uses *POTW* to refer to the legal entity that "owns" wastewater collection and treatment facilities (except satellite collection systems) and *WWTP* to refer specifically to the treatment facilities of the POTW. The manual avoids the use of WWTF, because this is not a regulatory term and may be confusing as to which "facilities" are included.
Reasonable engineering judgment	As a legal term of art, this is the statutory and regulatory standard for evaluating engineering practices.
Regulation	Rules, which have the force of law, issued by an administrative or executive agency of government, generally to enact the provisions of a statute.
Sanitary sewer overflow	An untreated or partially treated wastewater release from a sanitary sewer system (U.S. EPA, 2004).
Sanitary sewer system	A municipal wastewater collection system that conveys domestic, commercial, and industrial wastewater, and limited amounts of infiltrated groundwater and stormwater, to a POTW. Areas served by sanitary sewer systems often have a municipal separate storm sewer system to collect and convey runoff from rainfall and snowmelt (U.S. EPA, 2004).

TABLE 1.2 (Continued)

Term	Definition in this context
Satellite sewer system	Combined or separate sewer systems that convey flow to a POTW owned and operated by a separate entity (U.S. EPA, 2004).
Secondary treatment	Technology-based requirements for direct discharging from municipal wastewater treatment facilities. Standard is based on a combination of physical and biological processes for the treatment of pollutants in municipal wastewater. Standards are expressed as a minimum level of effluent quality in terms of 5-day BOD, suspended solids, and pH (except as provided for special considerations and treatment equivalent to secondary treatment) (U.S. EPA, 2004), [CWA 301(b) and 40 CFR 125.3].
Stakeholders	All parties with a direct interest, involvement, and investment, or that are likely affected by, wet weather-related effects and management programs. Stakeholders can include the public; environmental interest groups; municipal, county, state, and federal governments; recreation resource management agencies; public health agencies; landowners; point-source dischargers; ratepayers; local, state, regional, and federal regulators; and state and federal fish and wildlife agencies.
Statute	A written law passed by a legislative body of the government.
Technology-based effluent limit	Effluent limitations applicable to direct and indirect sources, which are developed on a category-by-category basis using statutory factors, not including water quality effects (U.S. EPA, 2004), [CWA §301(b)].
Total suspended solids (TSS)	A measure of the filterable solids present in a sample [of wastewater effluent or other liquid], as determined by the method specified in 40 CFR Part 136 (U.S. EPA, 1996; U.S. EPA, 2004).
Treatment works	Defined in section 212 of the CWA as "any devices and systems used in the storage, treatment, recycling, and reclamation of municipal wastewater or industrial wastes of a liquid nature to implement section 201 of this act, or necessary to recycle or reuse water at the most economical cost over the estimated life of the works, including intercepting sewers, outfall sewers, wastewater collection systems, pumping, power, and other equipment, and their appurtenances; extensions, improvements, remodeling, additions, and alterations thereof; elements essential to provide a reliable recycled supply such as standby treatment units and clear well facilities; and any works, including site acquisition of the land that will be an integral part of the treatment process (including land use for the storage of treated wastewater in land treatment systems prior to land application) or is used for ultimate disposal of residues resulting from such treatment" [CWA §212].

(*continued*)

Table 1.2 (Continued)

Term	Definition in this context
Upset	An exceptional incident in which there is unintentional and temporary noncompliance with technology-based permit effluent limitations because of factors beyond the reasonable control of the permittee. An upset does not include noncompliance to the extent caused by operational error, improperly designed treatment facilities, inadequate treatment facilities, lack of preventive maintenance, or careless or improper operation [40 CFR 122.42(n)].
U.S. EPA guidance document	A memorandum or other document written by U.S. EPA to provide information or advice on a regulatory matter. A guidance document may or may not have formal legal status, depending on factors such as the level of agency review, opportunity for public comment, or use of the guidance in compliance and enforcement actions.
U.S. EPA policy	Guidance which has been subject to a formal review process and published in the *Federal Register*, for example, the CSO control policy.
Virus	Smallest biological structure capable of reproduction; infects its host, producing disease. "Viruses belonging to the group known as enteric viruses infect the gastrointestinal tract of humans and animals and are excreted in their feces" (AWWA, 1999).
Wastewater treatment facility (WWTF)	Used by many in the industry interchangeably with *WWTP*, *wastewater reclamation facilities*, and, on occasion, *POTW*, to refer to the facilities owned, operated, and/or controlled by the POTW (i.e., the municipal or agency "owner"). This manual uses *POTW* to refer to the legal entity that "owns" wastewater collection and treatment facilities (except satellite collection systems) and *WWTP* to refer specifically to the treatment facilities of the POTW. The manual avoids the use of WWTF, because this is not a regulatory term and may be confusing as to which "facilities" are included.
Wastewater treatment plant (WWTP)	That portion of the POTW that is designed to provide treatment (including recycling and reclamation) of municipal wastewater and industrial waste (also, *POTW Treatment Plant*) [40 CFR 403.3(p)]. This manual uses *POTW* to refer to the legal entity that "owns" wastewater collection and treatment facilities (except satellite collection systems) and *WWTP* to refer specifically to the treatment facilities of the POTW. The manual avoids the use of WWTF, because this is not a regulatory term and may be confusing as to which "facilities" are included.

Table 1.2 (Continued)

Term	Definition in this context
Water quality	Legal term of art intended to describe the level of pollutants or other constraints within a receiving water.
Water quality standard	A law or regulation that consists of the beneficial use or uses of a water body, the numeric and narrative water quality criteria that are necessary to protect the use or uses of that particular waterbody, and an anti-degradation statement (U.S. EPA, 2004) [40 CFR 130.3].
Water-quality-based effluent limitations	Effluent limitations applied to dischargers when technology-based limitations are insufficient to result in the attainment of water quality standards (U.S. EPA, 2004) [CWA §302].
Waters of the United States	All waters that are currently used, were used in the past, or may be susceptible to use in interstate or foreign commerce, including all waters subject to the ebb and flow of the tide. *Waters of the United States* include, but are not limited to, all interstate waters and intrastate lakes, rivers, streams (including intermittent streams), mudflats, sand flats, wetlands, sloughs, prairie potholes, wet meadows, play lakes, or natural ponds (U.S. EPA, 2004). (See 40 CFR §122.2 for the complete definition).
Wet weather event	A discharge from a combined or sanitary sewer system that occurs in direct response to rainfall or snowmelt (U.S. EPA, 2004).

3.0 TARGET AUDIENCE

This manual provides a systematic overview of the technical, management, and regulatory issues regarding collection system overflows, for both dry and wet weather (CSOs and SSOs). Information is provided to help owners, managers, engineers, technicians, and operators understand and analyze overflow problems and to provide guidance in evaluating and selecting cost-effective strategies to reduce or eliminate them. It is also intended to serve as a planning guide in developing LTCPs for CSOs and overflow mitigation plans for SSOs. This manual is written to provide reference beyond the training professional engineers receive during their education or the training technicians and operators receive in their education or on-the-job training.

4.0 SEWER OVERFLOW ISSUE SUMMARY AND CONCERNS

The following two types of public collection systems are predominant in the United States: combined sewer systems (CSSs) and sanitary sewer systems (SSSs). Combined sewer systems, which were built until the first part of the 20th century, were among the earliest sewer systems constructed in the United States. They are designed to convey both sanitary flow (wastewater) and stormwater runoff. Sanitary sewer systems, which have primarily been constructed since the first part of the 20th century, are not designed to collect large amounts of stormwater runoff from precipitation events. Overflows occur from both of these systems. *Combined sewer overflows* are permitted discharge points that overflow by design. Nevertheless, unintentional discharges can occur during dry weather. Sanitary sewer overflows are currently not permitted by the U.S. Environmental Protection Agency (U.S. EPA), and discharges from sanitary systems are typically unintended during both dry and wet weather conditions (it should be noted that many systems do have intentionally constructed SSOs that function primarily to prevent wastewater from backing into private property). This section briefly summarizes issues and concerns with sewer overflows.

4.1 Definition of an Overflow

A *sanitary sewer overflow* is the intentional or unintentional diversion of flow from an SSS that occurs before the headworks facilities of a publicly owned treatment works (POTW). Overflows that occur after the headworks facilities are termed *bypasses*. Sanitary sewer overflows that reach "waters of the United States" (see Table 1.2 for a definition of this term) violate the Clean Water Act (CWA)(U.S. EPA, 1972) because they are unpermitted discharges. Sanitary sewer overflows that involve discharges to public or private property and the surrounding environment that do not reach receiving waters (e.g., basement flooding) are typically regulated at the state level. Additionally, they are often considered violations of standard National Pollutant Discharge Elimination System (NPDES) Conditions for Duty to Mitigate (40 CFR 122.41(d)) and Proper Operation and Maintenance (40 CFR 122.41(e)) and are often included in enforcement actions (e.g., consent order or decree). Both types of SSOs may pose significant public health and environmental challenges. A *wet weather SSO* is a wastewater overflow that results from the introduction of excessive infiltration and inflow into a sanitary sewer system, such that the total flow exceeds conveyance capacity (U.S. EPA, 2004). Dry weather SSOs occur during weather conditions when little or no precipitation has occurred. Dry weather SSOs typically do not occur due to capacity-related causes; rather, they occur from maintenance- and construction-related causes, system failures (electrical or mechanical), blockages, or vandalism.

A *combined sewer overflow* is the intentional or unintentional discharge of untreated sanitary wastewater mixed with stormwater runoff or snowmelt that occurs when the capacity of the conveyance system, a regulating structure, or the downstream wastewater treatment capacity is exceeded. Combined sewer overflow discharges during wet weather include a mix of domestic, commercial, and industrial wastewater and stormwater runoff. *Wet weather CSOs* are typically controlled overflows from the collection system that have been strategically placed to discharge to receiving streams when flows exceed, or nearly exceed, downstream capacities. A dry weather discharge from a CSS occurs during dry weather conditions and, as such, is not part of the design intent of a CSS. *Dry weather CSOs* typically do not occur from lack of pipe capacity, rather, from blockages, maintenance- or construction-related issues, mechanical or electrical failures, or vandalism.

4.2 Regulatory Perspective

Municipal engineers traditionally design sewers to protect public health and the environment within the restrictions of available budgets, materials, and regulations. With CWA and subsequent amendments, the legal framework governing sewer overflows has greatly expanded. The Clean Water Act, formally known as The Federal Water Pollution Control Act Amendments of 1972, and its subsequent amendments provide the basis for sewer overflow policy in the United States. The act specifically recognizes CSOs and SSOs. It requires permits for all point source discharges and, to address this, NPDES was established as a permitting framework. Current policy regards CSOs and SSOs as unlawful under the CWA unless specifically authorized by, and compliant with, a CWA NPDES permit. As noted earlier in this chapter, under current practice CSOs are permitted discharges under the NPDES framework, whereas SSOs are not (or are only rarely) permitted. Even if SSOs are included under a permit, there is almost always a general prohibition of them.

In the United States, regulations, laws, policies, and guidance differ from state to state. Although federal regulations, laws, policies, and guidance must be honored in all states, each state has the authority to pass regulations that are more restrictive than the federal version. Additionally, the legal framework for CSO and SSO regulatory policy is different.

For SSSs, because of the lack of specific regulations and guidance, current practices provide better measures for standardizing an SSO control approach. Chapter 3 of this manual provides a more detailed description of federal and state regulatory initiatives. More recently, U.S. EPA and WEF cooperated to develop *Guide to Managing Peak Wet Weather Flows in Municipal Wastewater Collection and Treatment Systems* (WEF, 2006) to document current practices and understanding in terms acceptable to both the wastewater profession and those enforcing applicable regulations.

The *Combined Sewer Overflow Policy* (U.S. EPA, 1994) provides a national approach for communities to address CSOs through the NPDES permitting program. Key assertions of the policy are that CSOs are point sources subject to NPDES permit requirements, including both technology-based and water quality-based requirements of the CWA, and that CSOs are not subject to secondary treatment requirements applicable to POTWs.

4.3 Environmental and Public Health Effects

While it is understood that sewer overflows contain pathogens, solids, debris, and other potentially toxic pollutants, the environmental and public health effects of CSOs and SSOs vary depending on the size and type of receiving water, duration of overflow, and time of year of discharge, among other factors. Any assessment of environmental or public health effects of sewer overflows should be understood in the context of overall pollution contributions to receiving waters. Watershed approaches for overflow control take this into consideration and attempt to document all potential sources of pollution, often applying modeling or other analytical techniques to assess the relative effect of different discharges on receiving water quality. Many recent studies have shown that receiving water quality impairment is often controlled by nonpoint-source pollution sources. With all of this in mind, this section summarizes commonly understood effects of overflows. Much of this discussion is adapted from *Report to Congress: Impacts and Control of CSOs and SSOs* (U.S. EPA, 2004).

Because CSOs contain raw wastewater with large volumes of stormwater, and contribute pathogens, solids, debris, and toxic pollutants to receiving waters, CSOs can create public health and water quality concerns. Combined sewer overflows have contributed to beach closures, shellfish bed closures, contamination of drinking water supplies, and other environmental and public health concerns. Similarly, because SSOs contain raw wastewater and can occur on land and in public spaces, SSOs can create public health and environmental concerns. Sanitary sewer overflows have contributed to beach closures, contamination of drinking water supplies, and other environmental and public health concerns.

The environmental effects of CSOs and SSOs are typically the result of large or recurrent discharges, and are most apparent at the local level (near the point of discharge). Data compiled by U.S. EPA indicate that CSOs were responsible for 1% of reported advisories and closings and 2% of advisories and closings that had a known cause during the 2002 swimming season. Sanitary sewer overflows were responsible for 6% of reported advisories and closings and 12% of advisories and closings having a known cause. Combined sewer overflows and SSOs have also been identified as a cause of shellfish harvesting prohibitions and restrictions in classified shellfish growing areas.

Although it is clear that CSOs and SSOs contain disease-causing pathogens and other pollutants, there is limited information on actual human health effects occurring as a result of CSO and SSO events. Furthermore, CSOs and wet weather SSOs also tend to occur at times when exposure potential may be lower (e.g., storm events). Identification and quantification of human health effects caused by CSOs and SSOs at the national level are difficult due to a number of factors, including under-reporting and incomplete tracking of waterborne illness, contributions of pollutants from other sources, and the lack of a comprehensive national data system for tracking the occurrence and effects of CSOs and SSOs. As an alternative to direct data on human health effects, U.S. EPA (2004) modeled the annual number of gastroenteritis cases potentially occurring as a result of exposure to water contaminated by CSOs and SSOs at Beaches Environmental Assessment and Coastal Health survey beaches, and estimated that CSOs and SSOs cause between 3400 and 5600 illnesses annually at the subset of recreational areas included in the analysis.

4.4 Hot Topics

This manual was developed to capture the state of current practice and to document changes and new technologies that have arisen in the last 10 years. There are several topics in the management of collection systems that are not necessarily emerging issues, but are in the early stages of application or are otherwise gaining broader acceptance and understanding. This section describes a few of these topics, most of which are addressed in some aspect in this manual.

4.4.1 Asset Management

Asset management is essentially about cost-effective service and efficient performance. It is a strategic approach to help prioritize investments, make choices for maintaining equipment and infrastructure, and deliver reliable service to customers for the long term. The asset management strategy can provide a framework for the collection systems manager to work effectively with internal utility staff (e.g., planning, engineering, finance) and external stakeholders (e.g., regulators, elected and appointed officials, public boards) to make the best, most defensible decisions related to the maintenance, rehabilitation, and replacement of infrastructure and equipment assets. Key components of the asset management decision framework include service levels, asset inventory, condition assessment, criticality (consequence) evaluation, and effective useful life and asset valuation. Chapter 4 provides a more detailed discussion of this issue and its application to prevention and control of sewer overflows.

4.4.2 Service Levels, Performance Metrics, and Benchmarking

The asset management framework begins with defined commitments to deliver a specified level of service to utility customers. Service level commitments typically cut across the following four categories: reliability, customer service, quality, and regulatory. Service levels focus on high-level measures, which link back to the utility's mission and strategic plan, and must align with, and respond to, external influences such as customer expectations and regulatory requirements.

Performance measures are internally focused on key activities, which provide the means to track and report progress against the service level commitments. Performance measures rely heavily on the utility's information management systems to support the necessary work practices and to provide the data management and reporting mechanisms.

Benchmarking allows utilities to compare their performance with other utilities to effect organizational change, although its strength is not in rating one utility against another. Benchmarking allows utilities to be able to identify their strengths and areas for improvement, and helps set achievable targets for how the utility can improve its current practices. Benchmarking can help utilities

- Identify, track, and measure already established and tested performance indicators;
- Compare their performance against their peers;
- Identify specific processes needing improvement; and
- Implement outstanding processes to improve performance.

Chapter 4 provides more detailed discussions of service levels, performance metrics, and benchmarking, with an emphasis on how utilities can structure their information systems to support these activities.

4.4.3 Green Infrastructure

Green infrastructure is a term used in wet weather management to describe a number of technologies that are viewed as more natural to the environment than traditional concrete and steel structures. Green infrastructure can include site-specific stormwater best management technology such as green roofs, rain gardens, and pervious pavement, and also larger, more regional facilities in scope such as bio-retention facilities. The key to green infrastructure's role in addressing wet weather is its ability to contain or retain stormwater (allowing for evaporation or infiltration), thereby reducing its effect on CSOs and SSOs. U.S. EPA has acknowledged the potential of green infrastructure and has promoted inclusion of these practices into wet weather programs, although the real benefits of green infrastructure continue to be researched. Chapter 7 describes some green infrastructure solutions and their potential use in the prevention and control of sewer system overflows.

4.4.4 Climate Change

The consensus of the scientific community, stemming from long-term international investment in the study of the earth's climate, has evolved to support the stance that climate change will have dramatic effects on water resources. According to *National Water Program Strategy: Response to Climate Change* (U.S. EPA, 2008), the potential consequences of climate change on water resources include

- Warming air and water,
- Changes in the location and amount of rain and snow,
- Increased storm intensity,
- Rising sea levels, and
- Changes in ocean characteristics.

The Water Environment Federation (2010) outlines several areas where climate change can affect wastewater utilities. It suggests that the industry needs to perform a comprehensive assessment of the potential range of effects and develop solutions that will allow utilities to adapt to changes in precipitation patterns, drought intensity and frequency, and sea level rise, and source water quality due to increasing temperatures. Some of these changes to the hydrological cycle are described in detail in this section.

Perhaps the most challenging effect of climate change on managing sewer overflows is the change in precipitation patterns. Precipitation, as a result of climate change, is expected to change in intensity, duration, and frequency. In some parts of the country, more frequent extreme rainfall events will alter boundaries of the 100-year flood plain, affecting development, transportation, and water and wastewater infrastructure systems, and further damaging the already fragile relationship between land and water ecologies. Sanitary sewer systems, storm sewer systems, and CSSs will frequently be overtaxed. Additionally, local streams and rivers will experience increased erosion and decreased capacity for biohabitat.

Rising sea levels, estimated by some scientists to be 1 m or more during the next century, will compound effects brought on by precipitation changes, increasing the severity in coastal zones. Many coastal water treatment facilities (for drinking water or wastewater) will require significant capital investment to protect against tidal surges. Combined sewer systems and MS4s in coastal communities that are designed to discharge into tidal rivers, streams, and bays may experience decreased discharge capacity and overflow as a result of increased "pressure" (or "head") from rising tides.

Because the effects of climate change on water are uncertain and will vary by region, wastewater utility managers face daunting challenges. Traditional approaches that relied on historical weather patterns to manage source water supplies, stormwater runoff, and wastewater conveyance and treatment will no longer work. Chapter 8 provides more information on how utilities can plan for climate change in the prevention and control of overflows.

5.0 ORGANIZATION OF THE MANUAL

The content of the manual is organized as follows:

- Chapter 1—introduction to, and overview of, the manual;
- Chapter 2—reviews the operation and history of sanitary and CSSs, and provides a definition of typical overflows and a description of causes of overflows;
- Chapter 3—explains the evolution of regulations governing sewer system management and overflows;

- Chapter 4—describes information management needs and requirements for the wastewater collection system manager responsible for prevention and control of sewer system overflows;
- Chapter 5—offers a systematic approach to existing wastewater collection system analysis and characterization;
- Chapter 6—describes how system maintenance and management are key to the prevention and control of sewer system overflows;
- Chapter 7—summarizes methods and technologies, both existing and emerging, for overflow control and mitigation; and
- Chapter 8—describes how to apply information in this manual toward developing an overflow mitigation plan.

6.0 REFERENCES

American Water Works Association (1999) *Water Quality and Treatment*, 5th ed.; McGraw-Hill: New York.

Code of Federal Regulations Title 40 (2010) Office of the Federal Register, National Archives and Records Administration, Washington D.C.

Random House (1968) *Dictionary of the English Language;* Random House: New York.

U.S. Environmental Protection Agency (1972) Clean Water Act, Public Law 92-500 (formerly the Water Pollution Control Act); U.S. Environmental Protection Agency: Washington, D.C.

U.S. Environmental Protection Agency (1994) *Combined Sewer Overflow (CSO) Control Policy;* EPA-830/94-001; U.S. Environmental Protection Agency: Washington, D.C.

U.S. Environmental Protection Agency (1995) *Combined Sewer Overflows: Guidance for Long-Term Control Plans;* EPA-832/B-95-002; U.S. Environmental Protection Agency, Office of Water: Washington, D.C.

U.S. Environmental Protection Agency (1996) *U.S. EPA NPDES Permit Writers' Manual;* EPA-833/B-96-003; U.S. Environmental Protection Agency, Office of Water: Washington, D.C.

U.S. Environmental Protection Agency (1999) *Combined Sewer Overflows: Guidance for Monitoring and Modeling;* EPA-832/B-99-002; U.S. Environmental Protection Agency, Office of Water: Washington, D.C.

U.S. Environmental Protection Agency (2004) *Report to Congress: Impacts and Control of CSOs and SSOs;* EPA-833/R-04-001; U.S. Environmental Protection Agency, Office of Water: Washington, D.C.

U.S. Environmental Protection Agency (2005) National Pollutant Discharge Elimination System (NPDES) Permit Requirements for Peak Wet Weather Discharges from Publicly Owned Treatment Plants Serving Separate Sanitary Sewer Collection Systems, *Fed. Regist.*,**70**, No. 245.

U.S. Environmental Protection Agency (2008) *National Water Program Strategy: Response to Climate Change;* EPA-800/R-08-001; U.S. Environmental Protection Agency: Washington, D.C.

Water Environment Federation (2006) *Guide to Managing Peak Wet Weather Flows in Municipal Wastewater Collection and Treatment Systems;* Water Environment Federation: Alexandria, Virginia.

Water Environment Federation (2010) Protecting Water Resources and Infrastructure from the Impacts of Climate Change, Position Statement. http://www.wef.org/GovernmentAffairs/page.aspx?id=6490 (accessed Sept 2010).

Water Environment Federation; American Society of Civil Engineers; Environmental and Water Resources Institute (2009) *Existing Sewer Evaluation and Rehabilitation*, 3rd ed.; WEF Manual of Practice No. FD-6; ASCE Manual and Report on Engineering Practice No. 62; McGraw-Hill: New York.

7.0 SUGGESTED READINGS

National Association of Clean Water Agencies; Association of Metropolitan Water Agencies (2009) *Confronting Climate Change: An Early Analysis of Water and Wastewater Adaptation Costs;* Association of Metropolitan Water Agencies: Washington, D.C.

Water Environment Research Foundation (2003) *Reducing Peak Rainfall-Derived Infiltration/Inflow Rates—Case Studies and Protocol;* Water Environment Research Foundation: Alexandria, Virginia.

Chapter 2

Definitions and Causes of Overflows

(continued)

1.0 INTRODUCTION

An understanding of the operation and history of sanitary sewer systems (SSSs) and combined sewer systems (CSSs) is a prerequisite for recognizing the causes of sewer overflows. For centuries, cities have provided drainage systems built to carry away stormwater runoff. The discharge of fecal matter and other wastes to these conduits was forbidden well into the 19th century, although no enforcement mechanism existed to facilitate compliance. Common practice in urban areas before this time was the accumulation of heaps of waste materials, including fecal matter, outside dwellings and businesses. This practice obviously created an unhealthy atmosphere, especially in areas where the poor were crowded into close living quarters.

The subsequent admission of these wastes into existing drainage systems offered the cities an inexpensive and easy solution to the aesthetic and hygienic predicament. The resulting "combined sewers" became the accepted mode of waste removal. At the time, most newly constructed conduits were also being designed to accept wastes. The result of this practice was to transfer the problems from street surfaces to nearby waterways (streams, rivers, lakes, and tidal estuaries). In general, these receiving bodies of water could not assimilate the increased organic loads being fed into them.

The next step in the evolutionary process of many of our "modern" CSSs was to include smaller-diameter sewers to convey dry weather flows from combined sewers to wastewater treatment plants that were being built. Wet weather flows would still be discharged to receiving waterbodies. The decision to collect and treat only dry weather flows was based on a combination of economics and an understanding of environmental factors at the time. Control structures such as weirs to divert flows during dry and wet weather were included in these CSSs. Screens were also added to keep large debris out of the smaller dry weather piping and the receiving treatment plants.

Separate sanitary sewers also became prevalent as the effect of wastewater on receiving waterbodies became more evident. The capacity of SSSs was typically much smaller than CSSs because the rain-related flow component was mostly excluded. Sanitary sewer systems were designed to carry sanitary flow with a peaking factor to account for limited groundwater and rainwater entering the system through defects. Because of poor construction practices, aging systems, illicit storm connections, and other factors, many SSSs became overwhelmed by the rainfall-derived, or wet weather, component. In many instances, the design peaking factor was not large enough. Controlled overflows from the sewer system were added to the systems as a reaction to unforeseen capacity problem areas. The controlled overflows prevented adverse effects on private property and human health that are associated with untreated wastewater entering streets and building basements. Transferring the excess flows to streams through controlled overflows was considered the prudent course of action to relieve the system. Again, as in CSSs, the prevailing belief was that wet weather overflows were diluted enough that they would have minimal or no effect on the receiving stream environment.

Although diluted, some combined and separate sewer overflows were found to have a negative effect on the environment. The overflows, in combination with wastewater treatment discharges and other nonpoint discharges, have caused impairment to receiving waterbodies. Because of the negative effect of overflows on the environment, the U.S. Environmental Protection Agency (U.S. EPA), with support from the Clean Water Act (CWA) and state agencies, has set goals to eliminate or reduce overflows from all sewer systems. For these reasons, it is important for water environment professionals to understand the causes of overflows.

2.0 SANITARY SEWER OVERFLOWS

2.1 Definitions

2.1.1 Sanitary Sewer System

A *sanitary sewer system* is defined as "a municipal wastewater collection system that conveys domestic, commercial, and industrial wastewater, and limited amounts of infiltrated groundwater and storm water, to a POTW [publicly owned treatment works]" (U.S. EPA, 2004). Sanitary sewer systems also include pumping stations and force mains because wastewater flows cannot always be conveyed by gravity. Separate storm sewer systems are typically used in SSS areas to collect and convey rainfall and snowmelt runoff, as shown in Figure 2.1 (U.S. EPA, 2004). There are estimated to be more than 15 000 municipal SSSs operating in the United States (U.S. EPA, 2003).

Although designed to collect and convey wastewater, separate SSSs were also designed to convey a limited quantity of extraneous clear water, that is, infiltration and inflow originating as either groundwater or surface runoff. The quantity of infiltration and inflow present is often dependent on the physical condition of sanitary sewers and the number of connections that can contribute additional flow to the sewer system.

2.1.2 *Sanitary Sewer Overflow*

There are many definitions of a *sanitary sewer overflow* (SSO). The U.S. Environmental Protection Agency (U.S. EPA) defines an *SSO* as an untreated or partially treated wastewater release from an SSS (U.S. EPA, 2004).

A more descriptive definition preferred by the author is the intentional or unintentional diversion of flow from a SSS that occurs before the headworks facilities of

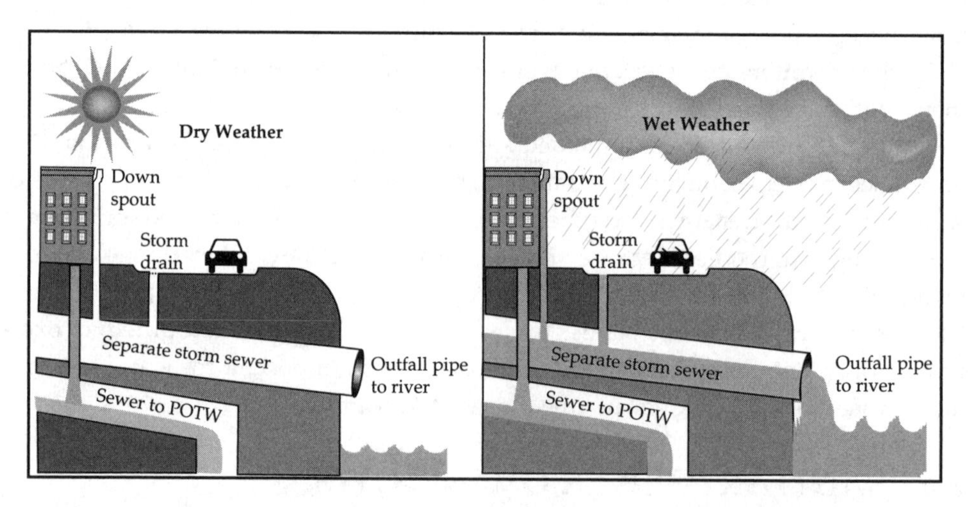

FIGURE 2.1 Typical separate sanitary and storm sewer systems. Sanitary sewer systems are designed to collect and convey wastewater mixed with limited amounts of infiltration and inflow to a treatment plant. A separate storm sewer system is used in many areas to collect and convey stormwater runoff directly to surface waterbodies (U.S. EPA, 2004).

a POTW. Overflows that occur beyond the headworks facilities are termed *bypasses*. Sanitary sewer overflows include discharges to receiving waters and diversions to public or private property and the surrounding environment that do not reach receiving waters (e.g., basement flooding) and, as a result, may pose significant public health and environmental challenges.

2.1.3 *Flow Components*

Flow components of typical wet weather SSOs include sanitary wastewater flow, infiltration, and inflow, as shown in Figure 2.2. These flow components are defined in the following subsections.

2.1.3.1 Sanitary Wastewater Flow

The sanitary wastewater flow component of an SSO includes liquid wastes produced in residences, commercial establishments, and institutions, and liquid wastes discharged from industries (McGhee, 1991). Sanitary wastewater flow may also be referred to as *wastewater production flow*, that is, wastewater without infiltration and inflow.

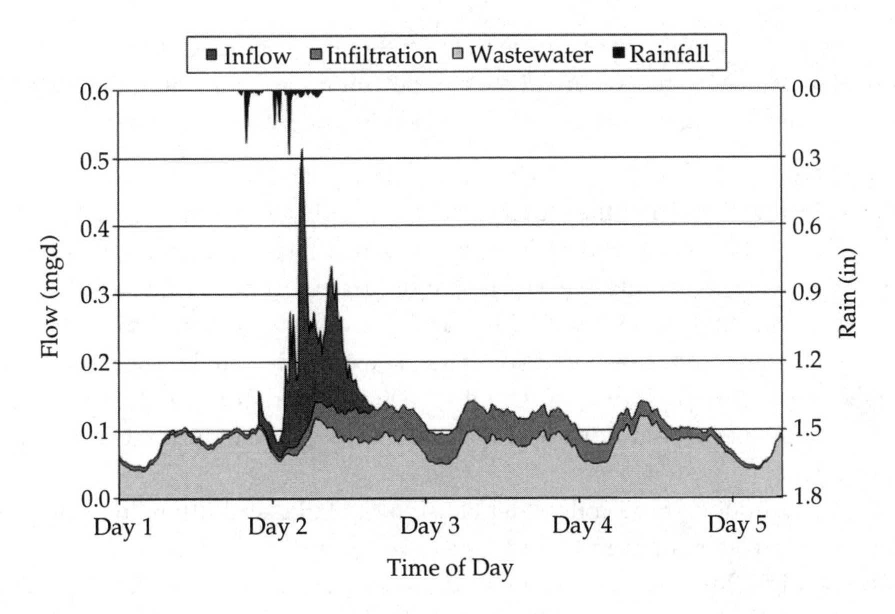

FIGURE 2.2 Peak wet weather flow components.

2.1.3.2 Infiltration

Sanitary sewer overflows also include infiltration and inflow flow components. *Infiltration* refers to stormwater and groundwater entering a sewer system (including service connections) from the ground through defective pipes, pipe joints, connections, manholes, or by other means (U.S. EPA, 2004). *Defective* is used to describe both poorly constructed and deteriorating structures.

When discussing causes of SSOs, it is important to differentiate between continuous (or base) infiltration and rainfall-derived (or stormwater) infiltration. Continuous infiltration is typically associated with leaks from average seasonal groundwater levels. Rainfall-derived infiltration is associated with leaks caused by higher-than-average groundwater levels as a result of wet weather event(s).

Infiltration rates are dependent on the depth of groundwater above the defects and the size of the defects (ASCE, 2004). Understandably, infiltration may become a significant capacity problem when the ground is saturated and groundwater levels are high and/or above system invert elevations. Infiltration can exceed the sanitary wastewater flow component by more than double and can reach higher rates in sewer systems near major waterbodies, where the groundwater approaches surface elevations.

Compared to the inflow component, infiltration contributes less to the peak flow component, but contributes more to the annual volume component. Infiltration typically occurs for days after a storm event, whereas inflow typically lasts for a few hours.

2.1.3.3 Inflow

Inflow is stormwater runoff that enters an SSS through such means as the following: roof leaders; cellar, yard, and area drains; crushed laterals; uncapped cleanouts; foundation drains; cooling water discharge; drains from springs and swampy areas; manhole covers, both dislodged and perforated; summit manhole plugs; cross-connections from storm and combined sewers; tide gate leakage; catch basin laterals; and street wash water (U.S. EPA, 2004). Many of the aforementioned sources can also be infiltration sources. Inflow sources are those that react rapidly to rain events.

The magnitude of peak inflow depends on rainfall distribution, intensity, antecedent groundwater conditions, and types and locations of inflow sources (ASCE, 2004). The high flow peaks caused by inflow, as shown in Figure 2.2, are a major cause of inadequate capacity in SSSs. Inflow levels can cause peak flows in separate sewer systems that are similar to peak flows in storm and CSSs.

2.2 Types

The terms, *dry weather SSOs* and *wet weather SSOs*, are used to distinguish overflows that are not caused by rainfall or snowmelt and those that are, respectively. Dry and wet weather SSOs have different characteristics and their causes and methods of control differ (U.S. EPA, 2004).

2.2.1 *Dry Weather Sanitary Sewer Overflows*

A *dry weather SSO* is an overflow of raw wastewater that occurs when inflow and rainfall-derived infiltration, two flow components caused by wet weather, are not the primary cause of the SSO. In other words, dry weather SSOs occur during weather conditions when little or no precipitation has occurred. Dry weather SSOs typically do not occur due to capacity-related causes (i.e., size of pipe), rather, they occur from operation-, maintenance-, and construction-related causes, system failures, or vandalism.

Because of the nature of dry weather SSOs, they can occur at any location of a sewer system regardless of the sewer capacity. Under gravity flow conditions, a dry weather SSO will occur at the lowest upstream relief point in the SSS, such as a broken pipe, low-lying manhole cover, or a basement floor drain at the lowest building upstream of a blockage. Under pressure flow conditions, a dry weather SSO may occur at any elevation that provides relief to the outside atmosphere. Dry weather SSOs generally have much higher concentrations of contaminants than wet weather SSOs and pose higher environmental and health impacts because they are not diluted by stormwater.

2.2.2 *Wet Weather Sanitary Sewer Overflows*

A *wet weather SSO* is a wastewater overflow that results from the introduction of excessive rainfall-derived infiltration and inflow into an SSS such that the total flow exceeds conveyance capacity (U.S. EPA, 2004). Wet weather SSOs occur during or immediately after rain events, and are typically diluted to a lesser strength with rain and groundwater compared to dry weather SSOs. A wet weather overflow typically occurs at the lowest discharge points within the overloaded area (i.e., areas were flow exceeds capacity) of the sewer system. A controlled wet weather overflow is typically a sewer pipe running from a manhole or pumping station wet well to the environment (typically a stream). Controlled wet weather overflows are typically strategically located at areas with known inadequate capacity, and are typically needed to protect residents from property damage and downstream facilities (e.g., pumping stations). Typical uncontrolled wet weather overflows occur at manhole covers just upstream of the overloaded sewers or the lowest building floor elevations connected to the overloaded system.

3.0 COMBINED SEWER OVERFLOWS

3.1 Definitions

3.1.1 Combined Sewer System

A *combined sewer system* is defined as "a wastewater collection system owned by a municipality (as defined by Section 502(4) of the Clean Water Act) that conveys domestic, commercial, and industrial wastewater and stormwater runoff though a single pipe system to a POTW", as shown in Figure 2.3 (U.S. EPA, 2004). Similar to SSSs, CSSs may also include pumping stations and force mains. Combined sewer systems are designed to discharge flows that exceed the system conveyance and/or treatment capacity directly to receiving waterbodies (U.S. EPA, 2004). There are estimated to be more than 800 active CSO permits in more than 700 communities (U.S. EPA, 2004).

Combined sewer systems are designed to convey large amounts of stormwater runoff. Because of varying amounts of clear water in the system, flowrates and pollutant concentrations in a CSS are highly variable.

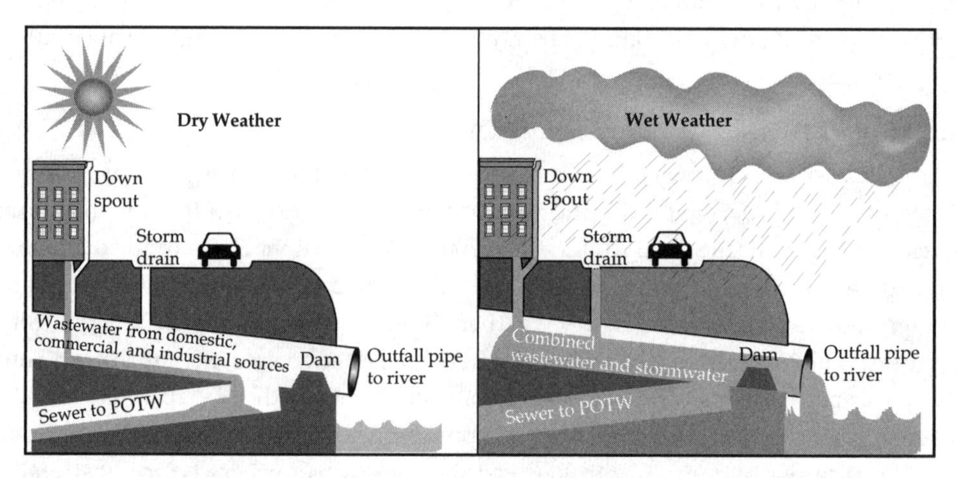

FIGURE 2.3 Typical combined sewer systems. Combined sewer systems are designed to discharge directly to surface waterbodies such as rivers, estuaries, and coastal waters during wet weather, when total flows exceed the capacity of the CSS or treatment plant (U.S. EPA, 2004).

3.1.2 *Combined Sewer Overflow*

Combined sewer overflow (CSO) is the intentional or unintentional discharge of untreated sanitary wastewater mixed with stormwater runoff or snow melt that occurs when the capacity of the conveyance system or treatment plant is exceeded. To control where CSOs occur, diversion structures are constructed at strategic locations within the CSS that discharge extraneous commingled flow to receiving waters. Combined sewer overflow discharges include a mix of domestic, commercial, and industrial wastewater; infiltration and inflow; and stormwater runoff.

3.1.3 *Flow Components*

3.1.3.1 *Sanitary Wastewater Flow*

The sanitary wastewater flow component within a CSS is the same as that in an SSS.

3.1.3.2 *Infiltration and Inflow*

Infiltration and inflow are components of CSSs that are similar to that of SSSs. The difference between stormwater flow and infiltration and inflow in CSSs is that infiltration and inflow represent rain and groundwater that are not intended to be collected by the CSS and that enter the system through defects such as broken pipes and illicit connections.

3.1.3.3 *Stormwater*

Stormwater is water caused by precipitation. Stormwater runoff that occurs during storm events is often conveyed to storm inlets to be transferred to conveyance systems for treatment or to be discharged to receiving waterbodies. For CSSs during or immediately after rainfall, stormwater is often the major flow component of CSOs. The stormwater runoff component of a CSO differentiates it from an SSO.

3.2 Types

Synonymous with SSOs, the terms, *dry weather* and *wet weather*, are used to distinguish sewer overflows that are rainfall- or snowmelt-derived from those that are not caused by rainfall or snowmelt.

3.2.1 *Dry Weather Combined Sewer Overflows*

A *dry weather CSO* is a discharge from a CSS that occurs during dry weather conditions. Dry weather CSOs are not part of the design intent of a CSS and typically do not occur from lack of pipe capacity, rather, from blockages, mechanical or electrical failures, and vandalism. Similar to dry weather SSOs, dry weather CSOs can occur at any location of a sewer system regardless of the sewer capacity.

3.2.2 *Wet Weather Combined Sewer Overflows*

Wet weather CSOs are typically controlled overflows in a system and have been strategically located to discharge to receiving waters when flows exceed or nearly exceed downstream capacities. Wet weather CSOs occur during or immediately after rain events and are typically diluted to a lesser contaminant strength than dry weather CSOs because of the stormwater flow component.

4.0 CAUSES OF OVERFLOWS

Recognizing causes of overflows is crucial for water environment professionals to prevent and control their occurrence. There are numerous causes of overflows in SSSs and CSSs. These causes can be grouped into the following three categories: operation-, maintenance-, and construction-related causes; capacity-related causes; and system failures and vandalism.

Although public health and environmental concerns associated with CSOs and SSOs are similar, their causes and predictability differ significantly. Combined sewer overflows are primarily caused by wet weather events when the combined volume of wastewater and stormwater entering the system exceeds the capacity of the CSS or treatment plant (U.S. EPA, 2004). Sanitary sewer overflows can be caused by excessive levels of infiltration and inflow entering the SSS during wet weather or by improper operation and maintenance of the sewer system (U.S. EPA, 2004). In a recent survey by U.S. EPA (2004), 48% of all SSOs with a known cause were the result of complete or partial blockage of a sewer line (dry weather), whereas only 26% were caused by wet weather and inflow and infiltration (U.S. EPA, 2004).

4.1 Operation-, Maintenance-, and Construction-Related Causes

Overflows caused by operation, maintenance, and construction practices are often caused by the growing age of the infrastructure. Many CSSs are 50 to 100 years old or more, whereas the age of SSSs can range from new to 100 years. The age of the older sewers brings into question the structural integrity and adequacy of conveyance capacity of these systems. Over time, as sewer systems continue to deteriorate and possibly become clogged, dry weather overflows can occur; these conditions worsen during wet weather due to reduced conveyance capacity (U.S. EPA, 2004).

4.1.1 *Operation-Related Causes*

Operation-related causes of overflows can occur because of numerous situations and operational settings within the complete sewer system (including pumps, gates, valves, and force mains) resulting in the system not operating as it was designed. For example, if pumping station settings and alarms are not properly set, overflows can occur at low-lying wet wells. Occasionally, pump controls are inadvertently turned off during routine pumping station maintenance and are not discovered until the pumps are needed for wet weather pumping, resulting in overflows. If power outages occur, pumping stations without standby power can result in overflows.

4.1.2 *Maintenance-Related Blockages*

Maintenance-related blockages are generally associated with inadequate sewer cleaning or inadequate frequency of cleaning. Surveillance of a system and work history records are required to identify the maintenance frequency needed throughout the system. Inadequate or infrequent maintenance can result in solids deposition; fats, oils, and grease (FOG) accumulation; and root intrusion that can create blockages and restrict flow (i.e., decreased capacity).

Fats, oil, and grease solidify and can reduce conveyance capacity and block flow. Grease from restaurants, homes, and industrial sources are the most common cause of reported sewer blockages (U.S. EPA, 2004). Accumulation of solids, such as grit and rocks, accounts for nearly one-third of reported blockages (U.S. EPA, 2004). Roots are problematic because they penetrate the pipe and cause blockages; approximately one-quarter of reported blockages is caused by root intrusion (U.S. EPA, 2004). Roots can enter the system through joints or can extend through service laterals and enter the main at service connections.

Lack of surveillance and maintenance can also result in structural failures. As described by the American Society of Civil Engineers (2004), minor sewer defects can lead to structural problems. As a pipe is subjected to surcharge and alternating exfiltration and infiltration cycles, soil particles may migrate into the sewer, resulting in voids around the outside of the sewer. This may then result in pipe cracking or collapse.

Lack of maintenance on pumping stations and other mechanical and electrical equipment can lead to overflows. Checking pump wear and motor operation and providing replacement prior to failure is imperative. Additionally, air entrainment in force mains can reduce the force main capacity, resulting in capacity problems upstream. Air entrainment can be avoided if air-release valves are checked and kept in operation. A proactive versus reactive maintenance program will prevent failures and resulting overflows through early detection.

4.1.3 *Construction-Related Causes*

Poor construction practices at the time of installation can result in overflows over time. Poor pipe and manhole installation can result in infiltration and inflow entering a newly constructed sewer system, resulting in a system with inadequate capacity for typical dry weather wastewater flows. Manhole construction problems, such as no joint sealing at the cone, wall sections, and pipe seals, can contribute to infiltration and inflow when subjected to water pressures. Additionally, manhole inverts and troughs that are poorly constructed and do not allow smooth transition of flow of single and multiple pipe connections can hinder flow and greatly reduce the overall capacity of the sewer system. Protruding lateral wye connections also reduce the intended capacity of sewers by reducing the pipe cross-sectional area. Sewers installed with offsets, sags, and flat or reverse slope reduce the capacity of the sewer due to the resulting solids deposition and FOG accumulation. Sags and breaks may be caused by poor compaction during construction and settlement over time (ASCE, 2004).

4.1.4 *Structural Pipe Failures*

Structural pipe failures can cause overflows through blockages that reduce capacity by reducing the cross-sectional area of the sewer pipe. Structural failures can be caused by corrosion, pipe deterioration, or excessive live and dead loads. In some instances, a complete blockage results in zero flow velocities, and upstream overflows result until the flow stream has been resumed. The probability of structural pipe failures can be a function of the age and type of pipe material.

Vitrified clay pipe (VCP) was one of the primary pipe materials used in gravity sewer construction prior to 1980, and is still prevalent throughout sewer systems in the United States. Although the pipe material is inert and has excellent corrosion-resistant properties, it is a rigid pipe and prone to breaking if not properly bedded. Since the 1980s, polyvinyl chloride (PVC) pipe has commonly been used for sewer systems. Like VCP, it also has corrosion-resistant properties, but has a flexible component that minimizes cracks and breaks. Therefore, structural pipe failures of gravity sewers are more prone to occur in older areas where VCP sewers were installed and not as prevalent in newer PVC installations. Additionally, VCP sewers have joints every 0.9 m (3 ft) on average compared to every 4.3 m (14 ft) for PVC. This increased joint frequency allows for additional avenues for roots to enter the sewer, resulting in decreased capacity. In gravity sewers, structural defects that cause overflows typically reduce the cross-sectional area of the conduit and restrict or stop the flow and/or result in excessive infiltration and inflow entering the system.

Reinforced concrete pipe (RCP) has commonly been used for larger sewers in both CSSs and SSSs since the 1930s and is still used for large-diameter installations such as interceptor sewers. Reinforced concrete pipe is known for its strength and durability, but does not have the corrosion-resistant properties of VCP or PVC. Hydrogen sulfide gas can develop in higher-strength wastewater or in long interceptors with long detention times. Corrosion of RCP can cause weakness in the pipe and its eventual collapse. Protective coatings and liners are used to prevent RCP corrosion.

Ferrous pipes are commonly used for pumping station force mains and in some gravity sewer installations such as aerial or buried stream crossings that require a high-strength material. Corrosion can affect ferrous pipes, especially those that are not protected with chemical-resistant coatings. Corrosion of ferrous pipes can be internal from sewer gases or external from corrosive soils. Tuberculation of ferrous pipe walls can restrict pipe openings and cause overflows due to the reduced cross-sectional area. As with RCP sewers, ferrous pipes under attack from corrosion can weaken, causing pressure breaks in force mains and pipe collapses in gravity sewers that may result in overflows.

Generally, pipe materials and bedding requirements are continuously improving in the sewer industry. For instance, joint materials for sewer pipe have greatly improved since the rubber gasket was introduced; indeed, it is far superior to older gaskets such as oakum or lead joint seals. However, continued surveillance of both old and new construction is needed.

4.2 Capacity-Related Causes

Sanitary sewer overflows and CSOs are caused by flow entering the sewer system that exceeds its capacity. Combined sewer systems are typically meant to overflow when the piping system capacity is exceeded. Excessive flow entering the system can occur from several sources, including excessive infiltration and inflow, wet weather events that result in the exceedance of the design capacity of the system, and changes in land use resulting in larger, unanticipated flows.

4.2.1 *Infiltration and Inflow*

Conveyance capacity in combined sewers and especially in separate sewers can be expended (i.e., used up) by infiltration and inflow from both public and private sources. Even in areas where new sanitary sewers were installed to separate the systems and eliminate CSOs, there is significant potential for SSOs if building foundation drains and/or private downspouts continue to be connected to the SSS.

Many SSSs that were designed according to industry standards experience wet weather SSOs because levels of infiltration and inflow exceeded levels originally anticipated, removal of infiltration and inflow has proven more difficult and costly than anticipated, or the capacity of the system has become inadequate due to an increase in service population without corresponding system upgrades (U.S. EPA, 2004). The rate and quantity of infiltration and inflow entering a sewer system is a function of many variables, including soil moisture and permeability, drainage conditions, sewer construction practices, groundwater elevation, and precipitation intensity, frequency, and duration. To stop overflows caused by infiltration and inflow requires either removal of the infiltration and inflow sources or increasing the capacity of the system.

4.2.2 *Inadequate Hydraulic Capacity*

Inadequate hydraulic capacity is a major cause of SSOs and CSOs. Lack of hydraulic capacity can result from many factors including lack of planning, addition of unplanned development, excessive flows caused by infiltration and inflow, and hydraulic design constraints.

Sanitary sewer systems are designed to collect and safely transfer wastewater flows with a safety factor to account for wet weather flows. Early design standards included a safety factor between 3 and 4. In many situations, this safety factor has been inadequate. Recent safety factors have moved toward designing for storm protection return periods (i.e., 10-year storm). A storm protection return period is often established for an existing system from flow monitoring and capacity modeling. Some areas of the sewer system may have a different storm protection return period than others. Because of aging infrastructure and outdated design practices, many existing systems have less than a 1-year storm protection, which means that sewers surcharge and potentially overflow the system on an annual basis. Generally, sanitary sewers today are designed to carry flows generated from a designated storm return period between 5 and 50 years.

The conveyance capacity of SSSs is important in that it affects the ability of an area to accommodate future growth. If the separate sewers in an area are already overloaded, assessment of the system capacity is important in understanding where the hydraulic constraints are and the extent to which additional sewer extensions will increase the overflow potential.

Combined sewer systems are designed to collect stormwater flow and sanitary flow and to prevent stormwater flooding. In contrast, SSSs are not designed to collect stormwater, but have a safety factor for rain and groundwater that enters the system through defects. Combined sewers are typically designed to overflow during lesser storm events than sanitary sewers through controlled overflows. Controlled overflows are typically needed to lower capacity requirements downstream and protect downstream properties and residents. Combined sewers, like storm sewers, have a maximum design capacity that can be associated with a storm return interval and is typically less than a 2-year storm protection. Although CSSs are generally not installed today, retrofit of CSSs has a typical design storm protection between 10 and 50 years. Much like storm sewer systems, CSSs are impacted by issues associated with urbanization that lower stormwater travel times due to increased impervious ground conditions. Upstream urbanization not only overwhelms the CSS, but also receiving streams, which then cause backups into the CSS. The result is that combined sewers often become stormwater management facilities by providing inline storage and detention of peak flow volumes. During smaller storms, CSSs may not provide an acceptable level of protection to prevent either basement or surface flooding. This is especially true for many older highly developed areas. Thus, in addition to concern for the receiving waterbody, there may also be a significant public health and safety concern from basement and street flooding.

Inadequate hydraulic capacity may also be caused by the hydraulic constraints in sewer system design. For example, hydraulic restrictions in a sewer system may result from severe bends, drop connections, plugged drop connections, free-fall discharges, and matching inverts of flows with significant differential velocities, thus creating the potential for surcharge, backup, and overflow during times of peak flow (ASCE, 2004).

4.2.3 *Excessive Wet Weather Events*

Excessive wet weather events can cause large quantities of infiltration and inflow that raise flow levels and exceed the design capacity of the sewer system. Excessive wet weather flows can overload sewer systems resulting in collection system surcharges, basement flooding, overflows, and treatment plant bypasses and effluent violations.

When a SSS experiences excessive wet weather flows, those flows are considered to be outside of the design capacity of the system. The cause of excessive flows is infiltration and inflow sources. Excessive flows are typically created by a storm event that is greater than the storm return interval for which the system was designed. The frequency of storm events, one or more large storms after another, often cause excessive flow even though each storm event may have been under the system's design storm protection.

Much like SSSs, design storm protection in CSSs can be exceeded. Excess flow may exceed a CSS's capacity and result in uncontrolled overflows that may include reverse flow and overflows from inlets and sewer backups for residents connected to the system. Multiple storms generally have less of an effect on a CSS upstream in the system because stormwater flows recede quickly after rainfalls. However, at receiving streams and large conduits downstream, the effect is higher because of longer times of concentration.

4.2.4 *Receiving Stream Flooding*

Flooding of receiving streams must be considered in CSS planning. Because CSSs represent a closed conduit system and not an open-channel one, inline detention structures are not commonly used to temporarily detain increased flows caused by urban development. Many problems associated with stormwater runoff can be attributed to land use, particularly insufficient attention to land drainage in urban planning and ineffective updating of existing stormwater control and conveyance systems. Combined sewers extend into drainage basins and often directly collect private storm flow such as downspouts and yard drains. The private sector flow typically is indirectly collected by street catch basins. Because stormwater has a more direct path to streams, combined sewers can stress a stream as higher flows scour flood channels during wet weather.

Land use markedly influences the runoff process and overflows from CSSs. The conversion of landscape from rural or open space to urban development with the ensuing increase of impervious land uses, such as roads, streets, parking lots, and rooftops, will increase the volume of stormwater runoff and ultimately contribute to CSOs both in terms of quality and quantity.

4.2.5 *Illicit Connections*

Illicit connections are direct stormwater flow connections to a SSS that are not permitted by law. Typical illicit connections to SSSs include storm sump pumps, downspouts, catch basins, and area drains. Illicit connections become a serious capacity issue when the receiving system is not designed to accommodate additional flows. Illicit connections can expend capacity of an SSS and overwhelm a system depending on the size and number of connections. Unfortunately, downstream low-lying houses and streams become overflow recipients due to their location, whereas those with illicit connections that contribute to overflows may be located at the top of the basin and out of harm's way.

Most CSS allow sump pumps, downspouts, area drains, and other private storm connections to be directly connected to the sewer system. As the country moves to separate sewer systems, connected private sources, illicit or non-illicit, must be accounted for to provide capacity for the resulting SSSs. Identifying and addressing private connections adds significant costs to sewer separation and often causes delays in eliminating controlled overflows.

4.3 System Failures and Vandalism

4.3.1 *Mechanical and Electrical Failures*

Mechanical and electrical failures of pumping stations and treatment plants may cause overflows by causing sewers to surcharge, back up, and then overflow. Mechanical failures are more likely to occur anytime regardless of weather conditions. Electrical failures occur more often during electrical storm events. The severity of the overflow caused by a pumping station mechanical or electrical failure depends on the size of the tributary area to the pumping station and the duration of failure. Excess flow basins (storage basins) and/or emergency overflows are often located near pumping stations to protect upstream residents from overflows.

Regulators are commonly located in CSSs to control the rate of flow and elevation of wastewater. Gates or dams are typically used to prevent overloading of downstream pipes and facilities. Overflows upstream may result from using the regulator to control storage. The duration and volume of overflow depends on flowrate, availability of storage, and influence of tide or water level in the receiving water.

4.3.2 *Vandalism, Sabotage, and Terrorism*

Overflows caused by vandalism, sabotage, and terrorism are more publicized and scrutinized today than in the past and can range from pranks, to disgruntled employees, to acts of war. Recent terrorist events and the precautionary acts of the U.S. Homeland Security Department have identified vulnerabilities of U.S. sewer systems. Sewer authorities are taking protective measures to lower the risk of illegal activities by developing security plans and incident remediation plans. Today, the occurrence of overflows caused by illegal activities is low; however, the potential for an overflow due to these sorts of activities has recently increased due to the growing knowledge and new tactics of terrorist groups.

4.3.3 Third-Party Damages

Overflows may also be caused by damages to sewers caused by third parties. A common example is other utility contractors installing conduits via open cut excavation or directional drilling that damage existing sewer pipes. Utility companies that commonly come in contact with CSSs and SSSs include water and gas utilities and telecommunication companies. Directional drilling can result in a gas or water line or telecommunication cable puncturing an existing sewer line, resulting in a broken or blocked pipe.

5.0 REFERENCES

American Society of Civil Engineers (2004) *Sanitary Sewer Overflow Solutions;* American Society of Civil Engineers: Reston, Virginia.

U.S. Environmental Protection Agency (2003) *2000 Clean Watershed Needs Survey Report to Congress;* EPA-832/R-03-001; U.S. Environmental Protection Agency: Washington, D.C.

U.S. Environmental Protection Agency (2004) *Report to Congress: Impacts and Control of CSOs and SSOs;* EPA-833/R-04-001; U.S. Environmental Protection Agency: Washington, D.C.

6.0 SUGGESTED READINGS

American Society of Civil Engineers (1999) *Optimization of Collection System Maintenance Frequencies and System Performance.* http://www.asce.org/pdf/finalreport.pdf (accessed May 2009).

American Society of Civil Engineers (2000) *Protocols for Identifying Sanitary Sewer Overflows;* American Society of Civil Engineers: Reston, Virginia.

California State University, Sacramento (2001) *Operation and Maintenance of Wastewater Collection Systems,* Volume I, 5th ed.; California State University, Sacramento Foundation: Sacramento, California.

California State University, Sacramento (2002) *Operation and Maintenance of Wastewater Collection Systems*, Volume II, 5th ed.; California State University, Sacramento Foundation: Sacramento, California.

McGhee, T. J. (1991) *Water Supply and Sewerage*, 6th ed.; McGraw-Hill: New York.

U.S. Environmental Protection Agency (1983) *Results of the Nationwide Urban Runoff Program: Volume I-Final Report;* EPA-832/R-83-112; U.S. Environmental Protection Agency: Washington, D.C.

U.S. Environmental Protection Agency (1994) *Combined Sewer Overflow (CSO) Control Policy;* EPA-830/94-001; U.S. Environmental Protection Agency: Washington, D.C.

U.S. Environmental Protection Agency (1996) *Setting Priorities for Addressing Discharges from Separate Sanitary Sewers.* http://www.epa.gov/compliance/resources/policies/civil/cwa/ssodoc.pdf (accessed May 2009).

Chapter 3

Regulatory Overview

(continued)

1.0 LEGAL AND REGULATORY BACKGROUND

Municipal engineers design sewers to protect public health and the environment within the restrictions of available budgets, materials, and regulations. With passage of the *Federal Water Pollution Control Act Amendments* of 1972 (U.S. EPA, 1972), now commonly known as the Clean Water Act (CWA), and subsequent amendments, the legal framework governing sewer overflows was greatly expanded. Under this legal framework, regulations have continued to evolve with attendant policies; additionally, guidance has been established to apply these to real world situations. The prevailing legal and regulatory framework for sewer overflows governs the design of contemporary sewers and improvements made to existing systems.

The legal and regulatory framework consists of laws, regulations, policies, and guidance. Each of these plays an important role in describing current requirements to address sewer overflows. They are defined as follows:

- *Law*—a written law passed by a legislative body of the government.
- *Regulation*—rules, which have the force of law, issued by an administrative or executive agency of government, generally to enact the provisions of a law or statute.
- *Policy*—documented interpretation of means to comply with a regulation, sometimes adopted and incorporated into regulation (e.g., Combined Sewer Overflow (CSO) Control Policy [U.S. EPA, 1994]). Generally, policy is only binding on the agency that issues the policy.
- *Guidance*—documents describing means to achieve objectives set forth in regulation, law, or policy, such as Combined Sewer Overflows: Guidance for Long-Term Control Plans (U.S. EPA, 1995b). In general, guidance does not have the force of law. However, conformity with guidance often facilitates review and approval because the reviewer can more readily verify conformance.

In the United States, law, regulations, policy, and guidance differ from state to state. Federal regulation, law, policy, and guidance must be honored in all states, but each state has the authority to pass laws and regulations that are more, but not less, restrictive than the federal version. The following sections in this chapter primarily address underlying federal regulations, with some examples of state regulations cited.

1.1 Clean Water Act of 1972 and Amendments

The CWA and its subsequent amendments provide the basis for sewer overflow policy in the United States. The act includes references to combined sewer overflows (CSOs) (sec 402(q)), but no specific reference to sanitary sewer overflows (SSOs). It requires permits for all point-source discharges and, to address this, the National Pollutant Discharge Elimination System (NPDES) was established as a permitting framework. Current policy views SSOs and CSOs (and any other discharge) as unlawful under CWA unless specifically authorized by, and in compliance with, a CWA NPDES permit. As the following section explains, the legal framework for SSO and CSO regulatory policy is different.

1.2 Sanitary Sewer Overflows

If they have the potential to reach nearby streams, overflows from sanitary sewers are subject to the provisions of CWA. As such, an SSO that meets this description is a violation of CWA unless it is authorized by a permit. Permits for discharges of sanitary wastewater overflows are difficult to justify because

- Any SSO may cause or contribute to a violation of water quality standards (WQS), particularly bacteria and narrative standards regarding objectionable materials, unless treated to appropriate standards; and
- Provisions of the CWA indicate that any discharge from a publicly owned treatment works (POTW), which, by the U.S. Environmental Protection Agency's (U.S.EPA's) current interpretation, includes the entire sanitary sewer system (SSS), must be based on secondary treatment standards. Current policy assumes that a properly designed SSS will collect sanitary waste only and convey these flows to the treatment facility. This logic has led to the position that SSSs are components of POTW and, therefore, are covered under the secondary treatment requirements of CWA, 33 U.S.C. § 1311(b)(1)(B). The result is that permit writers are tasked with drafting permits that are based on secondary treatment of SSOs.

1.3 Combined Sewer Overflows

Like SSOs, overflows from sanitary sewers are subject to the provisions of CWA if they have the potential to reach nearby streams. As such, a CSO that meets this description is a violation of CWA unless it is authorized by a permit. Combined sewer overflow regulation and policy has been determined on a different legal basis than SSOs. Combined sewer overflows are considered "point-source discharges" subject to NPDES permit requirements and to technology- and water quality-based requirements of CWA. The following court ruling is significant in that it provides a basis for U.S. EPA's current CSO policy direction:

> In *Montgomery Environmental Coalition v. Costle*, 646 F.2d 568 (1980), the District of Columbia (D.C.) Circuit Court of Appeals held that as a discharge point in the collection system, a CSO is not part of the 'treatment works' within the meaning of the statute, and thus the secondary treatment requirements of 33 U.S.C. § 1311(b)(1)(B) applicable to POTWs do not apply to CSO discharges. Instead, effluent limits for CSOs are to be

> set in accordance with 'best practicable technology' standards applying to 'point sources other than publicly-owned treatment works.' 33 U.S.C. § 1311(b)(1)(A) (AMSA, 2003).

Combined Sewer Overflow (CSO) Control Policy (U.S. EPA, 1994), detailed in Section 2.2, is based on this concept. This national policy received legal standing in 2000 when Section 402q was added to CWA as part of the Consolidated Appropriations Act for fiscal year 2001 (also known as the Wet Weather Water Quality Act of 2000). Section 402q requires that permits, orders, or decrees issued for a municipal combined sewer system (CSS), relative to CWA, shall be in conformance with the *Combined Sewer Overflow (CSO) Control Policy* (U.S. EPA, 1994).

1.4 Blending and Peak Wastewater Treatment Plant Flows

Blending is the practice of routing part of the peak wet weather flow at wastewater treatment plants around biological treatment units and combining it with effluent from all processes prior to discharge from a permitted outfall (WEF, 2006). Such rerouting where the capacity of the biological (or other advanced) treatment units is exceeded might be necessary to avoid damaging the treatment units. Traditionally, blending has been a common practice for POTWs. Effluent is typically required to meet dry weather concentration limits in the NPDES permit.

Officials in some U.S. EPA regions, however, have held the opinion that blending represents an "illegal bypass" subject to bypass regulations of 40 CFR § 122.41(m), and that permitting of blending practices at a POTW requires that substantial conditions be met.

1.4.1 Sanitary Sewer Systems

Recent controversy over blending at POTWs served by separate sewer systems includes views that are at odds with each other. For instance, sometimes blending is a necessary accommodation to prevent disruption of the treatment process. However, blending circumvents the requirement for full treatment of discharges from a POTW. There is a view that bypassing secondary biological treatment by a portion of the flow is a violation of CWA, even when POTW effluent pollutant concentrations typically contained in NPDES permits are met. In 2005, U.S. EPA proposed the "blending policy" (referred to as *peak wet weather discharges* in the document) to "provide a framework that (1) ensures appropriate management of the wet weather flows at the POTW consistent with generally accepted good engineering practices and criteria for long-term design, (2) clarifies technology-based requirements, (3) uses water quality-based effluent limitations to address residual site-specific health and environmental risks, and (4) provides appropriate safeguards, including comprehensive monitoring and protection for sensitive waters" (U.S. EPA, 2005). These proposed regulations have not been promulgated. In June and July of 2010, U.S. EPA held "listening sessions" and accepted written responses to their request for stakeholder input on a number of issues relative to SSSs, including peak wet weather discharges from POTWs serving separate sanitary sewer collection systems. These issues could be addressed in the near future as part of a U.S. EPA comprehensive national policy for SSSs.

If blending occurs at a POTW, specific conditions are set forth in wastewater treatment discharge permit conditions that commonly require the plant operator to justify that the conditions were met each time blending is initiated. Generally, permit conditions require that the blended effluent meet the discharge criteria listed in the governing NPDES permit.

The reader is referred to *National Pollutant Discharge Elimination System (NPDES) Permit Requirements for Peak Wet Weather Discharges from Publicly Owned Treatment Works Treatment Plants Serving Separate Sewer Collection Systems* (U.S. EPA, 2005) for further explanation of the history, provisions, and potential implementation of the blending policy.

1.4.2 Combined Sewer Systems

For CSS POTWs, *Combined Sewer Overflow (CSO) Control Policy* (U.S. EPA, 1994) recognizes that an NPDES permit may "...authorize a CSO-related bypass of the secondary treatment portion of the POTW treatment plant for combined sewer flows in certain identified circumstances (II.C.7)". The following excerpt, taken from *Combined Sewer Overflows Guidance for Long-Term Control Plans* (U.S. EPA, 1995b), frames the issue:

> *The regulatory basis for permitting a CSO-related bypass is included at 40 CFR 122.41(m), which defines a bypass as '... the intentional diversion of waste streams from any portion of a treatment facility.' At 40 CFR 122.41(m)(4), bypasses are prohibited except where unavoidable to prevent loss of life, personal injury, or severe property damage and where there were no feasible alternatives to the bypass. 'Severe property damage' is defined at 40 CFR 122.41(m)(1) to include '... damage to treatment facilities which causes them to become inoperable...'. Under CSO Policy, severe property damage could '... include situations where flows above a certain level wash out the POTW's secondary treatment system' (II.C.7).*

2.0 POLICIES AND GUIDANCE RELATIVE TO OVERFLOW MANAGEMENT

2.1 Sanitary Sewer Overflow Policy

U.S. EPA has developed comprehensive policy and guidance for CSOs, but not for SSOs. Although U.S. EPA has listed SSO compliance as a top enforcement priority, there is little established national regulation or guidance. Sanitary sewer overflow policies vary by U.S. EPA regions and individual states. In 2001, a national regulation was proposed when a draft SSO rule was prepared to be published in the *Federal Register*. However, it was withdrawn shortly thereafter when a new presidential administration took office. The draft rule and preamble were comprehensive, and some of their provisions, such as the capacity, management, operation, and maintenance (CMOM) program, are used extensively today. In contrast to U.S. EPA's standards approach to CSOs of best available technology economically achievable and best conventional control technology currently available, the regulatory position is that the SSS is meant to convey flow to the POTW and, therefore, discharge that does not receive secondary treatment is generally prohibited.

2.1.1 *History of Sanitary Sewer Overflow Policy Development*

As with any unpermitted discharge, unpermitted SSOs have been prohibited since passage of CWA. When and under what conditions SSOs were permitted has been handled differently throughout the country. Some important issues to be considered in development of SSO policy include sanitary collection system level of service, permitting satellite collection systems, and peak wet weather treatment facilities (i.e., treatment that may not achieve secondary standards or may not include biologic treatment).

Although SSOs have been discussed for more than a decade, a consistent national policy to address policy issues has been elusive. In administering the CWA Construction Grants Program, U.S. EPA sometimes used a "design storm" to describe what sanitary sewer capacity was allowable for funding. States and U.S. EPA regions sometimes had differing requirements for permitting SSOs. The following sections describe some important accomplishments in developing a national SSO policy.

2.1.1.1 Urban Wet Weather Flows Advisory Committee

The Urban Wet Weather Flows Advisory Committee was formed in 1995 by U.S. EPA "to address issues associated with water quality impacts from urban wet weather flows and other storm water discharges under Section 402(P) (6) of the Clean Water Act in accordance with the provisions of the Federal Advisory Committee Act

(FACA)" (Browner, 1995). The committee was formed of a broad array of stakeholders. A focus of the committee was cost–benefit, that is, examining the effect of wet weather flows and the cost of alternative methods of control. The development of recommendations for a national SSO policy was recognized as a potential function of the committee. An SSO subcommittee was formed and eventually agreed to a set of principles that formed the basis of a draft SSO rule. Although the rule was prepared for publication in the *Federal Register* and signed by the U.S. EPA administrator, it was withdrawn before actual publication took place.

2.1.1.2 Draft Sanitary Sewer Overflow Rule

In late 2000, U.S. EPA developed a draft SSO rule that was based on key principles of agreement of the SSO subcommittee of the Urban Wet Weather Flows Advisory Committee (U.S. EPA, 2001). The rule was prepared to be published in the *Federal Register* in early 2001, but was pulled back as the new presidential administration took office. While the draft rule was relatively short, it had an extensive preamble that discussed a number of important regulatory issues relative to SSOs, giving alternate regulatory positions and requesting comment from the public. The preamble is one component to understanding U.S. EPA's position and considerations involved in regulating SSOs. Because an SSO rule or other defining policy for SSOs has not been established, the draft SSO rule with preamble is the most comprehensive treatise on U.S. EPA's views on SSO regulation and control. Even though the draft rule was pulled back, U.S. EPA received extensive comments, which presumably play a role as the U.S. EPA continues to regulate SSOs. The following are important elements of the draft SSO rule:

- *Prohibition of overflows*—the CWA has a general prohibition of overflows from the SSS to surface waters. This part of the proposed rule describes limited protection for overflows from SSSs that occur due to circumstances beyond the reasonable control of the utility or severe weather conditions, provided there are no feasible alternatives.
- *Capacity, management, operation, and maintenance*—under the proposed rule, a CMOM program is required. The program is intended to ensure that the collection system has sufficient capacity and is operated in an efficient manner. The CMOM concept was initially practiced and promoted in U.S. EPA Region IV. Today, CMOM concepts form the backbone of many wastewater agencies' permits and operating procedures. U.S. EPA has recognized CMOM concepts as a significant component of programs to address SSO issues. U.S. EPA's FY08-FY10 Compliance and National Priority report states, "EPA and the States will continue

to address these problems using various derivatives of the capacity, management, operation, and maintenance [CMOM] concept ..." (U.S. EPA, 2007). The capacity component of the program typically includes systematic monitoring of the collection system and requires that chronic capacity problems are addressed. Management, operation, and maintenance requirements include developing and implementing a program that contains specific tasks for management, operation, and maintenance that are appropriate for the particular collection system. As a starting point, an audit of the collection system is often performed.

- *Satellite collection systems*—satellite collection systems are those where the collection system is owned and operated by someone other than the POTW. The proposed SSO rule would require that satellite collection systems not covered by the POTW's permit have their own NPDES permit.
- *Reporting, public notification, and recordkeeping*—the proposed SSO rule provides details on these requirements.
- *Peak flow treatment facilities*—peak flow treatment facilities typically use high-rate treatment technology, and would be placed in operation only under peak flow conditions. If allowed under certain conditions, these facilities would be recognized as temporary facilities unless discharges could meet all secondary and water quality-based requirements. The preamble to the proposed rule has extensive discussion of peak-flow treatment facilities.

2.1.1.3 *Most Recent Developments*

In 2010, U.S. EPA began efforts to develop a comprehensive national policy for SSSs. In June and July of 2010, U.S. EPA held "listening sessions" around the country and accepted written responses to their request for stakeholder input on NPDES permit requirements for municipal sanitary sewer collections systems, municipal satellite collection systems, SSOs, and peak wet weather discharges from POTWs serving separate sanitary sewer collection systems.

2.1.1.4 *State Initiatives*

In the absence of national guidance for collection system management (SSO rule), many states have used their existing authority under state law to develop programs for management and oversight of collection systems. These programs have had some success in parts of the country, but there remains no consistently applied national guidance upon which collection system managers can rely. States that have developed programs include California, Florida, Michigan, North Carolina, South

Carolina, Texas, and Wisconsin. Suggested insertion: The State of South Carolina Department of Health and Environmental Control (SCDHEC) issues the South Carolina general permit for satellite systems under R.61.9-610, June 2003. Individual permits may be issued, but most systems are covered by the general permit. The general permit broadly applies to all sewers serving more than one building as well as to industrial pretreatment systems. The general permit covers such items as notification, proper operations and maintenance, providing capacity, having a 24-hour telephone number, reporting releases to the environment, and obtaining SCDHEC approval of permit transfers. The State of California Water Resources Control Board (State Water Board) adopted statewide general waste discharge requirements for SSSs in May 2006. The Sanitary Sewer Order requires public agencies that own or operate SSSs to develop and implement sewer system management plans and report all SSOs to the State Water Board's online SSO database. Every enrollee is required to develop and implement a sewer system management plan (SSMP). The SSMP documents an enrollee's program to properly operate and maintain its sanitary sewer system. Each SSMP should address the following elements:

- Goal;
- Organization;
- Legal authority;
- Operation and maintenance program;
- Design and performance provisions;
- Overflow emergency response plan;
- Fats, oils, and grease (FOG) control program;
- System evaluation and capacity assurance plan;
- Monitoring, measurement, and program modifications;
- Sewer system management plan program audits; and
- Communication program.

The state of Florida established design, construction, and operation requirements for wastewater collection and transmission systems in November 2003. The state refers to U.S. EPA's Region IV's CMOM requirements. The state of Florida did not receive authorization for the NPDES program until May 1995; hence, U.S. EPA Region IV drafted permits and enforced SSO rules within the state prior to 1995.

The state of Michigan, Department of Environmental Quality, adopted an SSO control strategy in 2000. The strategy is based on timely reporting requirements of all SSOs and follow-up corrective actions.

The state of North Carolina, under the North Carolina Department of Environment and Natural Resources, has developed and implemented a system-wide collection system permit. The permit encompasses entire collection systems that were never permitted and have gone unmanaged since installation. The permit contains the following sections: performance standards, operation and maintenance, inspections, recordkeeping, and general conditions.

The Texas Commission on Environmental Quality developed an SSO initiative where collection system owners enroll and agree to develop a plan to control SSOs.

The state of Wisconsin issued a general permit to all owners and/or operators of collection systems that applies to those who do not also operate wastewater treatment plants. The conditions of these general permits require operation to enhance transport of wastewater to treatment and minimization of overflows. In 2000, the Wisconsin permitting authority, the Department of Natural Resources, developed an SSO strategy to identify collection systems at risk for SSOs; encourage development of capacity, management operation, and maintenance plans; and initiate enforcement in systems with chronic SSOs. Under this strategy, the department has recommended that closed-circuit television (CCTV) programs focus on those sewers most susceptible to failure and blockages, and that an analysis of system performance, maintenance history, age, and materials be used to prioritize sewers for CCTV inspection. The department has implemented improved systems for tracking and follow-up on reports of SSOs. The department also initiated an outreach program to ensure that all communities submit timely reports about SSOs from their sewer systems, as required by their NPDES permits, and that they become more aggressive in correcting the root causes of overflows, particularly excessive infiltration and inflow.

2.1.2 *Current Practices*

Because of the lack of specific regulations and guidance, it is important to describe current practices. Section 2.1.1.3 describes states' initiatives in standardizing an SSO approach. Additionally, U.S. EPA and Water Environment Federation (WEF) cooperated to develop the *Guide to Managing Peak Wet Weather Flows in Municipal Wastewater Collection and Treatment Systems* (WEF, 2006) to document current practices and understanding in terms acceptable to both the wastewater profession and those enforcing the applicable regulations.

2.1.2.1 *Industry Standards (Core Practices)*

In 2010, the American Public Works Association, American Society of Civil Engineers, National Association of Clean Water Agencies (NACWA), and WEF published a manual entitled *Core Attributes of Effectively Managed Wastewater Collection System* (2010), which was the result of a collaborative effort engaging a broad group of stakeholders that identified and developed good engineering practices and core attributes essential to managing and operating collection systems. These core attributes were designed to present key principles that support the core values. The core attributes are intended to support existing state programs and should not conflict with or preempt state efforts.

In the absence of clear federal guidance, the sponsors intend to embrace these baseline attributes as fundamental elements in the management of sanitary sewer collection systems. Many communities have developed programs and specific incorporating technologies or other tools that go above and beyond performance levels resulting from these core practices.

The core attributes are intended to provide guidance for utilities to evaluate their existing programs, either confirming they are performing according to good engineering practices or determining they have practices that are lacking and need enhancement. Implementation of a collection system management program incorporating these attributes will vary from one system to the next based on size, organizational structure, character of the waste stream, history of the system, needs, and availability of resources. Through development and implementation of a management program encompassing these attributes, collection system operators can provide efficient and effective maintenance and operation of the collection system while protecting public health and the environment.

2.1.2.2 *Dry Weather Overflows*

The majority of dry weather overflows are caused by blockages resulting from FOG buildup, root intrusion, debris, sewer breaks, and lack of sewer capacity. It is important to understand the cause of overflows in each community. Current practices for preventing overflows include inspection, cleaning, source control, condition assessment, planning, and pipe renewal. Evaluating the risk of overflow for each asset and sewer basin is crucial and will allow for the prioritization and optimization of the maintenance and overflow control activities.

2.1.2.3 *Wet Weather Overflows*

As described in Section 1.2, provisions of CWA require that all discharges from an SSS have a permit; the permit must require a level of treatment consistent with discharges from a POTW. The CWA makes no exception for wet weather discharges. Recognizing the impossibility of eliminating all wet weather overflows, U.S. EPA and WEF cooperated in development of a guidance document (WEF, 2006) to allow utilities and regulatory officials to use common principles to define locally appropriate performance objectives. The common principles recognize that elimination of all overflows cannot be guaranteed, and the appropriate level of wet weather performance must include an assessment and balance of benefits, risks, costs, and technologies to move toward the ideal of no discharges that cause or threaten receiving-water-quality standards violations.

Current practices for wet weather SSO control are listed here by category only because they are documented in other recent guides. The practices include

- Infiltration and inflow reduction, including monitoring to identify opportunities for flow exclusion;
- Conveyance capacity increases;
- Storage; and
- High-rate treatment.

Current practices for controlling wet weather overflows are detailed in *Guide to Managing Peak Wet Weather Flows in Municipal Wastewater Collection and Treatment Systems* (WEF, 2006) and in *Existing Sewer Evaluation and Rehabilitation* (WEF et al., 2009).

Recognizing that some SSOs are unavoidable, POTW permits sometimes include SSOs in the bypass provisions of the permits. Each time an SSO occurs in the POTW system, the POTW permit holder must report the SSO and document the reasons for the specific occurrence. If SSOs occur chronically, regulatory enforcement may require filing an enforceable plan for control of the SSOs, potentially resulting in a negotiated consent order.

Many SSSs are owned and operated by satellite utilities where treatment is provided in a regional treatment facility. Such satellite utilities often do not have a permit to discharge to waters of the state. In these instances, any SSO that reaches the waters of the state would be a violation of CWA and would be subject to enforcement as such. Where such SSOs are chronic, the regulatory enforcement agency may require filing an enforceable plan for control of the SSOs, potentially resulting in a negotiated consent order.

2.2 Combined Sewer Overflow Policy

2.2.1 *History of Policy Development*

Although CSO control is required by CWA, the first specific policy addressing CSOs was developed in 1989. *National Combined Sewer Overflow Control Strategy* (U.S. EPA, 1989) promoted the establishment of NPDES permitting for CSOs and described six minimum technology-based standards for CSSs. Despite this policy being established, except for a few exceptions, little was done nationally in either the development of local CSO control programs or enforcement of regulations or policies until 1994, with the development of the national *Combined Sewer Overflow (CSO) Control Policy* (U.S.EPA 1994). For the first time, this policy established specific requirements for allowable discharges for CSSs in terms of level of control of pollutants. Near-term, technology-based requirements, consisting of the original six controls of U.S. EPA's 1989 policy plus an additional three, are enumerated and known as the *nine minimum controls* in *Combined Sewer Overflow (CSO) Control Policy* (U.S. EPA, 1994). These were to be implemented by 1997. The policy also requires a plan be developed to address water quality-based requirements as part of a CSO long-term control plan (LTCP). Long-term control plans are to be developed and submitted to U.S. EPA as expeditiously as practicable.

Work to develop and implement CSO control programs increased dramatically following the establishment of the *Combined Sewer Overflow (CSO) Control Policy* in 1994. U.S. EPA conducted workshops around the country and developed a number of guidance documents intended to interpret and explain the intentions of the policy in real applications.

Combined sewer overflow communities around the country took notice and began the serious work of CSO control. Even with this impetus, many communities failed to meet the schedule goals for nine minimum controls and LTCP development planned by the original policy.

Combined sewer overflow control became an enforcement priority for U.S. EPA, and many communities found themselves in negotiations with government regulators. In 2000, Section 402q was added to CWA as part of the Consolidated Appropriations Act for fiscal year 2001(also known as the Wet Weather Water Quality Act of 2000). Section 402q requires that permits, orders, or decrees issued for a municipal CSS, relative to the CWA, shall be in conformance with the *Combined Sewer Overflow (CSO) Control Policy* (U.S. EPA, 1994).

A number of obstacles remain in the development and implementation of LTCPs for some agencies. Technical issues are aggravated by the enormous costs inherent in controls for some combined systems. Wet weather WQS were envisioned by the developers of *Combined Sewer Overflow (CSO) Control Policy*. Under the policy, the WQS authority (typically states) are to work with CSO communities in development of LTCPs and, if necessary, revise existing WQS. The process of establishing wet weather standards has proven to be difficult and few exist today. Only a few states have established wet weather standards. The state of Indiana has adopted an approach that permits a suspension of the recreational designated use of a waterbody provided a LTCP and a use attainability analysis (UAA) are performed. The issue of affordability and economic capability is intertwined with the ability to meet existing use designations at all times.

2.2.2 *Significant Elements of the 1994 Combined Sewer Overflow (CSO) Control Policy*

The *Combined Sewer Overflow (CSO) Control Policy* (U.S.EPA, 1994) provides a national approach for communities to address CSOs to be done through the NPDES permitting program. Key assertions of the policy are that CSOs are point sources subject to NPDES permit requirements, including both technology-based and water quality-based requirements of the CWA, and that CSOs are not subject to secondary treatment requirements applicable to POTWs.

Policy objectives are

- To ensure that, if CSOs occur, they are only as a result of wet weather;
- To bring all wet weather CSO discharge points into compliance with the technology- and water quality-based requirements of CWA; and
- To minimize water quality, aquatic biota, and human health effects from CSOs.

U.S. EPA's CSO policy intends to ensure "... municipalities, permitting authorities, and WQS authorities and the public engage in a comprehensive and coordinated planning effort to achieve cost-effective CSO controls that ultimately meet appropriate health and environmental objectives and requirements. The policy recognizes the site-specific nature of CSOs and their effects and provides the flexibility to tailor controls to local situations" (U.S. EPA, 1994).

The following key principles ensure that CSO controls are cost-effective and meet the objectives of CWA:

- Provide clear levels of control that would be presumed to meet appropriate health and environmental objectives;
- Provide sufficient flexibility to municipalities, especially those that are financially disadvantaged, to consider the site-specific nature of CSOs and to

determine the most cost-effective means of reducing pollutants and meeting CWA objectives and requirements;

- Allow a phased approach for implementation of CSO controls considering a community's financial capability; and
- Review and revise, as appropriate, WQS and their implementation procedures when developing long-term CSO control plans to reflect the site-specific wet weather effects of CSOs.

2.2.2.1 Nine Minimum Controls

A near-term CSO control program, which includes technology-based control standards, is described in the CSO policy (U.S. EPA, 1994). These are termed *nine minimum controls* and consist of the original six controls of the 1989 policy plus an additional three. The nine minimum controls are intended to be more operation-improvement types of controls with less emphasis on capital-intensive, significant structure-type improvements. The controls are as follows:

- Proper operation and regular maintenance programs for the sewer system and CSOs,
- Maximum use of the collection system for storage,
- Review and modification of pretreatment requirements to ensure CSO impacts are minimized,
- Maximization of flow to the POTW for treatment,
- Prohibition of CSOs during dry weather,
- Control of solid and floatable materials in CSOs,
- Pollution prevention,
- Public notification to ensure that the public receives adequate notification of CSO occurrences and CSO impacts, and
- Monitoring to effectively characterize CSO impacts and the efficacy of CSO controls.

2.2.2.2 Long-Term Control Plans

The CSO policy (U.S. EPA, 1994) describes the requirements of CSO LTCPs, which are intended to result in conformance with water quality provisions of CWA. Long-term control plans are to be conducted in coordination with WQS authorities, NPDES authorities, and the CSO community. The site-specific nature of CSOs is emphasized. Small systems (i.e., those serving populations under 75 000) may not need to perform all the requirements of long-term control planning based on a determination by the NPDES authority. The two general approaches for developing the LTCP, presumption and demonstration of attainment of WQS, are described. The presumption approach "presumes" that the water quality-based requirements of CWA are met by meeting prescribed levels of CSO control, providing there is reasonable belief by the permitting authority that water quality requirements can be met. The demonstration approach "demonstrates" through information available and developed in the LTCP planning process that water quality provisions will be met. Following either the presumption or the demonstration approach, implementation of the LTCP must result in attainment of water quality requirements or more work must be done.

Table 3.1 details the roles and responsibilities of the permittee, state WQS authorities, and NPDES permitting and enforcement authorities (U.S. EPA, 1995b).

2.2.3 *Combined Sewer Overflow Guidance Documents*

After the release of *Combined Sewer Overflow (CSO) Control Policy* in 1994, U.S. EPA published a number of guidance documents. These documents provide technical and policy information for use in developing plans to address CSOs. Since these documents have been written, progress has been made in the development of technology aids to understand the impacts of CSOs, scheduling and affordability, and refinements and interpretations of the CSO policy itself. However, these documents are considered relevant and valuable sources of information for the understanding of the CSO policy.

2.2.3.1 Guidance: Coordinating Combined Sewer Overflow Long-Term Planning with Water Quality Standards Reviews

One of the four key principles of the CSO policy is to "review and revise, as appropriate, [WQS] and their implementation procedures when developing long-term CSO control plans to reflect site-specific wet weather impacts of CSOs" (U.S. EPA, 1994). *Guidance: Coordinating Combined Sewer Overflow Long-Term Planning with Water Quality Standards Reviews* (U.S. EPA, 2001) provides a description of how WQS should be considered in the development of CSO LTCPs. It is an important source for providing detailed information on how existing standards should be evaluated in relation to CSO control alternatives and for determining the need for a UAA. A detailed process for water quality considerations in the development of CSO LTCPs is explained, including the roles of NPDES and WQS authorities and the CSO community (see Figure 3.1) (U.S. EPA, 2001).

TABLE 3.1 Roles and responsibilities (U.S. EPA, 1995b).

Permittee	NPDES Permitting Authority	NPDES Enforcement Authority	State WQS Authorities
Evaluate and implement NMC Submit documentation of NMC implementation by January 1, 1997 Support the review of WQS in CSO-impacted receiving water bodies Comply with permit conditions based on narrative WQS Implement selected CSO controls from LTCP Perform post-construction compliance monitoring Reassess overflows to sensitive areas Coordinate all activities with NPDES permitting authority, State WQS authority, and State watershed personnel	Reassess/revise CSO permitting strategy Incorporate into Phase I permits CSO-related conditions (e.g., NMC implementation and documentation and LTCP development) Review documentation of NMC implementation Coordinate review of LTCP components throughout the LTCP development process and accept/approve permittee's LTCP Coordinate the review and revision of WQS as appropriate Incorporate into Phase II permits CSO-related conditions (e.g., continued NMC implementation and LTCP implementation) Incorporate implementation schedule into an appropriate enforceable mechanism Review implementation activity reports (e.g., compliance schedule progress reports)	Ensure that CSO requirements and schedules for compliance are incorporated into appropriate enforceable mechanisms Monitor adherence to January 1, 1997, deadline for NMC implementation and documentation Take appropriate enforcement action against dry weather overflows Monitor compliance with Phase I, Phase II, and post-Phase II permits and take enforcement action as appropriate	Review WQS in CSO-impacted receiving water bodies Coordinate review with LTCP development Development of site-specific criteria Modification of designated use to Create partial use reflecting specific situations Define use more explicitly Temporary variance from WQS

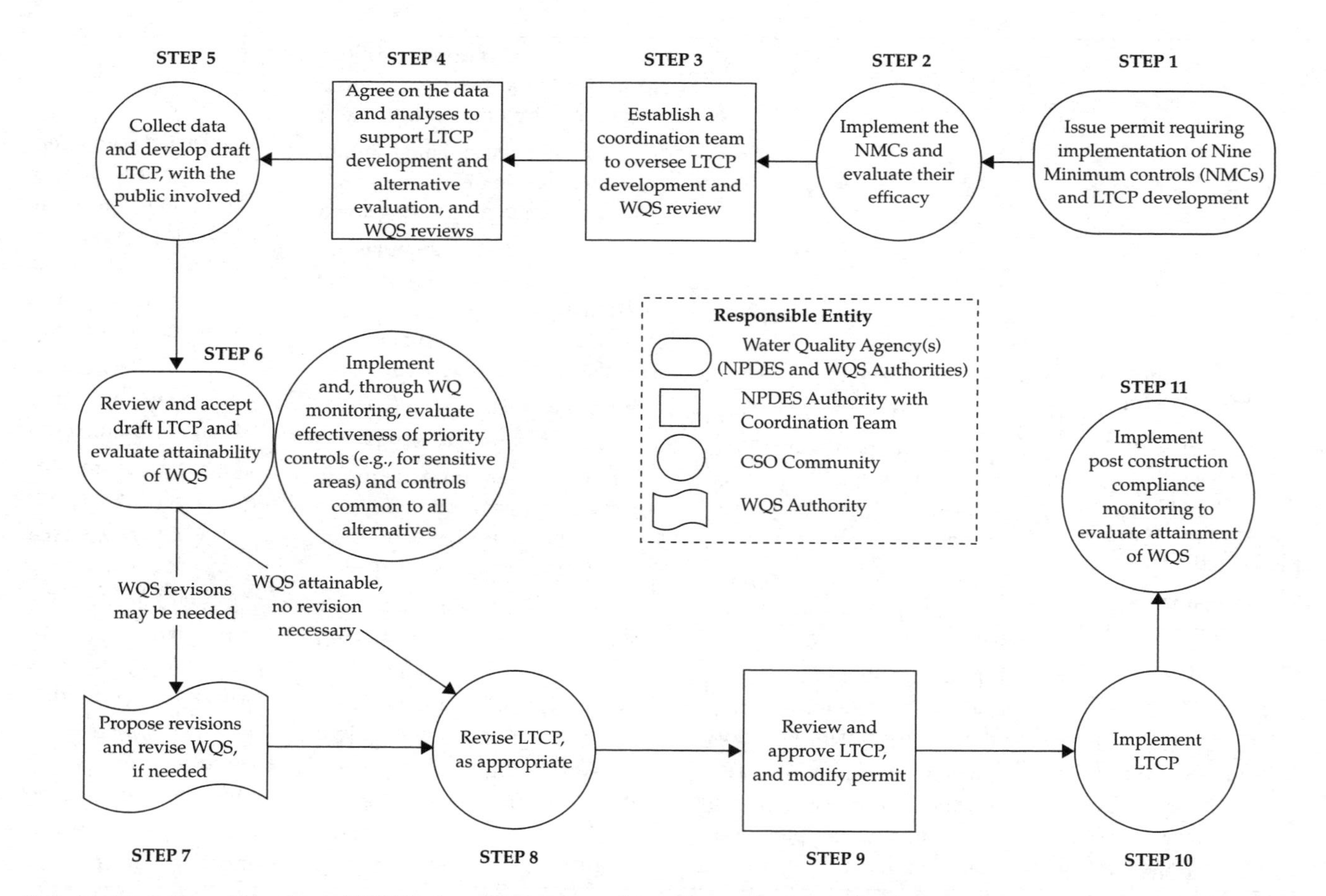

FIGURE 3.1 Coordination of LTCP development and water quality standards review and revision (WQ = water quality) (U.S. EPA, 2001).

2.2.3.2 *Combined Sewer Overflows: Guidance for Nine Minimum Control Measures*

A key element of the *Combined Sewer Overflow (CSO) Control Policy* is the expectation that "permittees should immediately implement the nine minimum controls (NMC), which are technology-based actions or measures designed to reduce CSOs and their effects on receiving water quality ..." (U.S. EPA, 1994). The purpose of *Combined Sewer Overflows: Guidance for Nine Minimum Control Measures* (U.S. EPA, 1995c) is to help CSO permittees and others understand the intent of these measures and to plan, implement, and document the appropriate measures. The document examines and discusses each of the nine minimum controls in depth and provides illustrated examples of potential alternatives to be considered.

2.2.3.3 *Combined Sewer Overflows: Guidance for Long-Term Control Plans*

Combined Sewer Overflows: Guidance for Long-Term Control Plans (U.S. EPA, 1995b) provides guidance for development of a LTCP in conformance with the CSO policy. The target audience for the document, which is written for the non-engineer, includes municipal officials who are developing long-term control plans. Components of the CSO LTCP development process are covered in more depth in other U.S. EPA guidance documents detailed in this section.

2.2.3.4 *Combined Sewer Overflows: Guidance for Monitoring and Modeling*

Because CSO policy is water quality based, it is, therefore, important to understand the impact of CSO discharges to the waterbody. *Combined Sewer Overflows: Guidance for Monitoring and Modeling* (U.S. EPA, 1999) presents guidance for monitoring both the CSO (source) and the receiving waterbody and mathematical modeling to characterize and understand the impacts of CSO discharges and predict the improvements to water quality to be achieved through various types and levels of control of CSO discharges. It is not a "how to" manual, rather, it discusses issues to consider and guidelines for developing site-specific monitoring and modeling programs to address a community's particular situation.

2.2.3.5 *Combined Sewer Overflows: Guidance for Permit Writers*

Combined Sewer Overflows: Guidance for Permit Writers (U.S. EPA, 1995d) provides guidance for NPDES permit writers in developing CSO permit conditions. The permit writer has a large role to play in conceptually facilitating and coordinating the development of CSO programs relative to attainment of WQS and compliance with CWA.

2.2.3.6 Combined Sewer Overflows: Guidance for Financial Capability Assessment and Schedule Development

The purpose of *Combined Sewer Overflows: Guidance for Financial Capability Assessment and Schedule Development* (U.S. EPA, 1997) is to provide a planning tool to evaluate available resources of a permittee to implement a CSO program, and to assist U.S. EPA and state NPDES authorities in determining an appropriate schedule for implementing a CSO program. This document can be especially important to small communities, where the cost of CSO control represents a significant investment for communities that have large combined systems and limited financial resources. The document presents a prescriptive approach to determining a permittee's financial capability that includes determination of a "residential indicator" and a "permittee financial capability indicator". With this information, the effect of total wastewater cost to the customer in terms of percent of median household income is determined and evaluated in terms of burden. In addition to the prescriptive approach described, the guidance has numerous references to the alternate or additional consideration of unique local conditions.

In 2010, U.S. EPA indicated interest in revisiting and revising this guidance. An important issue that may be considered includes additional flexibility in considering a community's unique economic condition.

2.2.3.7 Combined Sewer Overflows: Guidance for Funding Options

For some communities, the CSO LTCP represents a significant portion of the total wastewater budget. Funding for these programs can be a significant issue, especially in poor economic times for the community or the nation as a whole. *Combined Sewer Overflows: Guidance for Funding Options* (U.S. EPA, 1995a) provides an array of funding options with advantages and limitations of each.

2.2.3.8 Combined Sewer Overflows: Screening and Ranking Guidance

Combined Sewer Overflows: Screening and Ranking Guidance (U.S. EPA, 1995e) provides methods for screening and ranking CSSs by NPDES authorities to prioritize the issuance of NPDES permitting. It can also be used by CSO communities to prioritize individual CSOs within their systems, thereby addressing highest needs first. The guidance specifically states that it should not be used by U.S. EPA to prioritize enforcement actions. The value of the guidance may be reduced to communities experiencing enforcement actions, although it could be useful in determining the schedule sequence of addressing multiple CSOs.

2.2.3.9 *Interim Economic Guidance for Water Quality Standards: Workbook*

Interim Economic Guidance for Water Quality Standards: Workbook (U.S. EPA, 1995f) provides important information on economic considerations that may be considered in the change of a waterbody use designation or the granting of a variance from an existing standard. The concept of whether proposed measures would result in "substantial and widespread economic and social impacts" is considered. *Combined Sewer Overflows: Guidance for Financial Capability Assessment and Schedule Development* (U.S. EPA, 1997) was based, in part, on the 1995 document. *Interim Economic Guidance for Water Quality Standards: Workbook* (U.S. EPA, 1995f) also contains an evaluation process that considers socio-economic factors and median household income of the community.

2.3 Publicly Owned Treatment Works Blending Policies

The practice of "blending" or peak flow management at POTWs has traditionally consisted of routing a portion of the peak wet weather flow entering a treatment facility around the biological treatment component of the treatment process and re-combining the total flows before final discharge. The practice has been necessary to protect the biological treatment process and prevent overflows and backups during peak wet weather flow events. The treatment goal was to meet pollutant concentration requirements for final effluent as stated in the NPDES permit. This practice was common in the past and was recognized by treatment plant operators and regulators. The CWA grants program funded numerous treatment plant projects that included blending.

For POTWs serving CSSs, blending is addressed in the national CSO policy in terms of a "bypass" (see Section 1.4).

For sanitary system POTWs, however, there has been controversy over the use of blending. This is an important issue as many existing facilities use blending. U.S. EPA developed a draft policy to address blending requirements for NPDES permitting and, in 2003, published the document for public review and comment. The policy was based on permitting blending facilities provided six specific criteria were met. U.S. EPA received many comments on the proposed policy and, consequently, it was not promulgated. Representatives of NACWA and the Natural Resources Defense Council joined forces to develop a strategy referred to as "treatment facility peak flow strategy" that addressed issues of concern and was proposed as a new regulatory standard. U.S. EPA requested public comment on the peak flows policy on December 22, 2005. As of 2010, however, the policy has not been finalized. U.S. EPA is currently reviewing regulatory issues concerning peak wet weather discharges from POTWs as part of an effort to develop a comprehensive national policy for SSSs (see Section 2.1.1.3).

2.3.1 *Satellite Collection Systems*

Satellite collection systems are collection systems that discharge into a regional trunk sewer and treatment facility owned by others. Satellite systems are typically owned and operated by a community that has a formal agreement with the regional sewer authority. Wet weather regulations apply to satellite systems, just as any other system. Considerable discussion has occurred nationally on the issue of wet weather permitting for satellite systems. Permitting can take the form of a general permits issued to the regional authority or a separate permit for the satellite system. Complications arise with regard to the general permit because the regional authority often has limited control over the operation, maintenance, and performance of the satellite facilities. U.S. EPA is currently reviewing the role of satellite collection systems in an effort to develop a comprehensive national policy for SSSs (see Section 2.1.1.3).

2.4 Sewer System Overflow Regulatory and Enforcement History

Added interest in addressing SSOs occurred in the early 1990s. By this time, much work had been focused on providing reliable dry weather capacity for sanitary sewers and secondary treatment facilities for POTWs. *Combined Sewer Overflow (CSO) Control Policy* (U.S. EPA, 1994) provided a basis for enforcement of regulations for CSOs; additionally, the Urban Wet Weather Flows Advisory Committee, while not providing a solid regulation for SSOs, focused attention on the problem and, with it, brought increased enforcement. In addition to overflow remediation requirements being contained in NPDES permits, there was significant emphasis on consent decrees and other enforceable mechanisms. By 2007, U.S. EPA had entered into more than 50 judicial settlement agreements with municipalities to address SSOs and CSOs (Grumbles, 2007).

Sanitary sewer overflow and CSOs continue to be an enforcement priority for U.S. EPA and the states. U.S. EPA's Office of Enforcement and Compliance Assurance states in its *National Enforcement Initiatives for Fiscal Years 2011-2013,* "This enforcement initiative will focus on reducing discharges from combined sewer overflows ('CSOs'), sanitary sewer overflows ('SSOs'), and municipal separate storm sewer systems ('MS4s') in FY2011–13, by obtaining cities' commitments to implement timely, affordable solutions to these problems, including increased infrastructure and other innovative approaches" (U.S. EPA, 2010).

3.0 CONSENT DECREES AND OTHER ENFORCEMENT ORDERS

With the enforcement emphasis on consent decrees and other legal orders, many communities will need to participate in negotiations with regulators that will result in legal requirements to improve their wastewater systems so that overflows are reduced or eliminated to the extent feasible (see Section 3.2.1). Recent legal orders have been issued by federal (U.S. EPA) or state agencies. It is important to understand the legal and policy requirements for these issues and also the current regulatory positions that U.S. EPA and the states have taken with others in the negotiation process. In addition to the wastewater agency's in-house experts, outside engineers and legal experts should be included on the negotiation team. The National Association of Clean Water Agencies (formerly the Association of Metropolitan Sewerage Agencies) developed a handbook entitled, *Wet Weather Consent Decrees: Protecting POTWs in Negotiations* (AMSA, 2003), and subsequent updates to serve as a guide to communities faced with wet weather negotiations. This document, and its subsequent updates, describes current policy and guidance relative to wet weather negotiations, typical compliance programs, and negotiating strategies. Additionally, the handbook describes specific issues addressed in actual consent decrees of various wastewater agencies.

3.1 Typical Provisions

Along with an understanding of the legal and policy requirements for overflows, it is important to understand current regulatory positions that U.S. EPA and the states have taken with others in the negotiation process. This includes specific provisions for issues relating to SSOs and CSOs. This information can be obtained from references such as the NACWA handbook, *Wet Weather Consent Decrees: Protecting POTWs in Negotiations* (AMSA, 2003), described in the previous section, or by researching previous consent decrees from U.S. EPA. Networking with other wastewater agencies that have similar issues, either through a professional organization or individually, is another good source of information. Additionally, consultants are well versed in wet weather and overflow issues and can be a good resource for a wastewater agency.

3.1.1 Sanitary Sewer Overflow Program

Each consent decree or other official order is individually negotiated to address the deficiencies in performance of the particular sewer system. Most recent SSO-related consent decrees have implemented provisions consistent with SSO control recommendations of industry advisors of the Urban Wet Weather Flows Advisory Committee described in Sections 2.1.1.1 and 2.1.1.2. Industry advisors recommended pursuit of a defined CMOM program to define performance expectations for the SSS and development of system evaluation and capacity assurance plans (SECAP). Subsequent SSO-related consent decrees commonly require completion of a CMOM and SECAP and commitment to identification and construction of facilities necessary to achieve performance consistent with the CMOM. It is important to note that CMOM and SECAP should define and support a level of control appropriate to the specific collection system. Where CMOM and SECAP adequately support a reasonable level of control, subsequent overflows that do occur can be readily classified as either "failures" (i.e., could have been avoided) or "exceptions" (i.e., unavoidable due to forces beyond the control or expectations of the system design).

3.1.2 Combined Sewer Overflow Program

Combined Sewer Overflow (CSO) Control Policy (U.S. EPA, 1994) prescribes that implementation of CSO LTCPs should be done as soon as practicable and that the compliance schedule should be in an enforceable mechanism. For significant CSO systems, the policy states that the compliance schedule should be in a judicial order. Consent decrees for CSO programs often contain requirements to implement the nine minimum controls (if not already done) and develop and implement (or just implement if development has already occurred) a CSO LTCP. Primary elements of consent decrees for CSO implementation programs are defined projects and a fixed-date compliance schedule. The projects are intended to meet the provisions of law and regulation and the compliance schedule to reflect what is expeditious as practicable. Affordability, which is linked directly to the schedule, is sometimes an important issue in the development of a LTCP and schedule. U.S. EPA guidance describes a maximum schedule of 20 years. For some large programs, consent decrees contain a phased approach to scheduling, in which the initial set of projects is defined and scheduled along with a date for submittal of a schedule for projects in the next phase of work (see also Section 3.2.4, which discusses adaptive management). Civil penalties may be included in consent decrees; and, stipulated penalties for not meeting requirements of the decree are typically included.

3.2 Negotiating Strategies

Every consent decree has its own negotiating strategy. The strategy is typically based on unique agency-specific conditions. These conditions commonly include financial capability and affordability, scope and size of the overflow issues to be addressed, impact of the overflows, and whether the decree includes SSOs, CSOs, treatment facilities, or a combination of these.

3.2.1 *Level of Control and Cost–Benefit Analyses*

The goal for level of control for a consent decree-driven program can be a big issue, whether the system is an SSS or a CSS. As such, the following issues should be considered.

3.2.1.1 Sanitary Sewer Systems

As described previously, there is neither an SSO rule nor clearly defined SSO policy that establishes prescriptive control standards. The level of control for SSOs typically is infrequent events, often expressed in terms of a standard design storm. Level of control requirements for a sanitary system with few wet weather overflow problems and a comfortable financial picture may not be an issue. However, in many instances, especially for older systems in areas of moderate-to-heavy rainfall, overflow problems are widespread and funds are scarce. In these instances, cost–benefit analyses will play a necessary role in defining a reasonable level of control. The obvious benefit to consider in this instance would be reduction in overflow volume or frequency of occurrence. Other benefits that should be considered in identifying a meaningful level of control include public health (e.g., exposure to raw wastewater in basement backups, street flooding, or discharges) and water quality impacts to the receiving stream. These factors also play a part in any watershed management and adaptive management strategies as described in Sections 3.2.3 and 3.2.4. Note that the goal of zero overflows is typically unattainable; hence, levels of control that approach, but do not fully achieve, the zero-overflow goal must be selected. Typically, the level of control beyond which further expenditure does not result in further benefit is chosen. The eventual agreed-upon level of control must be contained in a discharge permit so that the utility will not be deemed non-compliant.

3.2.1.2 Combined Sewer Systems

Level of control is an important determination in a CSO control program. Level of control for CSOs is typically expressed as a percentage of annual volumetric control and/or annual frequency of overflows. Even though *Combined Sewer Overflow (CSO)*

Control Policy (U.S. EPA, 1994) is water quality-based, a cost-effectiveness evaluation of a number of types of overflow control alternatives at various levels of control is required to arrive at a final plan. After the cost–benefit analysis is performed and an initial level of control is determined, the plan must be assessed to ensure that water quality goals are expected to be met. If CSOs are not the only significant wet weather pollutant source, it should be shown that the resultant CSO contribution does not cause or contribute to the non-attainment of standards. If the wastewater agency decides to develop their LTCP based on the presumptive approach of meeting water quality, *Combined Sewer Overflow (CSO) Control Policy* (U.S. EPA, 1994) prescribes an attainment of a number of possible control criteria, one of which is a minimum 85% level of annual volumetric control of wet weather combined sewer flows.

3.2.2 *Affordability and Schedule*

There are three significant and interconnected components of every wet weather program: scope, cost, and schedule. For an implementable program, each of these components is dependent on the other, and, in some instances, determines the other. The wastewater agency and the government are interested in reasonable schedules that do not extend excessively into the future. The CSO policy (U.S. EPA, 1997) recommends a 15-year horizon, with 20 years considered for extreme cases. The practicality of excessive wet weather program schedules is questionable because it can tie up an agency for many years and leaves no flexibility to adjust to new requirements and needs as they arise. Some of the significant wet weather programs in the country have been confronted with affordability issues as they consider the potential scope of their wet weather program with associated cost and reasonable schedule limits.

Thus, in developing a wet weather program schedule, an agency must determine the scope and cost of the considered program and then determine a schedule. Part of the schedule determination will include the availability of funds to pay for the work and the burden on rate payers. This will include an analysis of current and future operating and capital costs in addition to the wet weather program. If the cost of the wet weather program is substantial, the wastewater agency may find that it will either take an extended time to implement the program or that it is not financially feasible to implement the program as described in a reasonable length of time. In the latter instance, a UAA may be appropriate. A UAA is used to determine whether attainment of an existing waterbody use is feasible considering such issues as economic effect. (See *Guidance: Coordinating CSO Long-Term Planning with Water Quality Standards Reviews* [U.S. EPA, 2001] for more information about UAAs.)

In determining what is affordable, it can be helpful to review *Economic Guidance for Water Quality Standards: Workbook* (U.S. EPA, 1995f) and *Combined Sewer Overflows: Guidance for Financial Capability Assessment and Schedule Development* (U.S. EPA, 1997). The regulatory water quality issue of showing that attainment of a use would result in "substantial and widespread economic and social impacts" (40 CFR 131.10(g)(6)) is an important consideration. It should be noted that a revision to an existing designated use of a waterbody that lowers the level of protection is serious and should be duly considered. Additionally, even if an existing use is revised through the UAA process, the reason for the revision must be revisited in the future and, if conditions change, the use may be restored accordingly.

In addition to describing a defined prescriptive approach, the two aforementioned financial capability and affordability documents both allow and encourage the review of unique local financial considerations. However, because of the subjective nature of every community's unique local circumstances, it is often more difficult to reach agreement about what is affordable by taking this subjective approach instead of the prescriptive approach. In addition to federal guidance, some states have their own financial capability and affordability policies that may be considered in either a federal or state decree.

U.S.EPA has expressed an interest in reviewing and possibly revising their present economic capability and affordability policy.

3.2.3 Watershed Management

Watershed management is a common-sense, holistic approach to addressing water issues in a comprehensive, best-bottom-line method to meeting established goals in a watershed. Watershed management can include many items outside of wet weather overflows from sewer systems. However, for this discussion, it means applying a structure for considering the effect of CSOs, SSOs, and stormwater and analyzing the impact these have on meeting watershed goals. By understanding these goals, a program can be developed that best uses available resources to make the most effect on the watershed. Although this approach is powerful, it can be difficult to implement due to the present paradigm of regulatory controls. The various specific and isolated regulations for the components (CSO, SSO, and stormwater) get in the way of a complete watershed approach. This is recognized by U.S. EPA, and progress is being made to advance this concept. For example, Clean Water Services in Oregon currently has a single permit covering NPDES and stormwater issues. The National Association of Clean Water Agencies has spent considerable effort in promoting this concept. U.S. EPA has recognized watershed management as of one their "four pillars" to sustainable infrastructure, which are recognized as representing a "fundamental change to viewing and managing water infrastructure" (Grumbles, 2007).

Presently, watershed concepts can be used as a negotiation strategy to describe and program the best allocation of available resources, especially if enforcement addresses multiple issues such as CSOs, SSOs, and POTW peak flows. Watershed considerations can be very useful in establishing priorities in implementation.

3.2.4 *Adaptive Management*

Adaptive management is a systematic, cost-effective approach to program development and implementation. An overall program concept is identified with a fixed overall structure such as performance goals, end date or schedule, and overall cost. Within the overall time period of the program, periodic steps are identified for review and analysis of what has been completed and what is proposed for the next period to meet program goals. These periodic reviews can ensure that critical program goals and assumptions are being realized and that new information or conditions are accounted for. (Sanitation District No. 1 of Northern Kentucky has a recent consent decree based on adaptive management.)

With an adaptive management approach, periodic reviews would take place on a given schedule (e.g., every 5 years), at which time the next phase of the program can be identified and scheduled. The new work should be programmed in light of current understanding of program needs and should include review of work done to date.

A comprehensive audit can be performed that includes the following:

- *Program costs*—actual costs of the program to date are compared to predicted costs. If actual costs are substantially different than predicted, an adjustment may be made to the program and schedule.
- *Project experiences*—successes and failures of implemented projects are reviewed and necessary adjustments will be made to future projects.
- *Priorities*—new priorities are considered as they appear and new projects are proposed.
- *New regulations*—new regulations are assessed and necessary adjustments are made to the program.
- *Rainfall-derived infiltration and inflow (RDII) programs*—results of RDII programs are assessed and additional projects for these areas will be proposed.
- *New developments*—any newly developed flow issues are addressed.

An adaptive management approach is practical for large wet weather programs, especially those that have extended schedules. However, all communities should consider this approach because their wet weather program may represent a large part of their capital program for years to come, and, it makes sense to adjust future projects based on the most current information and priorities. Even if the program is not totally structured around an adaptive management strategy, provisions in the consent decree and approved program that allows periodic reviews and adjustments is recommended.

3.2.5 Emerging Issues

Because of the high costs and long schedules associated with wet weather programs, it is important for wastewater agencies to understand and consider emerging issues in the wastewater industry as consent decrees and other enforcement orders are negotiated. Green infrastructure and asset management, described in the remaining sections of this chapter, are examples of emerging technologies and practices that may be advantageous to incorporate into the program. In addition to new technologies and practices, there are potential new regulatory requirements that should be considered, as they may be required over the time of the implementation of the consent decree program. The utility should understand what impact the costs of these new requirements could have on the ability to pay for the negotiated program and, consequently, make provisions in the consent decree to adjust the program if necessary based on the financial impact of new regulations.

3.2.5.1 Green Infrastructure

Green infrastructure is a term used in wet weather management to describe a number of technologies that are viewed as more natural to the environment than traditional concrete and steel structures, which are commonly termed *gray infrastructure.* Green infrastructure can include site-specific stormwater best management technology such as green roofs, rain gardens, and pervious pavement, and also larger, more regional facilities in scope such as bio-retention facilities. The key to green infrastructure's role in addressing wet weather is its ability to contain or retain stormwater, thereby reducing its effect on CSOs and SSOs. U.S. EPA has acknowledged and promoted the inclusion of green infrastructure into wet weather programs. A 2007 memorandum to U.S. EPA water division directors, regional counsel and enforcement coordinators, and state NPDES directors states, "… permitting authorities may structure their permits, as well as guidance or criteria for stormwater and CSO long-term control plans, to encourage permittees to utilize green infrastructure approaches, where appropriate, in lieu of or in addition to more traditional controls. U.S. EPA will also consider the feasibility of the use of green infrastructure as a water pollution control technology in its enforcement activities, and encourages state authorities to do likewise" (Boornazian and Pollins, 2007).

The effect of green infrastructure on wet weather programs continues to be researched. A number of communities and agencies including those in Philadelphia, Pennsylvania, Portland, Oregon, and Seattle, Washington, have implemented substantial green programs. Additionally, agencies including those serving Washington, D.C., Cincinnati, Ohio, Kansas City, Missouri, and St. Louis, Missouri, are including green infrastructure in their wet weather programs. Issues that are currently being addressed are actual quantities of rain to be controlled by the various green techniques and how these results can be reasonably portrayed in hydraulic models of the collection system. A significant amount of work has already been done in this regard. Most of the available research to date is directed at the ability of green infrastructures to reduce the impact of stormwater and CSOs. There is currently a lack of available information and documented performance on the impact of green infrastructure on SSOs.

Other green infrastructure issues that should be considered are life-cycle costing, including maintenance of decentralized facilities, and ensuring that voluntary best management practices on private property are maintained and remain in place.

3.2.5.2 Asset Management

Asset management is a term used to describe a structured, proactive system of maintaining and replacing existing assets to maximize benefit. It is imperative that wastewater utilities properly manage their assets and provide the proper resources to ensure that the day-to-day requirements of collection and treatment of wastewater are fulfilled. In negotiating consent decrees and other enforcement orders, it is important that the wastewater utility understands the resources needed to provide proper asset management and that these resources are accounted for in the affordability analyses.

4.0 REFERENCES

American Public Works Association; American Society of Civil Engineers; National Association of Clean Water Agencies; Water Environment Federation (2010) *Core Attributes of Effectively Managed Wastewater Collection Systems;* www.wef.org/CoreAttributesofWWCS/ (accessed October 2010); National Association of Clean Water Agencies: Washington, D.C.

Association of Metropolitan Sewer Agencies (2003) *Wet Weather Consent Decrees: Protecting POTWS in Negotiations;* Association of Metropolitan Sewer Agencies: Washington, D.C.

Boornazian, L.; Pollins, M. (2007) Memorandum—Use of Green Infrastructure in NPDES Permits and Enforcement. U.S. Environmental Protection Agency Memo to U.S. Environmental Protection Agency Water Division Directors, Regions 1-10; Regional Counsel/Enforcement Coordinators, Regions 1-10; State NPDES Directors.

Browner, C. M. (1995) U.S. Environmental Protection Agency, Advisory Committee Charter, Urban Wet Weather Flows. www.epa.gov/npdes/pubs/charter.txt (accessed Nov 2010).

Grumbles, B. H. (2007) Testimony before the Subcommittee on Water Resources and Environment, U.S. House of Representatives, Oct 16, 2007.

U.S. Environmental Protection Agency (1972) *Federal Water Pollution Control Act Amendments;* Public Law 92-500; 86 Stat. 816; 33 U.S.C.§1251 et seq.; U.S. Environmental Protection Agency: Washington, D.C.

U.S. Environmental Protection Agency (1989) *National Combined Sewer Overflow Control Strategy;* 54 *Fed. Regist.* 37371; U.S. Environmental Protection Agency: Washington, D.C.

U.S. Environmental Protection Agency (1994) *Combined Sewer Overflow (CSO) Control Policy;* EPA-830/94-001; U.S. Environmental Protection Agency: Washington, D.C.

U.S. Environmental Protection Agency (1995a) *Combined Sewer Overflows: Guidance for Funding Options;* EPA-832/B-95-007; U.S. Environmental Protection Agency, Office of Water: Washington, D.C.

U.S. Environmental Protection Agency (1995b) *Combined Sewer Overflows: Guidance for Long-Term Control Plans;* EPA-832/B-95-002; U.S. Environmental Protection Agency, Office of Water: Washington, D.C.

U.S. Environmental Protection Agency (1995c) *Combined Sewer Overflows: Guidance for Nine Minimum Controls;* EPA-832/B-95-003; U.S. Environmental Protection Agency, Office of Water: Washington, D.C.

U.S. Environmental Protection Agency (1995d) *Combined Sewer Overflows: Guidance for Permit Writers;* EPA-832/B-95-008; U.S. Environmental Protection Agency, Office of Water: Washington, D.C.

U.S. Environmental Protection Agency (1995e) *Combined Sewer Overflows: Guidance for Screening and Ranking;* EPA-832/B-95-004; U.S. Environmental Protection Agency, Office of Water: Washington, D.C.

U.S. Environmental Protection Agency (1995f) *Interim Economic Guidance for Water Quality Standards: Workbook;* EPA-823/B-95-002; U.S. Environmental Protection Agency, Office of Water: Washington, D.C.

U.S. Environmental Protection Agency (1997) *Combined Sewer Overflows: Guidance for Financial Capability Assessment and Schedule Development;* EPA-832/B-97-004; U.S. Environmental Protection Agency, Office of Water: Washington, D.C.

U.S. Environmental Protection Agency (1999) *Combined Sewer Overflows: Guidance for Monitoring and Modeling;* EPA-832/B-99-002; U.S. Environmental Protection Agency, Office of Water: Washington, D.C.

U.S. Environmental Protection Agency (2001) *Guidance: Coordinating CSO Long-Term Planning with Water Quality Standards Reviews;* EPA-833/R-01-002; U.S. Environmental Protection Agency, Office of Water: Washington, D.C.

U.S. Environmental Protection Agency (2005) *National Pollutant Discharge Elimination System (NPDES) Permit Requirements for Peak Wet Weather Discharges from Publicly Owned Treatment Works Treatment Plants Serving Separate Sewer Collection Systems;* 70 *Fed. Regist.* 76013-76018; U.S. Environmental Protection Agency: Washington, D.C.

U.S. Environmental Protection Agency (2006) *Proposed Policy, Peak Wet Weather Discharges from Municipal Sewage Treatment Facilities* Web Site. http://cfpub.epa.gov/npdes/wetweather.cfm?program_id=0 (accessed June 2010).

U.S. Environmental Protection Agency (2007) U.S. Environmental Protection Agency -October 2007 FY08-FY10 Compliance and Enforcement National Priority: Clean Water Act, Wet Weather, Sanitary Sewer Overflows.

U.S. Environmental Protection Agency (2010) *National Enforcement Initiatives for Fiscal Years 2011-2013* Web Site. http://www.epa.gov/compliance/data/planning/initiatives/initiatives.html (accessed June 2010).

Water Environment Federation (2006) *Guide to Managing Peak Wet Weather Flows in Municipal Wastewater Collection and Treatment Systems;* Water Environment Federation: Alexandria, Virginia.

Water Environment Federation; American Society of Civil Engineers; Environmental and Water Resources Institute (2009) *Existing Sewer Evaluation and Rehabilitation,* 3rd ed.; WEF Manual of Practice No. FD-6; ASCE Manual and Report on Engineering Practice No. 62; McGraw-Hill: New York.

Chapter 4

Information Management

(*continued*)

1.0 INTRODUCTION

This chapter focuses on information management needs and requirements for the wastewater collection system manager responsible for preventing and controlling sewer system overflows.

The following forces are compelling collection system managers to evolve away from paper-based (or non-existent) recordkeeping to computer-based systems: environmental regulators, such as U.S. Environmental Protection Agency (U.S. EPA); asset management requirements promulgated in the Governmental Accounting Standards Board (GASB) Statement 34 ("GASB 34") (U.S. EPA, 2002); and overall industry improvements (e.g., technical certification programs). In theory, paper-based systems could fulfill the analysis and reporting requirements of these three demand areas; however, the cost of manually collecting and reporting the data is not financially acceptable in light of the availability of computer-based systems. In June 1999, GASB 34 was published, requiring state and local governments to begin reporting all financial transactions, including the value of their infrastructure assets, roads, bridges, water and sewer facilities, and dams, in their annual financial reports on an accrual accounting basis. (For more information, search http://www.epa.gov for "GASB 34").

Wastewater managers make decisions every day that are aimed at improving system reliability and compliance with regulatory requirements. As time passes, however, systems change, people retire, the knowledge base is lost, and the probability of costly failures increases as infrastructure ages. Against this backdrop, a new generation of managers is finding that old tactics to replace aging assets in broad strokes or fixing them after they break are not cost-effective solutions. To address these issues, today's managers seek a more strategic approach that improves their knowledge base as it instills confidence that the decisions made are the most cost-effective (Lovely, 2008).

Following the behavioral science theory of management, mainly developed at Carnegie Mellon University (Pittsburgh, Pennsylvania), most of what goes on in service organizations is information and decision-making processes. The crucial factor in the information and decision-making process is, thus, an individual's limited ability to process information and to make decisions under these limitations.

However, the evolution toward computerization has not been without its own problems. For example, computerization can create a flood of raw data from multiple sources that can result in an incoherent picture, thus becoming a source of frustration and difficulty, rather than assistance, for a collection system manager. However, when properly filtered and sorted, the data can become *knowledge*; indeed, with a collection system manager and staff's experience and intuition, it can yield information that leads to effective resource decisions. The objective of this chapter is to provide the reader with a sufficient framework with which to make those decisions in a balanced, professional manner.

1.1 Definitions

Wikipedia, the free Web encyclopedia, defines *information management* as the collection and management of information from one or more sources and the distribution of that information to one or more audiences (Wikipedia, 2009). *Management* means the organization of, and control over, structure, processing, and delivery of information. Information management, therefore, reflects an organization's ability to process information.

An *information system,* or an *information management system* (IMS), is a sequence of operations that begins with planning and collection of data, continues with storage and analysis of the data, and ends with the use of the derived information in some decision-making process (Shamsi, 2002b). *Information management system* also refers to a database and transaction management system that was first introduced by IBM Corporation (Armonk, New York) in 1968.

A *management information system* (MIS), not the same as an IMS, is a subset of the overall internal controls of a business covering the application of people, documents, technologies, and procedures by management accountants to solving business problems such as costing a product, service or a business-wide strategy. Management information systems are distinct from regular information systems in that they are used to analyze other information systems applied to operational activities in the organization (O'Brien, 1999).

Information technology, as defined by the Information Technology Association of America, is "the study, design, development, implementation, support or management of computer-based information systems, particularly software applications and computer hardware." Information technology deals with the use of electronic computers and computer software to convert, store, protect, process, transmit, and securely retrieve information.

1.2 Chicago Information Management System Case Study

In January 2007, the Metropolitan Water Reclamation District of Greater Chicago (MWRD) commissioned an information technology strategic plan to align information systems with operations, facilitate best business practices, address business and operational drivers, and support MWRD's strategic plan. Through a formal planning process, MWRD developed an integrated, organization-wide information technology strategy and consensus among managers on information technology priorities. Metropolitan Water Reclamation District of Greater Chicago expected that the plan would be the catalyst for continuous business performance improvements in asset management, regulatory management, financial management, customer service, facilities engineering and construction, integrated budgeting, planning, and more. The foundation of MWRD's information technology strategic plan was a set of interrelated programs and projects that addressed business needs. The defined information technology projects are organized within a series of nine programs that match up with industry best practices. The projects comprising the information technology strategic plan were directed to the design, implementation, and deployment of a district-wide information management and business process automation solution. This regional solution was based on an information management concept that is now considered an industry standard (see Figure 4.1). The following are the significant aspects of the information management concept (Lanyon et al., 2009):

- All applications and databases can be accessed via a Web-based information portal through secure links within MWRD's intranet; based on authorization levels, district staff can retrieve the information they need, when they need it, to streamline business processes.
- Using Web services and "enterprise integration bus" technologies, MWRD can integrate applications and databases without having to deploy any of the inflexible, expensive, high-maintenance adaptors associated with integration approaches of older systems.
- The solution provides online links with external parties, including customers and suppliers, thus enabling such e-business capabilities as e-commerce and electronic bill presentment and payment.

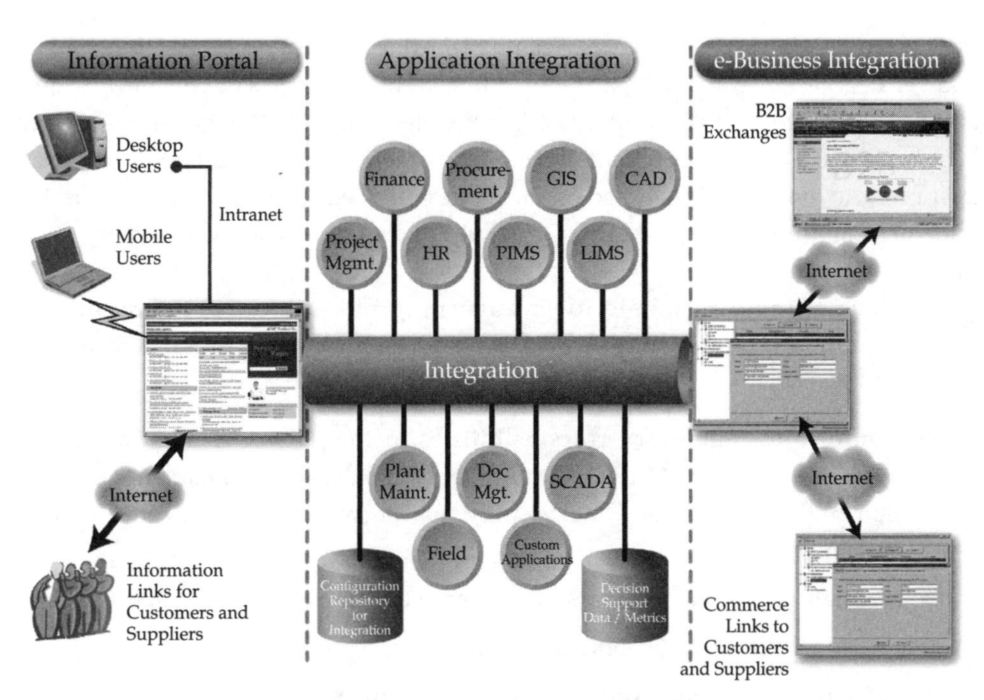

FIGURE 4.1 Modern information management concept (courtesy of Westin Engineering, Inc.). GIS = geographic information systems. CAD = computer-aided design. HR = human resources. PIMS = project information management system. LIMS = laboratory information management system. SCADA = supervisory control and data acquisition. B2B = business to business.

2.0 BENEFITS OF ESTABLISHING AN INFORMATION MANAGEMENT SYSTEM

The collection system industry has evolved at a deliberate, measured pace because it is a publicly funded infrastructure. Decisions that affect public infrastructure often are a product of careful consideration and much deliberation. Collection systems have historically also been low-tech and viewed by the public and elected officials as somewhat simple in nature. Therefore, establishing an electronic IMS can be more complicated for the manager than, for example, purchasing a sewer cleaning machine, which is a much more routine, familiar proposition. Establishing and maintaining an

IMS also requires a noticeable investment, including the cost of utility labor, and, to a lesser degree, software and hardware. Such an investment will be a new, added, and permanent cost for the utility. These factors tend to draw the scrutiny of elected boards. To make the establishment of an IMS palatable to governing boards, the manager needs to develop a case based on the highest possible goals and rationales rather than on technical merits, emphasizing the public benefit of an IMS to engender much broader support.

In developing a rationale for a governing board, the following are the most common and defensible principles upon which to establish a collection system IMS:

- Conform to regulatory responsibilities, principally U.S. EPA's national CSO policy (U.S. EPA, 1994) requirements and capacity, management, operation, and maintenance (CMOM) program for sanitary sewer overflow (SSO) systems;
- Maintain perpetual calendars for preventive maintenance and inspections;
- Provide justification for operations budgets;
- Track prioritized work order and repair schedules;
- Organize capital rehabilitation and replacement plans based on asset management data;
- Preserve the utility's corporate memory;
- Perform timely and consistent information searches;
- Provide real-time, user-friendly information to field staff; and
- Provide useful and timely maintenance trend and productivity reports.

2.1 Preservation of Institutional Knowledge

Of all the principles upon which to establish and maintain a collection system IMS, preservation of corporate memory is perhaps the most important. A collection system is a long-lived public asset and the knowledge of how the system was built, operated and maintained, and rehabilitated or replaced is a knowledge base that will outlive current and future utility staff.

Too often, the retirement of a knowledgeable employee has meant that a substantial part of the utility's corporate memory was lost, to the detriment of those still working there and the ratepayers who support the system. For this reason alone, an IMS that can maintain this knowledge across generations of employees is a valuable asset for a manager.

2.2 Compliance with Regulatory Requirements

Since the mid-1990s, collection system agencies have been subject to varying degrees of regulatory requirements embodied in CMOM. An IMS is an integral part of CMOM because of the accuracy and consistency that can be achieved with well-structured and executed computerized management programs.

As described in the *Guide for Evaluating Capacity, Management, Operation, and Maintenance (CMOM) Programs at Sanitary Sewer Collection Systems*, CMOM programs incorporate many of the standard operation and maintenance activities that are routinely implemented by utilities (U.S. EPA, 2005). However, the document does create a new set of information management requirements to

- Better manage, operate, and maintain collection systems;
- Investigate capacity-constrained areas of the collection system;
- Proactively prevent SSOs; and
- Respond to SSO events.

The IMS requirement has been incorporated by state and regional regulatory bodies into their own versions of CMOM; these requirement(s) form the development and use of an IMS within the collection system utility.

2.3 Timely and Consistent Information Retrieval

The ability to retrieve information from the Internet in microseconds has raised both the public's and media's expectations for local governments' Web sites. This has obvious implications for collection system managers who are subject to the same expectations for speed. The sheer speed of information retrieval and sorting by electronic systems is an asset that collection system agencies cannot do without. Poring over reams of paper records to retrieve and distribute information is a grinding, problem-laden task that does not speak to the professionalism of a collection system utility. The following are the most frequent demands for information retrieval:

- Field crews need to be more efficient by accessing much of the information regarding a work order electronically rather than poring through paper records.
- Customer and media inquiries are almost always related to complaints or problems because there is almost no interest in the normal operations of a collection system. The ability to provide an immediate reply using information retrieved electronically leads to high marks awarded for customer service.
- Elected officials expect to be kept informed on the performance of the collection system and its engineering and operations staff, especially during rate-increase requests. A well-organized database can yield compact, uniform reports that bolster the confidence of elected officials and provide a sound basis for making resource decisions.
- In GASB 34, requirements were promulgated to measure the degree to which the infrastructure of the United States was being maintained. It sought to establish its life span and when it would need to be replaced or rehabilitated. Although GASB is not an infrastructure management organization, it is responsible for setting the standards by which the financial health of cities and agencies is measured because they must be able to keep their infrastructures intact and functioning.
- An IMS that is designed to yield current and projected infrastructure data can be used to satisfy GASB 34 reporting requirements. Engineering data, asset management data, inspection records, and maintenance histories compose the data set that forms the foundation of GASB 34.
- When a collection system utility is faced with a damage or liability claim, gathering and arranging historical data is difficult without an IMS. Attorneys who represent agencies in litigation are at a distinct disadvantage when records and reports are not digitally organized. Additionally, a utility can depend on its IMS to alert them to repetitive conditions such as flooding patterns that can leave them exposed to damage claims.

2.4 Inspection and Maintenance Schedules

The repetitive nature of preventive maintenance and inspections can create a mind-numbing experience when trying to organize, assign, and report activities without an IMS. Using an IMS to keep track of commitments made over long periods of time helps ensure reliability and continuity. In fact, preventive maintenance and inspections that have long intervals between activities are the ones most likely to be forgotten without the infinite calendars built into an IMS. Paper reminders or other manual scheduling methods are more likely to become lost or forgotten.

The GASB 34 has specific timetables for inspection intervals, which can be loaded into the IMS. Because pipe inspections tend to occur at long intervals, the IMS becomes invaluable for this task.

2.5 Prioritization of Work Orders and Repairs

Prioritization of work orders is a dynamic process because collection systems are hydraulically and structurally dynamic. Additionally, environmental factors such as wet weather add to this. Because an IMS operates in real time, the priority shifts that will occur also can be reported in real time. This provides an accurate picture of the demands on a utility at any given moment. This is especially important during emergencies when a manager will be called on to explain the service status of the organization. An IMS-based report will quickly be able to display both the specifics and the global status of the situation. Additionally, linking the status reporting of an IMS to the Internet would provide the broadest information distribution to the public. Another benefit is that Internet linkage may reassure the public that the service they are expecting is not lost in the "bureaucracy" of local government.

2.6 Justification for Operations Budgets

Operations budgets are typically calculated on the basis of resources needed for each of the following four areas: preventive maintenance, inspections, complaint servicing, and spot repairs. Without an IMS, many collection system managers are left to define their budgets based on historical experience or intuition, neither of which has solid professional footing.

An IMS that is properly loaded with data can easily create a profile of the correct budget by examining resource requirements that are defined for each of the four areas of operations. For example, a manager can use IMS to sum the total labor hours required for a year's worth of preventive maintenance to determine if the budget is sufficient to accomplish all the tasks. If the manager has 10 000 labor hours of preventive maintenance work to do in a year, but has a budget for only 8000 labor hours, then it is easy to see that the utility has an accumulating deficit of work. These types of calculations are readily available in an IMS and provide elected officials supporting documents to prove the accuracy of proposed budgets.

2.7 Capital Rehabilitation and Replacement Plans

Prioritized capital replacement and rehabilitation plans that are linked to asset management information are similar to prioritized work order and repair schedules, as previously described. However, the sheer scale and expense of doing capital work requires a separate prioritization scheme that is much more complex than the relatively simple variables associated with work orders.

The following collection system attributes can cause asset management to be complex: structural materials and performance, maintenance and repair history, anticipated future flow, and the ability of the customer base to finance capital work. Capturing this information and producing a rational capital management plan, however, is only achievable with an IMS. The rise of asset management as a refined science is primarily due to the availability of data analysis capabilities associated with computer systems. Until recently, the complexity of this information has been too overwhelming to cost-effectively analyze using other means.

3.0 TYPES OF INFORMATION MANAGEMENT SYSTEMS

Computer-assisted design, relational database management system (RDBMS), geographic information system (GIS), and asset management system (AMS) software can offer convenient solutions for storing and maintaining the map and related tabular data compiled for key conduits and facilities within the study area. Compared to paper-based methods, these technologies provide additional functionality in the form of increased analytical, graphical display, and quality assurance and quality control (QA/QC) capabilities. The function of each of the aforementioned systems and their role in a combined sewer overflow (CSO) and SSO project are described in logical succession in this section.

3.1 Computer-Aided Design and Drafting

Computer-aided design and drafting (CADD) software is used for designing and drafting new objects and making engineering design drawings. Computer-aided design and drafting allows for automated mapping of the primary facilities composing the collection and treatment system to be represented as a geographic entity (pipe segment, plant, manhole, etc.) registered to a common base map. Additionally, delineation of drainage areas, land use, soil, impervious cover, and layering of other spatial data sets can be readily accomplished. Automated mapping is a special application of CADD technology to produce digital maps. Automated mapping is a computerized alternative to traditional, manual, cartographic map making. Data are organized in layers that are conceptually like registered film overlays. Layers organize data by theme (streams vs roads) and type (linework vs text). There are no spatial relations (topology) among data elements except orientation. Automated mapping is display-oriented; it produces plots of selected layers of point and line features. Graphic information systems, on the other hand, are analytically oriented; they analyze relationships among point, line, and polygon features. One cannot query features between layers using automated mapping. For example, the user cannot select sewer pipes in one layer that are located inside a sewershed in another layer.

3.2 Facility Management Systems

Facilities management is a CADD technology for managing utility systems data. It is used for inventory, inspection, and maintenance work. Facilities management includes an infrastructure management database. Facility management systems are used for maintaining records of the infrastructure owned and maintained by the utility and managing periodic maintenance. Some also assist in prioritizing capital improvement projects.

Compared to automated mapping, there is less emphasis on graphics detail or precision and more emphasis on data storage, analysis, and reporting. Relationships among utility system components are explicitly defined as networks and stored in the database in advance compared to a GIS, which can compute these relationships on demand (or on the fly). Because facility management systems have network definitions, a facility management system knows the pipes connected upstream or downstream of a given pipe. Generally, facility management systems do not have full topology; rather, they offer connectivity and orientation only. Where a richly attributed GIS is in place, a utility may not use a facility management system. In these instances, all facility data are often stored in the GIS and work-related activity is managed in a work order system.

Facility management systems have been used for the maintenance of linear network features. Linear networks are the infrastructure of water distribution and wastewater collection systems, such as pipes, manholes, catch basins, diversion chambers, and outfalls. Facilities management databases store detailed characteristics of the features, such as size, type of material, lining, date installed, work history, and location. The location is typically recorded as descriptive text, identifying the structure or pipe it is connected to, the street it is located on, and so on, rather than as coordinates in CADD and GISs. Facility management systems for wastewater utilities will commonly be associated with video inspections of pipe, so that users of the system can jump to videotape showing the respective reach of pipe. Cost data are also often included and, depending on the complexity of the system, may contain estimates, actual installation costs, and historical information. Presently, many large utilities are not using standalone facility management applications because they are not powerful enough to do effective work management, planning and scheduling, backlog management, failure reporting and analysis, and performance management. The trend is toward using computerized maintenance management systems (CMMSs) and a combination of automated mapping and facilities management (AM/FM) and GIS applications, as described in the following sections.

3.3 Computerized Maintenance Management Systems

Computerized maintenance management systems are relational database management systems (RDBMS) that provide management of tabular data sets and serve as excellent engines for storing, consolidating, and analyzing tabular maintenance data of a sewer system. Computerized maintenance management systems provide enhanced capabilities for managing upkeep of system changes, storage of monitoring and inspection data, and similar data relevant to day-to-day operations (e.g., work orders and maintenance management) of wastewater collection facilities. Each system change, work task performed, and other descriptive attribute data can be maintained in a user-friendly environment, typically eliminating the need for many separate data sets. Generally, it is best to implement these systems based on a complete understanding of the transfer of field-to-office information so that accuracy is preserved in the updating process.

Primary users of CMMSs are supervisors and crew chiefs responding to sewer problems and investigating complaints or performing routine maintenance work, such as cleaning catch basins or televising sewers. Managers and engineers use CMMSs to evaluate the network for prioritizing and budgeting for pipe replacement, extension, and rehabilitation projects.

Infor Hansen Asset Management (Infor, 2010) and CASSWORKS (RJN Group, 2010), which have evolved from facility management systems, are examples of CMMS software.

3.4 Geographic Information Systems

More than 80% of a wastewater utility's data are geographically referenced, that is, they have geographic coordinates. Examples of a geographic data are treatment plants, pumping stations, storage tanks, sewer pipes, manholes, catch basins, diversion chambers, and outfalls. Given wastewater utilities' large investment in distributed and buried assets, geographic information management is of special significance. Additionally, geospatial information is associated with the location of important activities such as maintenance work, construction projects, and water quality sampling (Lanyon et al., 2009).

A GIS is an information system to manage geographically referenced data. Simply stated, GISs can be defined by the following equation (Shamsi, 2002a):

$$\text{GIS} = \text{Geography} + \text{Information System} \tag{4.1}$$

A GIS takes numbers and words from the rows and columns of tables in databases or spreadsheets and displays them on a map. A GIS allows users to view, understand, question, interpret, and visualize data in ways that are simply not possible in the rows and columns of a spreadsheet. A GIS stores information about a map's features in a database table and maintains a link between the features and their stored information. Thus, a GIS can be defined as a link between a map and a database (Shamsi, 2009). Figure 4.2 shows a sewer system GIS map and GIS database table for sewer pipes. The map-to-table link allows a selected pipe in the map to automatically select corresponding pipe data in the table.

A GIS provides tools for interacting with computerized data by its location on planet earth. A GIS can be described as an intelligent map that can serve as an efficient tool for analysis, enhanced decision-making through data visualization, and improved access to information; it is the underlying principle for interacting with "spatial data".

A GIS provides additional functionality in that descriptive or attribute data can be stored as data linked to each geographic entity. The GIS then allows the analytical capabilities of the RDBMS to be executed using geographically based inquiries. Extensive documentation exists for the use of GISs in this type of application (Shamsi, 2002b, 2005; WEF, 2005).

Computerized data within a GIS has an organizational structure or format known as a *database,* which is used to describe physical features located on earth. However, databases are not always bound to physical features. For instance, financial databases are used extensively to store information about business transactions or customer history, which represents "non-spatial" data. In any database, data can be organized in rows and columns, where columns represent unique topics (fields) of information and rows represent individual units (records). An example of this structure is

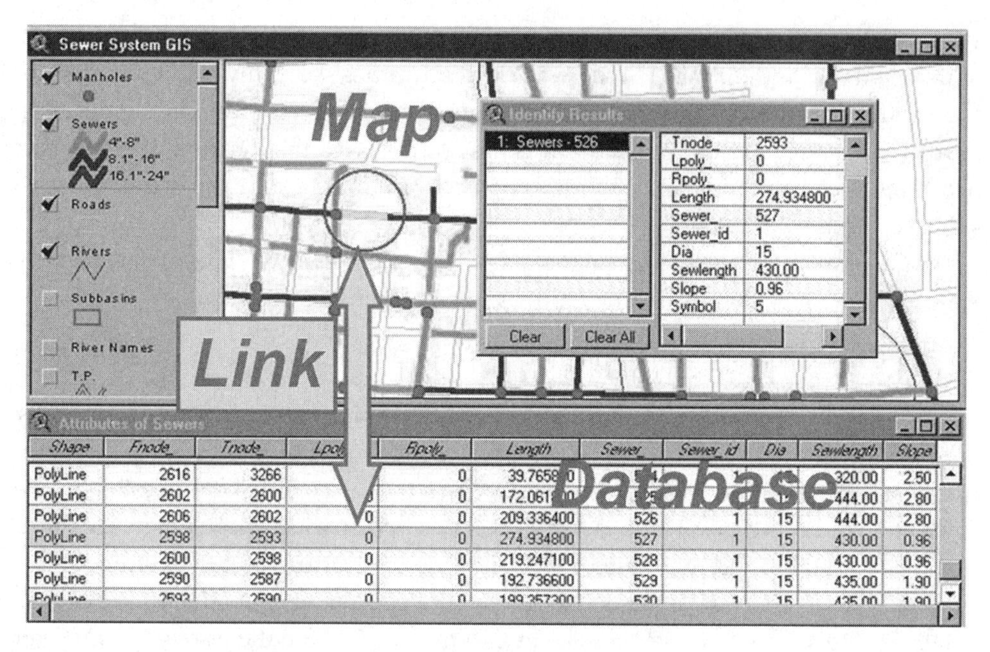

FIGURE 4.2 Geographic information system linking map features to database records (Shamsi, 2002a). Reprinted with permission from ASCE.

a database that describes trees. In this example, fields would include such things as the trunk diameter, species of tree, or an indication on the health of the tree. Every time a tree is cataloged, a new record is created that describes each tree. In this manner, earthly features such as trees, roads, and lakes can be described and recorded in records and fields of information. Geographic information systems represent features on a map using the following three graphical elements: point (nodes), lines (arcs), and polygons. In this example, trees would be represented on a GIS map by nodes, roads would be represented by arcs, and lakes would be represented by polygons.

A single database has always been an efficient tool for managing information storage and retrieval. However, database technology's most productive use is its ability to link several databases together for queries and analyses. Without GIS, databases of information have to be linked to each other through the use of established relationships. Relationships between non-spatial databases are established using identical information within fields of each database. Database management software allows for querying and analyses through several databases by using these established relationships or links. This limits analyses to only those databases that were designed to be integrated. With GIS, interacting with individual databases is made possible through relating these data sets based on their spatial components. In this way, databases that have been assigned to an earthly coordinate system can be related to one another as geographic "layers".

If a new mapping job is required, whether computerized or manual, addressing the following issues may assist in defining the specific scope of work for optimum results:

- Who will be the users? Policy decisions are required as to how broad the network of users will be; in a small town, a system may be completely centralized so that base maps created will be accessed by entities representing water and sewer, streets, parks, planning and development, and utilities. Larger cities may have large departments with independent systems and procedures for maintaining maps, but networked within the department.
- What will be on the maps? A comprehensive map of a sewer system should include all sewers, regulators, diversion chambers, pumping stations, interceptors, and outfalls on the base map. Separate sewers, water supply lines, and industrial connections can be added as applicable. These maps can then be used by design sections, operation and maintenance crews, and industrial waste control pretreatment regulatory programs or spill response teams.
- What is the selected system? If a decision is made to use a computerized system, knowing what features are available (and the answers to the aforementioned questions) should guide the selection. Hardware and software compatibility, user training, continuous vendor support, number of networked users, detail of resolution, features for menus, file retrieval, data storage capacity, enlargement and reduction, use of colors and overlays, and production time are some of the factors to evaluate in system selection.
- How will updated map information be incorporated to the mapping system?
- Where do information documenting system changes come from?

The cost of hardware and software equipment to achieve the benefits of information systems discussed is typically minimal (approximately 10% to 20%) compared to the cost of the data collection effort itself. Because of the size of data collection efforts to support complex planning objectives typical of CSO and SSO abatement programs, it is prudent to undertake implementation of these systems consistent with the information system plan for operation of the utility or department as a whole. In other words, the decision to convert record drawings to digital format for use in abatement planning should be made in close consideration with long-term strategies for record preservation and CADD, RDBMS, GIS, and AMS development

3.5 Automated Mapping, Facilities Management, and Geographic Information Systems

Automated mapping and facilities management represent a combination of AM/FM technologies. Automated mapping and facilities management software is used to automate maintenance. Although GISs and AM/FM are different, they both have their own advantages and applications (Shamsi and Fletcher, 2000). A GIS can help users locate their worst pipe. Automated mapping and facilities management can help them prioritize the work required to bring their worst pipe up to a minimum operating standard. For many years, people have used both the GIS and AM/FM systems separately. However, developing and maintaining two different systems is expensive and inefficient. Mainly because of advances in computer hardware and software, integrated AM/FM and GIS systems known as "AM/FM/GIS" systems are now available. These systems are particularly useful for asset inventory, inspection and maintenance, and work management. They are also popular among map-savvy users who like to "click" on a manhole in a GIS map to initiate a work order rather than locating it by querying a database. Cityworks software (Azteca Systems, 2010) and GBA Sewer Master (GBA Master Series, 2010) are examples of AM/FM/GIS software.

3.6 Differences among Computer-Aided Design and Drafting, Automated Mapping and Facilities Management, and Geographic Information Systems

Non-GIS software, such as spreadsheets (e.g., Microsoft Excel; Microsoft Corporation, Redmond, Washington), RDBMS (e.g., Microsoft Access; Microsoft Corporation), and drafting packages (e.g., AutoCAD; Autodesk, Inc., San Rafael, California) can handle simple spatial data. However, these programs cannot perform such spatial queries as how many residents are located within 100 m of a proposed water main.

The database concept is central to a GIS and is the main difference between a GIS and a simple drafting or computer mapping system, which can only produce high-quality graphic output (ESRI, 1992). By the same token, a GIS is not complete without map creation and editing functions. In fact, some GIS vendors are implementing

"CADD-like" editing functions in their products. Therefore, essential elements of a GIS are (1) the ability to perform spatial operations on the database and (2) map creation and editing functions. All GIS packages use some type of database for storing and manipulating data. Thus, a database management system (DBMS) is an integral part of typical GIS software. The unique method of storing and manipulating data differentiates GIS from other drafting or cartographic software (U.S. EPA, 2000). The GIS might best be understood as the intersection of CADD and DBMS technologies (Shamsi, 2002a).

Automated mapping and facilities management is CADD technology applied to manage utility system data. Compared to a GIS, AM/FM (1) does not have topology and (2) places less emphasis on the detail and precision of its graphics and more emphasis on data storage, analysis, and reporting. In AM/FM, relationships among utility system components are defined as *networks*. Simply stated, GIS emphasizes location and topology, whereas AM/FM emphasizes database and connectivity.

Table 4.1 presents a comparison of CADD, facilities management, GIS, and AM/FM/GIS systems. Figure 4.3 shows the difference between automated mapping, facilities management, AM/FM, and GIS and how they are combined to create an integrated AM/FM/GIS system.

In distinguishing between various systems, certain characteristics should be kept in mind. For instance, if a system can identify all the catch basins flowing into a combined sewer outfall, or the location of the worst sewer pipe, then it is a GIS; otherwise, it is probably a CADD or automated mapping system. Additionally, a system

Table 4.1 Comparison of CADD, facilities management, GIS, and AM/FM/GIS*.

Feature	CADD	FM	GIS	AM/FM/GIS
Map layers	Y	N	Y	Y
Topology (spatial relations)	N	N	Y	Y
Network connectivity	N	Y	Y	Y
Data attributes	N	Y	Y	Y
Positional accuracy	Y/N	N	Y	Y
Map intelligence	N	N	Y	Y
Asset management	N	Y	N	Y

* CADD = computer-aided design and drafting; FM = facilities management; GIS = geographic information system; AM/FM/GIS = automated mapping/facilities management/geographic information system; Y = yes; N = no; and Y/N = can be both yes and no.

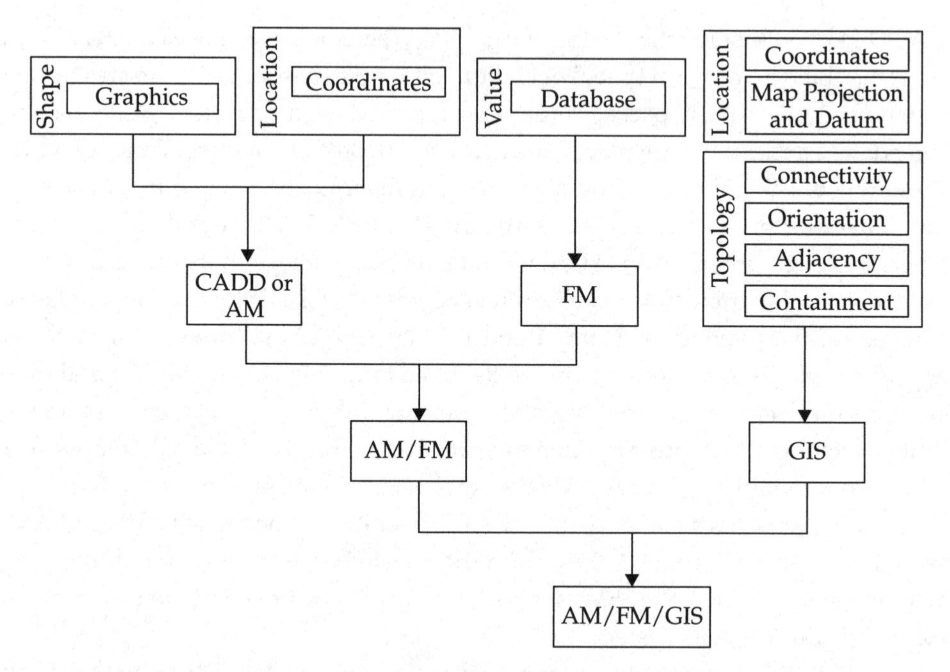

FIGURE 4.3 Comparison of CADD, facilities management, and GIS and evolution of AM/FM/GIS as the integration of CADD, facilities management, and GIS (Shamsi, 2002a). Reprinted with permission from ASCE.

has facility management capability if it can prioritize the work required to bring the worst pipe up to a minimum operating standard.

3.7 Customer Information Systems

Customer information systems (CISs) manage information about the services or laterals serving a property and the customers associated with the service. There may be multiple services to the property, each with separate accounts, use types, and responsible parties. In addition to information about the property being served, such as the address and type of use (residential, commercial, industrial, etc.) and the names of owners and tenants, a CIS may include information on the lateral or service pipe (i.e., size, material, and date installed), the main it is connected to, and the location of the connection, curb stop, and meter. Also included in the CIS is information on the utility's work history at the site and customer contact history. Typically, there will be feature identification numbers assigned to one or more appurtenances, such as the tap or connection, and to all meters on the property. Employees using the system can access the correct record for a customer by entering the identification or the address, or by cross-referencing billing records. Information is somewhat transient, with numerous changes occurring annually due to replacement of old taps or service pipes, new installations, and changes in property ownership or tenancy.

Primary users of CISs are customer service representatives because they manage accounts. New accounts need to be established when a new service is requested. All changes in the account or pipe must be entered, such as changes in property ownership or tenancy, replacement of the tap, and change in the type of service. The system will keep track of all call history, work orders, and comments related to the service, customer, or property. Customer service representatives generate a large share of work orders or service requests. Therefore, if the work order processing is also automated, there likely is a link between the two systems to allow representatives to automatically generate work orders or service requests. The CIS is also used by field service personnel when installing or repairing service pipes, marking the facilities in response to "one-call" or "call-before-you-dig" notices, meter setting or replacing, and by the supervisors managing those activities. Both the customer service representative and field personnel need maps of pertinent call or work locations. Therefore, linking GISs and CISs is a common integration.

4.0 BUILDING AN INFORMATION MANAGEMENT SYSTEM

Cooperation among various departments is needed to establish an IMS because every department has different needs. For example, mapping needs CADD/GIS to draw maps, engineering needs CADD for design, maintenance needs maps and work orders to plan work, asset management needs a register of assets and history of maintenance, and finance needs the CIS to send out utility bills. All these information sources are often kept in separate departments, with different systems and different levels of detail. The key step in establishing an IMS is getting these different departments to work together.

Once the premise and rationale for an IMS has been established and approved, the process of building it can begin. This process involves an understanding of the types of data associated with collection systems, attributes that need to be captured, and reports that are generated.

4.1 Types of Collection System Information

Building IMS for a collection system involves an understanding of the different types of data found in a typical system. In analyzing the information management needs of a utility, the manager should be familiar with basic types of data that are typically associated with collection systems. Collection system information can be categorized into the following principal types:

- *System data* are the heart of collection system information, upon which all other data are dependent. System attributes consist of an inventory of a collection system's assets, its maintenance, repair, inspection, and condition assessment. Customer service records and the history of complaint resolution are also components of system data. (Failure to maintain customer service data may invite litigation as a result of claims for damages.)
- *Spatial data* describe the location of system structures with respect to real world geographical references, typically survey coordinates, and are often derived from global positioning satellite (GPS) readings. Spatial data are the backbone of GISs, which are used to create electronic maps of collection systems and to link data to them.
- *Dynamic data* are the reporting of real-time data that describes how the collection system is functioning. The most common dynamic data are pumping station activity and alarms reported by supervisory control and data acquisition (SCADA) systems, flow metering results, and rain gauge readings.
- *Modeling input data* (e.g., loadings and pump curves) and *output data* (e.g., depth, velocity, and overflow location) are used to determine existing and future hydraulic capacity of a collection system. Input data are derived from GIS, CADD, and/or CMMS programs that combine inventory parameters, such as pipe size, with topographic data, such as slope, with flow monitoring and rainfall data and modeling results to profile the system's ability to handle flow during dry and wet weather events.
- *Asset value data* represent calculation of the monetary value of the collection system, the cost to rehabilitate or replace it over its lifetime, and the revenue demands necessary to support those capital improvements. There is a close link between how the system is maintained and the length of its life expectancy.

4.2 Collection System Data Types

There are two types of collection system data; (1) spatial data and (2) attribute data.

4.2.1 *Spatial Data*

At a minimum, the following spatial data sets should be compiled into the IMS for a sewer overflow study:

- Basin topography, including watershed and sewershed boundaries;
- Land use and land cover;
- Facility and outfall locations;
- Hydrologic features;
- Watershed data (population, land use, and impervious cover);
- Industrial user inventory; and
- Field screening locations.

Sources for supplying necessary data must first be obtained and then scrutinized for their suitability to needs of the study. Knowing the age and growth pattern of the area under investigation will aid the search for map source data. Any existing reports describing drainage areas or results of prior compilation of drawings should be reviewed for possible incorporation to the mapping and information management strategy. Wherever possible, duplication of effort should be minimized.

City and regional planning commissions are typically good sources for locating land-use data. As such, these data are readily incorporated to a GIS-based IMS. These data can be used to develop estimates of impervious cover for a project area. Photo-interpretation of aerial photography can provide similar results, but requires higher levels of effort to digitize individual land uses from the photographs. Satellite imagery is also available at several resolutions, and costs of the technology have steadily declined. Multispectral scanners used in conjunction with aerial photography or satellite images can take most of the guesswork out of generating impervious cover layers through photo-interpretation. More advanced digital methods can be used repeatedly throughout facility planning (e.g., for site selection). If no, or aged, photographs exist, the additional expense of procuring new aerial photographs or satellite imagery is recommended.

4.2.2 *Attribute Data*

Table 4.2 summarizes some of the most common types of "system data", as defined in Section 4.1, along with the frequency of change and the typical systems of record where the data would be managed.

A collection system manager should carefully consider how resources should be best used to capture, store, manage, and report various data attributes and related collection system information. There is always a level of effort and cost involved in the initial data conversion and population of the IMS along with the ongoing cost of maintaining the information. Physical attributes, which change less frequently, typically carry a comparatively higher initial data conversion cost and a lower long-term cost. Physical data for vertical assets, such as pumping stations, will generally have higher long-term costs than physical data for the linear pipe assets because the pumping equipment is shorter-lived and, therefore, is rehabilitated or replaced more often, resulting in the need for data updates. Data related to condition assessment, work orders, and customer complaints are typically much larger in volume and change more frequently than physical attribute data. The total cost of this large volume of data is almost entirely the result of update processes. Overall implementation costs for an IMS, which include data conversion and loading efforts, are typically three to

TABLE 4.2 Common system data types.

System data types	Typical examples	Frequency of change	System of record
Linear asset physical attributes	Pipe material, length, invert, manhole rim elevation, depth	Very infrequently	GIS
Vertical asset physical attributes	Pump capacity, type, motor horsepower	Infrequently	CMMS
Condition data	Condition assessment results from *Pipeline Assessment Certification Program* (NASSCO, 2001) and *Manhole Assessment Certification Program* (NASSCO, 2006)	Varies with inspection frequency	CCTV software or CMMS
Operations and maintenance data	Work order history	Frequently	CMMS
Customer service	Complaint history	Frequently	CIS or CMMS

five times the actual software cost. Therefore, the goal should be to gather a sufficient, but not overwhelming, amount of information to allow a reliable understanding of the collection system to meet operational and regulatory goals. The following subsections outline key issues to consider, which will vary by utility size and complexity.

4.2.2.1 Reporting Needs

A careful review of current reporting needs is the best place to begin defining the attribute data required from the various information systems. Documenting data requirements and sources for all ongoing daily, weekly, monthly, and annual reports will establish the minimum input requirements for a successful implementation and will focus initial efforts on gathering the highest-value information while eliminating information of questionable value. Attention should be paid to both internal and external reporting. Often, especially in larger organizations, the volume of internal reporting is much higher and the reports are more data-intensive than reports produced for external purposes (e.g., regulatory reporting). Regardless of utility size, the collection system manager should consider involving other departments, such as planning and engineering departments, in the review process such that the data needs of the entire organization are considered at the earliest possible stage in IMS development.

4.2.2.2 Initial Data Population

Requirements for all initial data population should be carefully evaluated and documented based on minimum data requirements from the reporting review. For example, successful GIS implementations require high-quality spatial and attribute data population. Similarly, successful CMMS implementations depend on a well-defined asset hierarchy along with the population of numerous look-up tables related to work order types, inspection codes, failure codes, and so on. Planning for the initial data population should consider data sources, conversion requirements, loading, QA/QC review, and final acceptance. Resource planning and scheduling are also important at this early stage.

4.2.2.3 Standard Operating Procedures

Maintenance, inspection, and condition data all rely on effective work practices to keep information current and accurate. Therefore, identification and early review of standard operating procedures (SOPs) for crucial work practices such as planning and scheduling, closed-circuit television (CCTV) inspection, general work order management, and failure analysis for overflow reporting should be performed to identify any gaps in data collection that need to be addressed as part of IMS implementation. For smaller systems, written SOPs are probably sufficient to document the full work processes, including all data inputs and outputs. For larger organizations, however, review of written SOPs may not be sufficient to fully document all the data flow processes (e.g., inputs, outputs, and review and approval) associated with key work practices, particularly if those practices span multiple departments. In this instance, use of business process modeling should be considered. In addition to documenting data flow, results of the initial business process modeling can be used later in the configuration and training stages of IMS implementation.

4.2.2.4 Physical Attribute Data

Physical attribute data include all the information necessary to describe and characterize the various collection system components, including all linear and vertical assets. Collection system linear assets include all the buried infrastructure components such as gravity mains, gravity main fittings (tees, junctions, wyes, bends, etc.), manholes, overflow structures, cleanouts, service laterals, catch basins, stormwater inlets, septic tanks, force mains, and force-main components (e.g., air-release valves and other buried valves). Attribute data for linear assets are generally managed in the GIS. Table 4.3 summarizes some of the most common collection system asset components and associated attribute data.

A useful tool during the planning process is developing a simple spreadsheet that lists all the known types of components present in the collection system, the quantity of those components, and the required attribute data and the potential sources for

Table 4.3 Typical collection system assets and attributes.

Gravity main	Force main	Manhole
Install date and cost	Install date and cost	Install date and cost
Size, material, and length	Size, material, and length	Material, size, and depth
Upstream and downstream inverts	Flow direction	Rim and floor elevations
Slope and flow direction	Valve type, size	Type (sanitary, combined, etc.)
Type (sanitary, combined, etc.)	Cleanout-type, size	Connection sizes and locations
Cross-section shape, height, width		

the data (electronic CADD files, paper record drawings, sewer atlas sheets, previous sewer system evaluation survey reports, work order history, inspection reports, etc.). The spreadsheet can then be used to estimate key metrics for planning the data conversion and update process, including (1) the percentage data gaps for each component by attribute type, (2) the potential level of effort to fill the data gaps from available sources, and (3) the potential magnitude of a field data collection program if data are not available from current sources. The overall data conversion effort can then be scheduled and prioritized to best meet the utility's needs and budget. A 3-to-5-year, or longer, data conversion process is not uncommon depending on the size and complexity of the system.

In addition to typical physical attribute data, all collection systems have unique characteristics related to such issues as access restrictions, known problem areas, or safety, which can influence the overall operations and maintenance of the system. For example, a collection system manager should ask the following questions:

- Does a structure have an exceptionally hazardous aspect, such as a nearby industrial waste discharger?
- Is it a gravity main of relatively flat slope, which is more subject to grease or other buildup, thereby requiring more frequent cleaning?
- Is it a work-order location that is known to be barricaded and, therefore, more hazardous to the public?
- Is it a structure that is privately owned or maintained and is, therefore, not the utility's responsibility?
- Is unusual traffic control, special equipment, or other special actions required for maintenance?
- Is a structure's construction guarantee still in effect when a repair is required?

Modern GISs, with the use of a relational database, provide almost unlimited opportunity to capture the unique and high-value institutional knowledge about the collection system, which serves to make day-to-day operations and maintenance more effective and efficient. Therefore, the planning process should consider interviews or workshops with a cross section of staff to capture, categorize, and prioritize this type of information for inclusion in overall IMS implementation. Such discussion at the earliest stages in the IMS development process also provides the additional benefit of building buy-in from staff, which is crucial for successful implementation.

Collection system vertical assets include items such as pumping stations, retention basins, regulator structures, and so on. The vertical asset locations ("pumping station no.1", "regulator structure ABC", etc.) are generally represented as point features in the GIS, with associated assets (pumps, gates, valves, etc.) and attribute data managed in the CMMS. Exceptions are sometimes made for smaller collections systems, wherein an individual asset, such as a pump, is listed in the GIS with a minimal amount of attribution. However, vertical assets are most efficiently represented within a CMMS using a standard parent–child asset hierarchy. The design for the vertical asset hierarchy and attribution of the collection system is recommended to follow any standards and conventions in place for the wastewater treatment plant assets to maintain consistency across the utility.

4.2.2.5 Financial Data

Financial data include all information related to the installed cost, calculation of depreciation, and the estimate of replacement cost. The installation date and installation cost are essential financial data attributes in support of critical asset management functions related to valuation, depreciation, and capital planning. The decision of where to store this financial data is not always straightforward, with the most common choices being the CMMS or financial information system . The approach can vary with factors related to utility size, type, and associated management policies. Access to this data is important for effective collection system renewal and replacement planning, which is a key process in the control of system overflows. Therefore, management of this data should be discussed and a policy developed early in the IMS planning stage. The preferred, most straightforward approach for collection system management would be to store the install date and cost in the CMMS with support for data access and reporting directly from the CMMS, and also via the CMMS/GIS integration.

4.2.2.6 Condition Data

Condition data include the scoring results of all inspections, which follow a defined standard for scoring of defects. Examples of standards for scoring defects include the National Association of Sewer Service Companies' *Pipeline Assessment Certification Program* (NASSCO, 2001) and *Manhole Assessment Certification Program* (NASSCO, 2006), which outline standards for gravity main- and manhole-condition assessments. The assessments are generated from an inspection process involving CCTV, or a similar technique, combined with software conforming to the assessment scoring standard.

The final results of a condition assessment are often transferred, in whole or in part, from the assessment software to the CMMS to enable further analysis and planning. Integration of the CMMS and GIS further supports the analysis process such that mapping of system-wide conditions and specific problem areas within the system is readily available. Deciding on how much condition data to transfer to the CMMS is critical, as the volume of data can be quite large. For example, the current *Pipeline Assessment Certification Process* can produce assessment data for a single gravity main in up to 12 structural defect groups, five operational and maintenance defect groups, four construction feature groups, and several other related scores. The data transfer from CCTV software to CMMS software is generally dependent on the following two factors: (1) both software vendors must support the most current *Pipeline Assessment Certification Process* data standard and (2) a contractual relationship often exists between the two software vendors such that they agree to mutually support data export and import functions between the two products. Smaller utilities may find the added level of complexity involved in the CMMS condition data import to be a burden; consequently, they may be able to perform acceptable sewer planning using the data and video viewing tools built in to most CCTV software products. Conversely, larger utilities will generally benefit from importing the inspection data to the CMMS, which creates the opportunity to simultaneously report and analyze condition data integrated with work history and the added advantage of GIS-based analysis via the GIS and CMMS integration.

In addition to the structural, operational, and construction defects rated by the *Pipeline Assessment and Certification Process* and the *Manhole Assessment Certification Process*, a utility seeking to make the most of collection system management, including minimizing overflows, will benefit from detailed knowledge of hydraulic performance, or hydraulic condition, which is available from the utility's collection system hydraulic model. It is recommended that the planning stages for an IMS include provisions to capture hydraulic performance. Similar to the approach for capturing unique physical data about the collection system, modern GIS architecture can readily incorporate data fields relating the normal, peak, and wet weather flow capabilities of the collection system.

4.2.2.7 Criticality or Consequence Data

Criticality or consequence data include results of evaluations to rank the consequence of failure for the particular collection system asset. The analysis often involves a complex GIS-based analysis, with a variety of factors related to the economic, social, and environmental consequences of failure, generally referred to as the "triple bottom line" approach. The results of the consequence evaluation for the collection system linear assets can be retained in the GIS to support maintenance optimization and capital planning.

4.2.2.8 Operation and Maintenance Data

Operation and maintenance data include all scheduled maintenance activities along with the complete work history (predictive, preventive, corrective, and emergency) for all assets comprising the collection system. Generally, the CMMS will be the repository for the maintenance and work history related to the collection system. Typical examples of this include

- Scheduled preventive maintenance (cleaning, flushing, roots, etc.),
- Scheduled corrective (repair) work orders,
- Scheduled inspections (condition-assessment activities), and
- Unscheduled and emergency work orders.

Depending on business practices in place for the utility, the CIS may also contain relevant work history generated in response to customer calls.

The CMMS uses the combination of attributes of physical data and work history data to provide reporting on key performance indicators relating to maintenance activities and effectiveness. Examples of this include

- Weekly, monthly, and yearly maintenance and crew productivity reports for all work order types (miles of pipe inspected, miles of pipe cleaned, etc.) and
- Planned versus unplanned maintenance ratios.

4.3 Guidelines for Establishing an Information Management System

Developing and implementing an IMS requires careful planning based on an in-depth analysis of user needs. Some useful strategies to accomplish this are presented in the following section.

4.3.1 Information Management Strategies

Because of complexities involved in studying even small collection systems, establishing a means for ongoing tracking of the physical and operational status of the collection system is an important objective in cost-effective sewer overflow pollution abatement planning. With advances in capabilities of personal computers over the past decade, low-cost, user-friendly tools are now available for consolidating information available from existing geographical, tabular, and field sources. Modern information systems offer a convenient means of organizing the multitude of data sources into a readily accessible file management environment, creating a tool for the capture and management of the following essential information:

- Maps and associated descriptive data for key facilities and sewer system,
- Input data for hydraulic and hydrologic models,
- Tracking of record drawings describing current status of collection and treatment systems,
- Tracking of system modifications made through operational and capital improvement projects,
- Archiving and management of historical operating data describing key facilities,
- Quantity and quality of industrial pretreatment data for industrial users,
- Tracking of sampling data for industrial pretreatment programs,
- Maintenance log data for key facilities,
- Baseline system data for maintenance management systems, and
- Regulatory report data.

4.3.2 *System Selection and Implementation*

Like all organizations, utilities face a daunting challenge in selecting and implementing the best, most cost-effective technology solutions to support their objectives. The focus of this manual is specific to one objective, that is, the prevention and control of sewer system overflows. However, it is neither practical nor cost-effective to implement an IMS solely to control overflow. Rather, overflow prevention and control will be one of many outcomes realized through sound collection system management practices, which, in turn, are supported by an effective IMS. Therefore, a holistic approach, as illustrated in Figure 4.4, wherein all the potential benefits, risks, and cost factors are identified, evaluated, and prioritized is critical to achieving the highest value and return on investment from the implementation of an IMS.

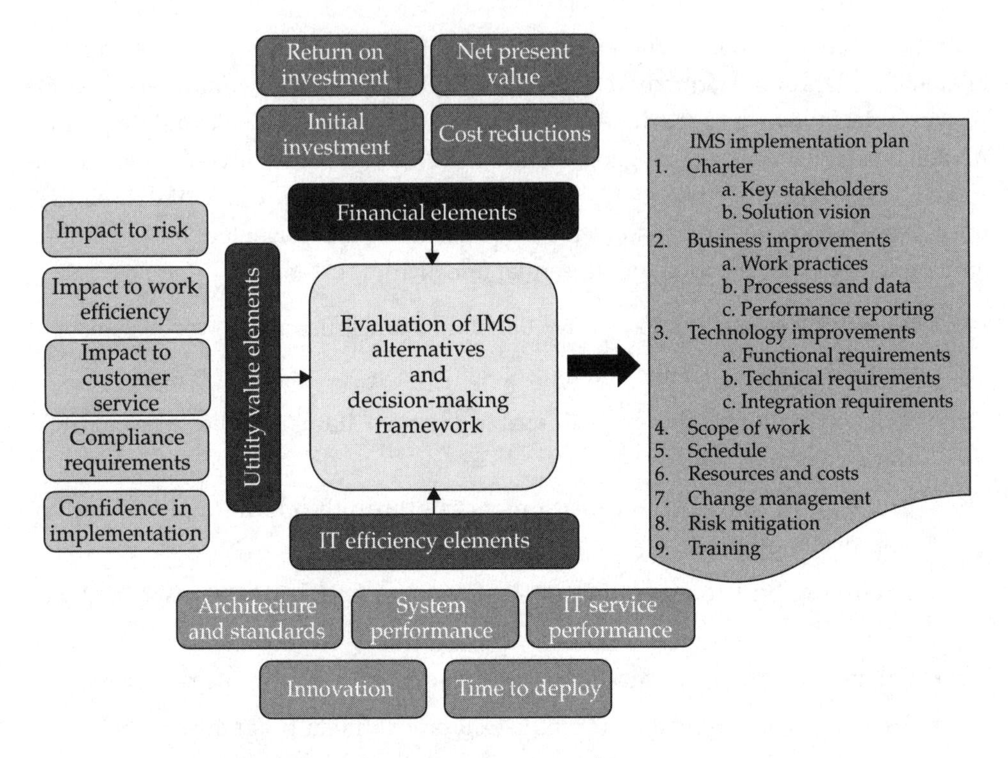

FIGURE 4.4 Holistic IMS implementation approach.

For example, year 1 of the IMS implementation may focus on overflow prevention, year 2 on maintenance optimization, and year 3 on capital planning support. Utilities that focus on the long term will be the best positioned to receive immediate and continuous benefit from their IMS investment.

4.3.2.1 Conducting a Technology Needs Assessment

4.3.2.1.1 Initial Planning and Information Management System Charter

Successful IMS implementation begins with comprehensive upfront planning, which is designed to identify and prepare for organizational, business, and technology challenges. A team approach, which involves the appropriate people within the utility, can effectively deliver an IMS from planning through implementation, thereby providing value and aligning with the utility's mission. Regardless of utility size, successful initial planning should result in an "IMS charter" that documents findings and conclusions in the following key areas:

- *Goals, objectives, and measures* define the following: (1) "what" the utility needs to accomplish with the IMS, (2) "why" it is necessary to do so, and (3) the criteria to be used for measuring success. Prioritizing the utility's goals and objectives in two or three phases (e.g., immediate, short term, and long term), with specific performance measures, can be a significant benefit in the selection and implementation of an IMS. For example, prioritizing goals and objectives allows the IMS evaluation and selection process to place a higher weighting on those specific system features and functions that are the most beneficial to the utility. Similarly, the implementation planning process can identify specific data collection and system configuration (work process, reports, etc.) that must be accomplished early in the overall implementation.
- *User groups* include all the people who will use the IMS. With larger utilities, user groups can be very diverse, including, for example, operations, maintenance, planning, engineering, customer service, finance, and information technology. Utilization typically ranges from "power users", who interact with the IMS every day, to "casual users", who may only receive monthly or quarterly reports. The role of each user group in meeting the utility's goals and objectives for the IMS should be defined. Additionally, the role of each user group in the evaluation, selection, and implementation of the IMS should also be defined. The overall IMS implementation team can then be defined in the charter, including the appropriate representation from the various user groups.
- *System ownership and administration* defines the roles for IMS ownership, including data, user management, system integration, and overall maintenance. The larger the organization, the more complex these roles could become. For example, with a large collection system, different groups may be in charge of pumping station maintenance versus pipeline maintenance. In this situation, ownership of attribute data for the vertical assets could be different than for the linear assets. Regardless of their size, wastewater utilities that are part of a larger utility department or municipal government may need to consider questions such as system administration and data sharing with one or multiple information technology departments. Establishing ownership of key system integrations, particularly with GIS and any existing legacy systems, is critical to successful IMS implementation for wastewater collection system management.

- *Work practices* include all the functions performed by the user groups that are necessary to accomplish day-to-day activities and meet the utility's goals and objectives. At this stage in the IMS planning process as part of the charter, it is beneficial to evaluate work practices in terms of the following: (1) frequency, which defines how often the activity is performed; (2) resource requirements, which defines the magnitude of resources (e.g., people) involved in the activity; and (3) data intensity, which defines the type and quantity of data necessary for the activity and the data generated as a result of the activity. Work practices can then be further defined in the needs assessment process.

Smaller utilities may be able to complete upfront IMS planning in less than a month, while larger utilities may need to devote several months.

4.3.2.1.2 Matching Technology to Collection System Management Goals

Many different technology solutions can be involved in day-to-day functions, depending on the size and complexity of the utility's operations.

The most common IMS technology solutions involved in collection system management for the prevention and control of overflows include

- *Geographic information systems* for spatial data management, analysis, and planning;
- *Computerized maintenance management systems* for work order management and data management related to asset condition and useful life;
- *Supervisory control and data acquisition* for monitoring and reporting of flow data and alarms;
- *Closed-circuit television* software for field data collection and scoring of gravity main and manhole conditions; and
- *Hydraulic capacity model* software for predicting and investigating collection system performance under different flow conditions.

The role and importance of each IMS technology solution can vary depending on the size, complexity, and specific needs of the utility. Other systems, which may have a role in the control and prevention of sewer system overflows, include the following:

- *Customer information system* for responding to customer calls and complaints (e.g., water in the basement);
- *Laboratory information managements system (LIMS)* for recording, managing, and reporting on water quality or similar regulatory requirements related to overflow monitoring;

- *Pretreatment information management system (PIMS)* for managing flow and discharge information from permitted industrial users. A PIMS is sometimes a stand-alone application or can be part of the LIMS software package; and
- *Electronic document management systems* for managing content related to the sewer system including standard operating procedures, as-built drawings, vendor equipment manuals, and so on.

The role of these secondary systems also varies greatly with such factors as utility size, complexity, and needs. For example, if the reduction of "water-in-basement" (WIB) events is a high priority, the CIS may play a significant role. Similarly, if water quality monitoring and reporting is a requirement, the LIMS and/or PIMS will have a significant role. Generally, the IMS technology solutions summarized in Table 4.4 will meet the majority of needs involved in collection system management in support of the control and prevention of overflows.

4.3.2.1.3 Defining Functional and Technical Requirements

The software marketplace offers an array of products with wide ranges in capabilities and costs. The field can be quickly narrowed to the most viable alternatives if care is taken upfront to define the scope and breadth of the utility's needs and then match those needs to software function. For example, the CMMS tends to act

Table 4.4 Computerized maintenance management system requirements for collection system management.

Example CMMS requirements – Collection system management		
CMMS requirements	**Priority**	**Comment**
Functional requirements		
Asset inventory	High	Support the full attribute information necessary for vertical and buried assets; support assignment of criticality to assets
Condition assessment	High	Support condition scoring for vertical assets; support for use of condition standard (e.g., *Pipeline Assessment Certification Program* [NASSCO, 2001] and *Manhole Assessment Certification Program* [NASSCO, 2006]) for pipe/manhole inspections

(continued)

TABLE 4.4 (Continued)

CMMS requirements	Priority	Comment
Work orders	High	Work order classification, scheduling, tracking, and reporting
Failure modes	High	RCFA capabilities
Asset cost tracking	Varies	Assignment of work costs (labor, materials, etc.) to assets
Remaining useful life	Varies	Required for asset management and for depreciation
Customer call-ins	Varies	Tracking of work orders created as a result of a customer call-in
Inventory management	Varies	Inventory management of parts and consumables
Purchasing	Varies	Generating and tracking of purchasing orders for internal cost accounting
Fleet management	Varies	Support for trucks, equipment, and other mobile assets
Technical requirements		
Open architecture	High	Non-proprietary database and published API for integration
Vendor strength	High	Training and ongoing support capabilities
GIS integration	High	The CMMS must support integration with market-leading GIS software
Performance reporting	High	Predefined reports and ad-hoc reports are required
Application security	High	Specify varying levels of access by user and by user group
User-defined fields	Varies	Asset and work order data fields to capture additional information
CCTV integration	Varies	Integration with the most popular CCTV inspection packages
SCADA integration	Varies	Integration with flow and alarm data for work order generation
Other integration	Varies	Support for integration with utility systems such as CIS or FIS
Web-based access	Varies	Interface functionality (e.g., view only vs edit) should be considered
Mobile computing	Varies	Typical options include wireless vs "dock and synch"

as the focal point of the IMS solution supporting collection system management. A capable and well-implemented CMMS not only tracks the day-to-day activities and changes to a utility's assets, but coordinates those activities through preventive maintenance schedules and resource allocation tools in addition to supporting capital planning.

The requirements for CMMS, or any software solution, can be divided into two fundamental categories, functional and technical, which can then be used to develop a "system scorecard" for use in the evaluation of both new and existing technology solutions. The following list summarizes considerations for software functional requirements, technical requirements, and overall cost:

- *Functional requirements* are defined as the ability of the CMMS package to support the utility's staff and overall work practices for maintenance, asset management and capital planning. Depending on utility size and complexity, these work practices may only include wastewater collection or they may extend to wastewater treatment. In the case of municipal utilities, the ultimate needs may extend beyond wastewater to include other utilities and potentially other city departments (customer service, fleet maintenance, parks, streets, etc.) should a city decide to expand use of CMMS in the future.
- *Technical requirements* are defined as the technical quality of the product (database design, user interface, GIS integration, reporting capabilities, mobile access, etc.) and the overall quality of the software vendor in terms of market strength, support, and training services.
- *Cost* is defined as the initial software licensing and implementation costs (e.g., installation, configuration, data population, and integration) plus the costs of training and ongoing support. Large, phased implementations should also consider the potential future cost of additional modules, as required.

The core functions required by a CMMS in support of collection system management can be generally summarized as follows:

- Vertical asset inventory and attribute data management for pumping stations and related installations;
- Condition assessment data management and reporting for all linear assets (e.g., pipes and manholes) and all vertical assets;

- Work order management including scheduling, tracking, cost accounting, and reporting;
- Failure tracking, analysis, and reporting; and
- Performance reporting against defined service levels for productivity, efficiency, and cost.

Computerized maintenance management system products are capable of performing a variety of functions related to maintenance, customer service, and asset management. Some of the more common functional and technical requirements typically considered in the selection of a CMMS are illustrated in Table 4.4.

Core requirements considered essential to collection system management are indicated as "high priority" in Table 4.4. Smaller utilities may only need to focus on these core needs, while larger utilities may need to consider some of the typical supporting functions, including expanding the list as required to define any additional needs. For example, large utilities, with a significant investment in trucks and other mobile equipment, may realize an advantage by also incorporating fleet management into the overall CMMS implementation.

Investing the time upfront and with the appropriate representation from all stakeholders in the utility is critical to defining and prioritizing all the functional and technical requirements for the CMMS.

4.3.2.2 *Evaluating New and Legacy Information Management Systems*

The first crucial step in the evaluation and selection process is developing a system scorecard from the technical and functional requirements previously discussed. The scorecard can then be used to facilitate the initial product screening and the final selection. This process, illustrated in Figure 4.5, which always ties directly back to the utility's prioritized needs, helps to achieve consensus among all stakeholders as to the best overall CMMS selection.

4.3.2.2.1 *Developing a System Scorecard*

The system scorecard for a CMMS evaluation can be quickly developed from the prioritized list of functional and technical requirements (refer to Figure 4.5) by first assigning a weight to each requirement. For example, high priority requirements could be weighted a "3", medium priority a "2", and low priority a "1". A simple scoring process can then be used with stakeholders involved in the evaluation and selection process as follows: 1 (does not meet the requirement),

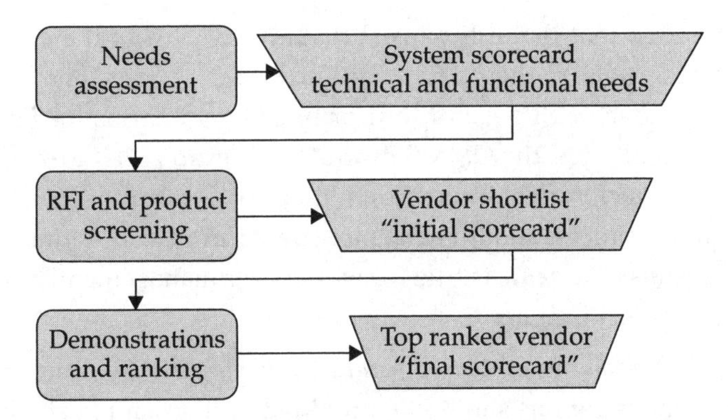

FIGURE 4.5 Computerized maintenance modeling system selection process.

2 (partially meets the requirement), or 3 (fully meets the requirement). The total score in each requirement is obtained by first multiplying the weight by the score from each stakeholder and then summing the results. Each candidate system can then be given two final scores, one for technical and one for functional. The scorecard can then be used to rank candidate CMMS products for consideration and final selection.

4.3.2.2.2 *Request for Information and Product Screening*

The competitive procurement process, which most utilities must follow under public bidding laws, typically begins with a request for information (RFI) or request for qualification, followed by a review and scoring of responses conducted by a selection committee. The system scorecard, which can be used by the selection committee during the RFI evaluation, fits very well into this process, with the outcome being a shortlist of qualified vendors.

4.3.2.2.3 *Demonstration and Final Ranking*

The software demonstration is the critical final step in the selection of a CMMS product. The following are recommendations for a successful demonstration process:

- *Consistent stakeholder involvement*—The same selection committee, which scored the initial short list, should also score vendor demonstrations.
- *Utility specific-use demonstrations*—Software vendors take great pride in demonstrating the "bells and whistles" of their particular product. However, it is

recommended that the utility insist that at least 50% of the product demonstration be specific to their data, work processes, reports, and so on. Vendors can be provided with a list of four or five key scenarios for how the utility plans to actually use the CMMS product. For example, if *Pipeline Assessment Certification Program* data import is a high priority need for the utility, then the actual import process should be demonstrated from start to finish and not just display results. The same is true for work order management, mobile computing, or any prioritized need.

- *Geographic information system integration demonstrations*—Vendors invited for CMMS demonstrations should be provided with a sample geodatabase for a portion of the utility's collection system service area so they can demonstrate the GIS integration capabilities for their product with live utility data. Using the utilities' actual GIS will make it much easier for the selection committee to evaluate as they will be familiar with the assets and underlying data. It is important that the functionality to add, edit, and delete assets (pipes, manholes, etc.) be clearly shown and defined for where and how each will occur. Additionally, it is recommended that the common routines for splitting and combining pipe assets be clearly demonstrated as this is often a stumbling block for CMMS/GIS integration, which can result in "orphan" records, particularly when work order history is involved.

The use of the system scorecard approach, linked back to prioritized needs, combined with product demonstrations specific to those needs, and using actual utility data, will provide high confidence in the final CMMS selection.

4.3.2.2.4 Evaluating Legacy Systems

In addition to the need to implement a new CMMS, utilities are often faced with the need to evaluate an existing, or "legacy", CMMS to determine if it is cost-effective to upgrade the product or if replacement with a new product is the better option. The aforementioned system scorecard approach works the same way for a legacy system, that is, by identifying and ranking key deficiencies in the areas of functional and technical performance. A legacy CMMS scoring well on technical requirements, but poorly on functional requirements, is an indication of misalignment between the CMMS and the business processes required by the utility. For example, the CMMS product may have the ability to manage asset condition data or failure-mode data, but neither is being done as part of the inspection or work order close-out processes, which result in limited ability to optimize maintenance and capital planning decisions. This could be addressed by improving those work practices, including the development of standard operating procedures and additional user training. Conversely, if those capabilities were not available in the legacy CMMS configuration, one solution could be the creation of user-defined fields to hold the required data.

Technical deficiencies determined in the evaluation of older CMMS products can involve key integration needs such as GIS. This is often the case when a utility desires to start using a CMMS currently in place at a wastewater treatment facility to also perform collection system management functions. If the legacy CMMS product does not support GIS integration, there are three basic choices: (1) perform a custom GIS integration, (2) replace the product such that there will be one CMMS product for the wastewater treatment facility and one for the collection system, or (3) implement a new product for sole use in the collection system, thereby running two separate CMMS packages. These options need to be carefully evaluated in terms of cost–benefit considering both technology costs and any potential changes in established work practices.

4.3.2.3 Managing a Successful Implementation

Companies that do not perform their own implementations typically will have business partners certified to do the software implementation. It is important for the utility to know if the software will be implemented by an entity other than the software company as this introduces another party to the contractual relationship. The method of implementation, along with specific personnel and qualifications for any third parties, should be part of the disclosure required during the RFI process.

Common options for managing system implementation include the following: (1) the utility can "self-perform" management and oversight of the software vendor or system integrator or (2) the utility can hire an independent consultant to perform that function. Factors in the choice include the size and complexity of the implementation along with the availability of staffing and technical resources. Regardless of the management approach, the key factor is a comprehensive implementation plan and schedule with key milestones for installation, initial configuration, data loading, initial testing, final data loading, integration, and testing. The roles and responsibilities of all parties involved in the implementation should be clearly delineated. Key components of the implementation plan should include the following:

- *Testing*—defining the requirements and expected outcomes of each phase of testing. Incremental testing during implementation is recommended to occur following initial installation and after each successive data load or integration. System-wide testing should occur over a period of several weeks with defined "scripts" for various users to test. Feedback sheets describing the scenarios and key success factors for each test are essential. Retests to address bug fixes and other issues should involve the same set of users for consistency.

- *Communication and change management*—identifying how the utility will modify or adopt any new business practices required for use of the CMMS. This is often the case with work management and inspection activities involved with collection systems. The needs assessment process should have identified where the changes need to occur. The communication and change management plan needs to define the steps and people involved. For example, the transition to import of CCTV data to the new CMMS could require new responsibilities for export from CCTV software, import to the CMMS, QA/QC, and generation of maps and reports. Any effects on data collection activities on the truck need to be defined and any required training needs to be identified and delivered before the new process is implemented.
- *Training and support*—identifying all the levels of training (e.g., power user, general user, and system administrator) required for the new CMMS. The vendor should be required to submit all training materials and credentials of the training professional for approval. The ongoing support agreement with the vendor (e.g., phone support and on-site assistance) should also include the option for refresher training, particularly in larger utility organizations.

No system will ever be perfect because it is impossible to account for all the possible variables, especially in a competitive procurement environment where the utility has limited control over the field of software candidates. Therefore, a postproject-implementation assessment is recommended after 3 to 6 months of use to document any outstanding issues (missing data, work process definition, reporting, etc.) that should be addressed going forward. This assessment process also helps sustain momentum and support for the new CMMS from utility staff.

4.3.3 System Integration

The two most common implementation options available in the market include "best-of-breed" and "enterprise". Best-of-breed refers to the approach whereby the best systems of a certain type (e.g., breed) are implemented separately and then integrated as necessary to achieve the defined needs. The term, *best-of-breed*, is typically used to describe general software solutions with wide applicability across many different markets (utilities, manufacturing, etc.), which perform key functions such as maintenance management, customer service, billing, GIS mapping, and so on. Specialized software tools within an industry, such as hydraulic models and CCTV, in the case of the utility industry, are generally not referred to as best-of-breed. The best-of-breed approach scales well to utilities of all sizes for the implementation of technology solutions.

Enterprise, sometimes called *enterprise resource planning* or *enterprise asset management* (EAM), refers to the approach where a single product or suite of products from the same vendor is implemented to meet multiple needs. The enterprise approach originated with the market for large corporate software suites, whereby critical business functions such as inventory, purchasing, sales, human resources, and, most recently, maintenance were integrated into a single software product or group of related products (e.g., modules). The enterprise approach, which can deliver greater economies of scale, is generally more viable for larger utilities or for utilities that are part of a larger entity, such as city or county municipal governments.

Regardless of the approach selected, a successful GIS integration is important to effectively managing collection-system buried assets. The approach to GIS integration varies among the EAM and best-of-breed CMMS vendors, and should be carefully considered when making the final software selection.

Effective collection system management also benefits from highly specialized information sources such as CCTV and hydraulic models. Therefore, some level of data integration is required to bring over condition and capacity information. Supervisory control and data acquisition is another major data source used for collection system management. Similar to CCTV and hydraulic models, some level of integration is required to bring over the flow and level data. Collection, storage, and reporting of flow data from remote monitoring locations represent one area where companies offering application service provider (ASP)-based solutions have established some presence in the utility software solution marketplace. In the ASP approach, the utility pays to access software via the Internet, which is remotely hosted and managed by a third party.

The final decision on EAM versus best-of-breed often cannot be based on a single need, such as the control and reduction of sewer system overflows, or even collection system management in general. The best decision will be arrived at when considering the broader information management needs across the utility. Often, an evaluation of available software offerings from EAM and best-of-breed vendors will determine that the smaller, more task-focused best-of-breed solutions actually perform key functions for collection system management, such as CCTV data import, more efficiently than larger, more complex EAM solutions. Conversely, a large utility seeking to maximize economies of scale with its overall information technology approach could benefit from the more comprehensive EAM solution.

5.0 INFORMATION MANAGEMENT SYSTEM APPLICATIONS

An IMS can be effectively used for performing various sewer overflow management tasks. This section provides the IMS applications for the following sewer overflow-related activities:

- System inventory and characterization,
- Hydraulic characterization and modeling,
- Asset management,
- Work order management, and
- Regulatory compliance.

5.1 System Inventory and Characterization

U.S. EPA's national CSO policy (U.S. EPA, 1994) requires that CSO dischargers perform a system inventory and hydraulic characterization as part of their National Pollution Discharge Elimination System permit compliance programs. Detailed hydrologic and hydraulic (H&H) models of the sewer system and tributary drainage areas are typically developed to help understand the existing system, to characterize CSO frequency and volume, and to evaluate the effectiveness of alternative interceptor and wastewater treatment plant operating scenarios. The models also are used in later phases of the compliance program to document the nine minimum controls and develop a long-term control plan. U.S. EPA's CMOM program for SSO systems, described previously in this chapter, has similar system inventory and characterization requirements.

There are several factors to consider in using and managing descriptive information. The cataloging procedure should be decided before starting. Establish a base map and adopt a uniform numbering and plotting system. Progress made in locating drawings should be clearly evident from viewing the centralized map. As a result, gaps in locating information will become readily apparent. Interviews with existing and former personnel are warranted to determine if gaps exist. For example, someone remembers there was flooding in a basement used for storage, thereby destroying some as-builts, or a former municipal engineer or consulting engineer has copies of records. Care should be taken to ensure the accuracy and completeness of the information and descriptive data collected. Having to re-collect invalid data that was poorly screened at the time of collection can have serious cost repercussions if erroneous data are allowed to propagate through subsequent phases of the project.

Differentiation between design drawings and as-builts is essential. Dates on drawings or surveyor's notes must be clear. Datums used in drawings, especially vertical datum, should be documented. There will be several sets of drawings per construction job. It is essential that the latest modification and most recent record drawing be located for an accurate description. The procedure for accounting and tracking this information should be detailed as part of the information management strategy. Preservation of record drawings is recommended. Once it is agreed to incur the expense of a comprehensive search, documenting findings and preserving drawings should be included in the scope of the work. Both manual methods and computerized methods are available for accomplishing this objective. With steadily improving scanning, automatic digitizing, and computer technology, compilation of geographic and tabular data in digital format is rapidly becoming standard practice. Additionally, computer-based methods can be stored on media that have lifetimes far exceeding that of manually prepared media. This greatly reduces storage space required, and the inherent reproducibility of digital media greatly diminishes the likelihood of catastrophic loss.

Budgetary and time constraints and preference may dictate that manual methods be used to perform system mapping. In this instance, some relative tradeoffs are provided. Typically, if only original tracings or faded blueprints are available, production and use of full- and half-size Mylar copies will guard against fading, wear and aging of paper, and damage caused during handling and printing. Use of microfilm or microfiche is common, and available commercial services should be consulted to determine what best suits the particular needs of a facility. Microfilm is better suited for drawings and has better resolution. Appropriate equipment for viewing and printing should be provided at each location where copies are stored for use. A detailed index with correspondingly labeled aperture cards or multifilm jackets enables easy access. Understanding design terminology in use during different eras will expedite interpretation of drawings and eliminate misunderstandings during subsequent phases of facility planning.

Using the aforementioned guidelines, a system characterization and summary report can be prepared to describe the existing system and network of separate, combined, and intercepting sewers in relation to treatment plants and other key collection facilities. This reference can serve as a valuable tool throughout CSO and SSO abatement facility planning. Functional schematics, including key system components (e.g., regulators, pumping stations, and hydraulic control points), will uncover misconceptions about how a system functions hydraulically. Computer-based strategies offer great flexibility in presenting the data. For example, schematics can be centered around a specific treatment plant's service area or a watershed. The report would contain these schematics, showing the network at a glance and tabulation of available information (drawings, reports, and data), indicating how information was prepared and where it can be retrieved.

The extent of a summary report varies with the size of the drainage area and level of accurate documentation already available. This comprehensive review will highlight where field investigations need to be focused and define the hydraulic and geographic nature of the project.

A GIS can be effectively used to store, manage, and display spatial information for creating an inventory of the sewer service area. Most common features included in the inventory are sewer pipes, outfall pipes, manholes, catch basins, inlets, regulators, and pumping stations. These features are overlaid on a base map that typically shows aerial photographs (orthophotos), political boundaries, watershed and sewershed boundaries, receiving waters (streams, rivers, lakes, and oceans), streets, and parcels.

Before in-depth monitoring and modeling efforts begin, however, the permittee should assemble as much information as possible from existing data sources and preliminary field investigations. Such preliminary activities will contribute to a baseline characterization of the sewer system and its receiving waters and help focus the monitoring and modeling plan. Physical characterization of a sewer system includes identification and description of all functional elements of the system and sources discharging into the system, delineation of the system drainage areas, analysis of rainfall data throughout the drainage area, identification of all outfalls, and preliminary system hydraulic analyses (U.S. EPA, 1999).

System inventory can be conducted by digitizing the as-built drawings and/or conducting a field survey. For example, the following data are collected for a manhole:

- Pipe invert(s) elevation into the manhole (two elevations are needed for drop connection pipes),

- Pipe invert elevation out of the manhole,
- Rim elevation,
- Pipe diameter into the manhole,
- Pipe diameter out of the manhole, and
- Manhole construction material.

The New York City Department of Environmental Protection (NYCDEP) developed enterprise GIS for the entire New York City five-borough sanitary and storm sewer system (Michael Baker Corporation, 2008). The project involved GIS database development that included scanning approximately 175 000 as-built or contract drawing sheets, converting the data into a seamless, city-wide enterprise geodatabase, field-investigating attributes for 20 000 manholes, and providing GIS applications and training for NYCDEP personnel.

The original scanned sewer maps were indexed to the appropriate city centerline segments and used as a source document for the creation and placement of digital infrastructure (see Figure 4.6). During the scrubbing process, attribute information such as sewer type, pipe shape, pipe diameter, manhole location, and catch basin location are determined from the best possible source document indexed along that street. In instances where a street segment had no associated source document or conflicting information was shown, field survey crews were used to collect the necessary information to fill in the gaps.

A custom QA/QC application was developed to verify point, linear, and attribute integrity of the geodatabase. This application is responsible for checking the NYCDEP network data to identify various potential errors resulting from the mass data conversion process in the geodatabase environment. The application currently consists of Visual Basic for Applications code that is encapsulated in an ArcObjects ".dll" file. The ArcMap interface displays all the data that are currently in use in the geodatabase. A new button is added to the toolbar that displays the QA/QC interface form. The following are brief explanations of each of the functions:

- *Junction connectivity check*—checks features to ensure that it is connected to the geometric network.
- *Pipe directionality check*—checks that network facility features have system correct pipe flow entering or exiting.

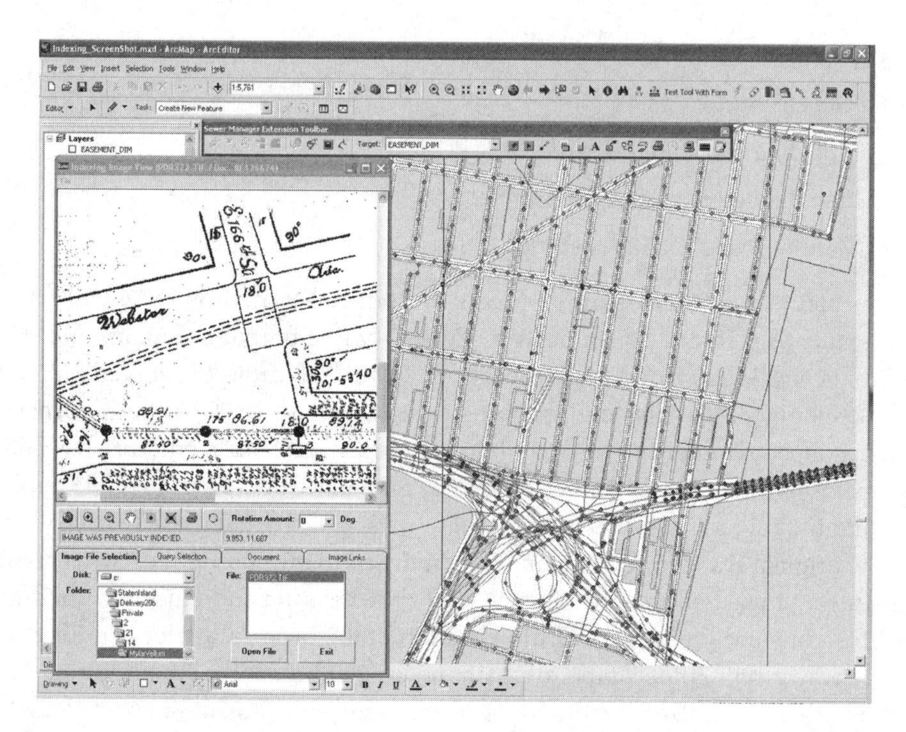

FIGURE 4.6 Indexing application screenshot from NYCDEP (courtesy of Michael Baker Jr., Inc., Moon Township, Pennsylvania).

- *Pipe invert check*—checks pipe features to ensure that invert values correspond with directionality. Also checks pipe inverts to manhole inverts to ensure that there is not a conflict in inflowing and outflowing inverts and to ensure that pipe inverts are not lower than the manhole invert.
- *Attribute check*—(always enabled)—checks each feature attribute and compares them to the populated attribute values in the data entry setup in the geodatabase and logs inconsistent values.
- *Domain check*—checks each feature domain field and compares them to the default domains setup in the geodatabase and logs inconsistent values.
- *Null check*—checks domain and certain other attribute columns for null values and logs them as an error.

- *Facility identification check*—checks each feature's facility identification against the database to ensure there is no error, and checks to ensure that facility identification and unit identification match for each feature. This function also checks to ensure that the facility identification matches the grid number that has been populated.
- *Unit identification check*—checks each feature's unit identification against the database to ensure there is no duplication, and checks to ensure that facility identification and unit identification match for each feature. This function also checks to ensure that the unit identification matches the grid number that has been populated.
- *Related documents check*—checks for the presence of related documents.
- *Node secondary document and pipe secondary document check*—checks secondary documents to confirm what was recorded for each feature in the data entry tables.
- *Spatial attribute check*—checks to ensure spatially that a feature lies within the correct grid, checking against the populated grid number field in the features attribute. This function also checks to ensure that features lie within the correct drainage district polygon, maintenance yard polygons, and treatment plant polygons to the populated domain fields in the features attributes.
- *Confidence-level attribute check*—checks to ensure that the correct confidence level for a feature based on the rating for the linked primary related document is populated.
- *Pipe-shape material attribute consistency check*—the NYCDEP provided a list of possible pipe-shape and material combinations that could or can exist in the city's sewer systems. This check runs against the database to check all sewers for possible combinations and flags any that are not found from that provided list.

Five applications were developed to help in the upkeep of the digital sewer map as new or rehabilitated projects occur. Included in these applications is the ability to perform data maintenance, view the data via an ArcIMS Web browser, spatially represent customer service requests, and produce profile sheets from plan data. As shown in Figure 4.7, a sewer hydraulic capacity calculation tool was also developed that allowed selecting a pipe in the GIS to calculate its flow velocity and

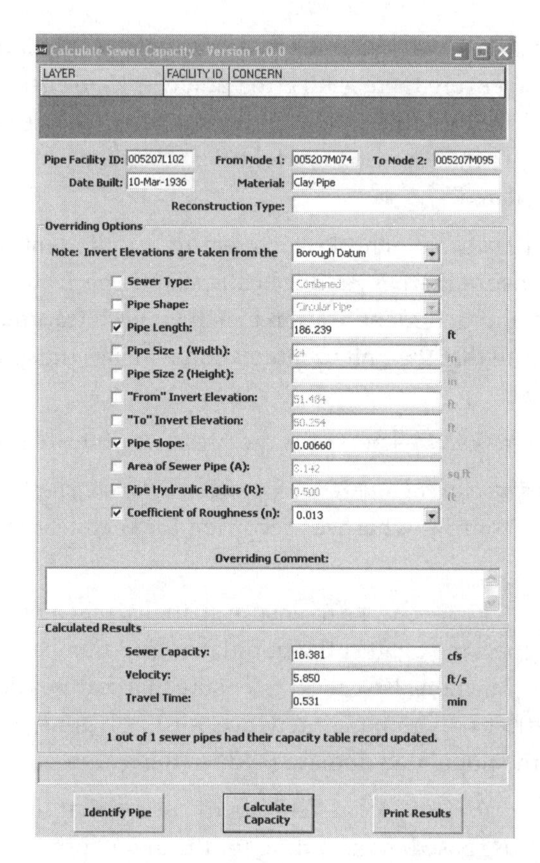

FIGURE 4.7 Geographic information system-based sewer capacity calculation tool for NYCDEP (courtesy of Michael Baker Jr., Inc., Moon Township, Pennsylvania).

hydraulic capacity using Manning's equation. Figure 4.8 shows a page from the sewer inventory map book.

5.2 Hydraulic Characterization and Modeling

Geographic information systems are ideally suited to conduct system hydraulic characterization which generally requires H&H modeling. Land use data, often developed from spectral analysis of satellite imagery, and slopes developed from digital elevation model (DEM) data are used to further characterize the sewer service area. These are some of the typical geographic data available to begin H&H modeling of a combined sewer system (CSS).

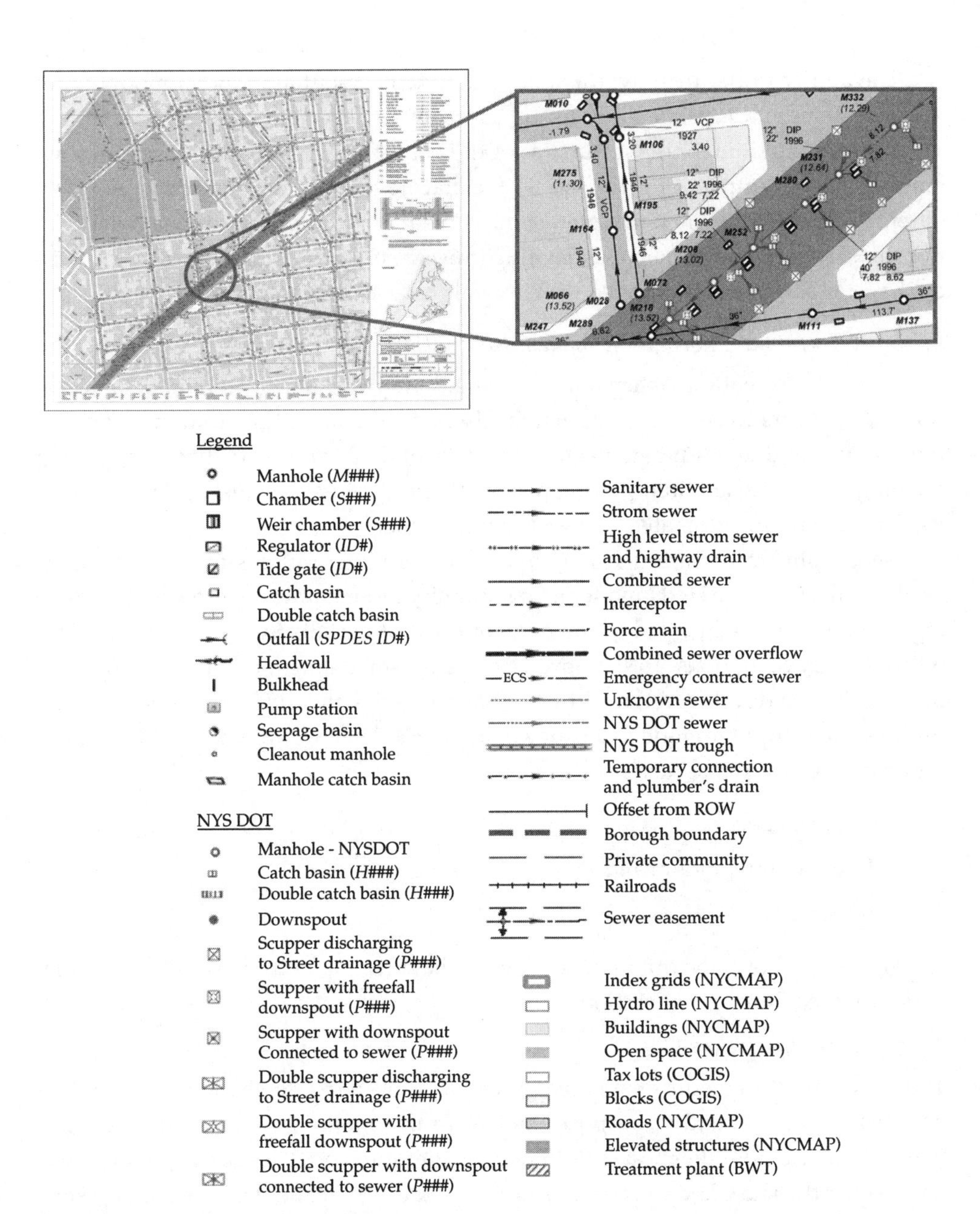

FIGURE 4.8 Geographic information system-based sewer system inventory from NYCDEP (courtesy of Michael Baker Jr., Inc., Moon Township, Pennsylvania).

The ability to perform overlays of such data is essential in H&H modeling of sewer systems and is particularly useful as a comparison method to estimate impervious cover. Providing initial estimates of the percentage of imperviousness of sewersheds is an example of the GIS easily facilitating an otherwise burdensome task. Often, these estimates are performed using standard land use–impervious cover relationships. Another approach to estimating impervious cover with a GIS uses planimetric-vector base maps developed from low-altitude aerial photography.

5.2.1 *Geographic Information System Integration*

Geographic information system and hydrologic and hydraulic (H&H) model integration allows users to be more productive. Users devote more time to understanding the problem and less time on mechanical tasks of data input and checking, getting the program to run, and interpreting reams of output. More than just tabular outputs, models become an automated system evaluation tool.

Geographic information system and modeling integration started in the mid-1990s. At the time, the H&H modeling community did not always agree on the meaning of integration, which led to the development of various integration techniques with arbitrary definitions. To standardize these GIS-modeling integration techniques, in 1998, Shamsi developed the first taxonomy to define the different ways a GIS can be linked to computer models (Heaney et al., 1999). The three methods of GIS application defined by Shamsi are

- Interchange,
- Interface integration, and
- Integration.

Figure 4.9 shows the differences between the three methods. Each method is discussed in the following subsections.

5.2.1.1 *Interchange Method*

The interchange method uses a batch process approach to interchange (i.e., transfer) data between a GIS and a computer model. In this method, there is no direct link between the GIS and the model. Both the GIS and the model are run separately and independently. The GIS database is pre-processed to extract model input parameters, which are manually copied into the model input file. Similarly, model output data are manually copied in a GIS as a new spatial layer for presentation mapping purposes. Although script programming is not necessary for this method, it may be done to

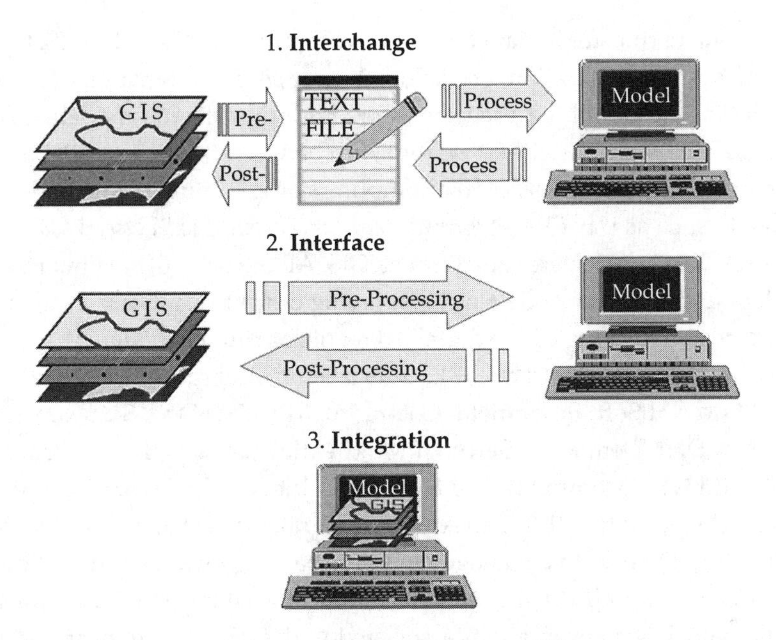

FIGURE 4.9 Modeling integration methods (Shamsi, 2005).

automate some manual operations, such as derivation of runoff curve numbers for calculating stormwater runoff. This was the most practiced method in the 1990s. In this method, GIS is essentially used to generate model input files and display model output data. Any GIS software can be used for the interchange method.

5.2.1.2 Interface Method

The interface method provides a direct link to transfer information between the GIS and the model. *Pre-processing* is defined as transfer of data from the GIS to the model. *Post-processing* is defined as transfer of data from the model to the GIS. The interface method consists of at least the following two components: (a) a pre-processor that analyzes and exports GIS data to create model input files and (b) a post-processor that imports the model output and displays it as a GIS layer. The interface is designed by adding model-specific menus or buttons to the GIS software. The model is executed independently from the GIS; however, the input file is created in the GIS.

5.2.1.3 Integration Method

In the interface method, options for data editing and launching the model from within the GIS software are not available. An interface simply adds new menus or buttons to a GIS to automate transfer of data between a computer model and a GIS. Geographic information system integration, on the other hand, is a combination of a model and a GIS such that the combined program offers both the GIS and modeling functions. This method represents the closest relationship between the GIS and the model. This approach generally uses the available computer codes of public domain legacy computer models, such as U.S. EPA's *Storm Water Management Model User's Manual, Version 5.0* (U.S. EPA, 2010), which are called from a GIS. All the tasks of creating model input, editing data, running the model, and displaying output results are available in GIS. There is no need to exit the GIS to edit the data file or run the model.

Software programs that have GIS interface or integration capabilities include InfoSWMM (MWH Soft, Broomfield, Colorado), XPSWMM (XP Software, Australia), MIKE Urban (DHI, Denmark), SewerGEMS (Bentley Systems, Exton, Pennsylvania), and PCSWMM.NET (Computational Hydraulics Int., Guelph, Ontario). For example, InfoSWMM (MWH Soft, 2010) is an ArcGIS-integrated H&H and water quality simulation model for the effective management of urban stormwater and wastewater collection systems. InfoSWMM and the other programs cited can be used for CSO and SSO studies and system evaluations associated with U.S. EPA's regulations including NDPES permits, CMOM, and total maximum daily load. Built atop Environmental System Research Institute (ESRI) ArcGIS using the latest Microsoft .NET and ESRI ArcObjects-component technologies, InfoSWMM integrates sewer collection systems modeling with ArcGIS. It allows users to conduct efficient GIS analysis and hydraulic modeling in a single environment using a single dataset. Users can create, edit, modify, run, map, analyze, design, and optimize their sewer network models and instantly review, query, and display simulation results from within ArcGIS. InfoSWMM allows selected facilities in the GIS to be exported as a standalong to U.S. EPA's *Storm Water Management Model User's Manual, Version 5.0* (U.S. EPA, 2010). It also allows importing several U.S. EPA models into one InfoSWMM project. Computations can be performed using U.S. EPA's *Storm Water Management Model User's Manual, Version 5.0* engine or other computational engines provided by the vendor. Figure 4.10 shows a screenshot of InfoSWMM running inside ArcGIS software, in which manholes of a collection system trunk sewer are color coded by modeled water depth and sewer pipes are color coded by modeled flow. While ESRI products have the dominant share of the GIS software market in the United States, there are other products and platforms, such as Oracle Spatial (Oracle, Redwood Shores, California), Bentley Map and Bentley Wastewater (Bentley Systems, Exton, Pennsylvania), and Small World (GE Energy, Atlanta, Georgia), which are more popular outside of the United States.

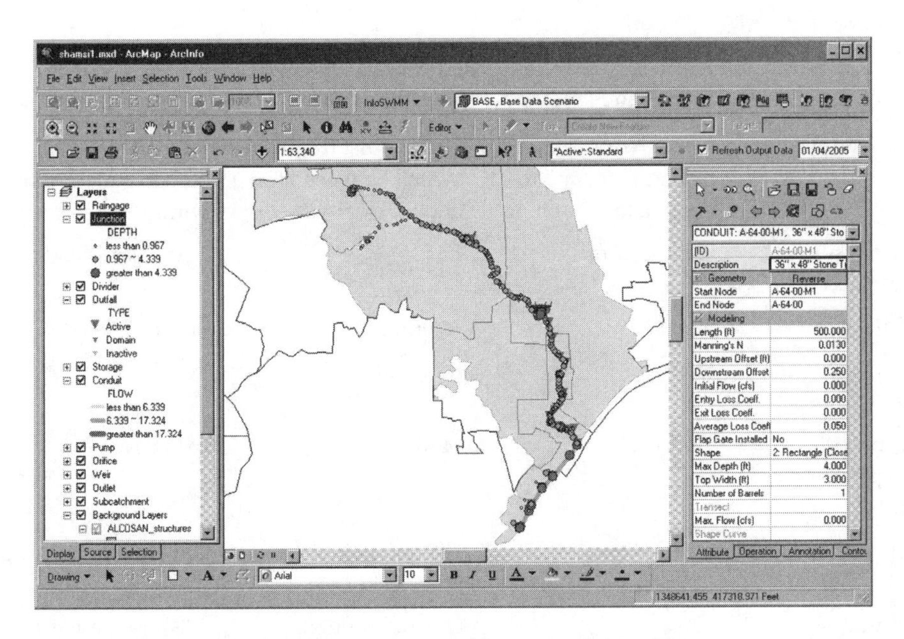

FIGURE 4.10 Screenshot showing a sewer system model in GIS.

5.2.1.4 Integration Method Selection

Table 4.5 lists the advantages and disadvantages of the three methods of GIS integration. Each method has its pros and cons. For example the table indicates that the integration method is easy to learn. However, model integration with GIS is not without some disadvantages. The simplistic modeling approach and user-friendly tools may encourage inexperienced users to become instant modelers, which could be risky. The GIS can easily convert the reams of computer output into colorful thematic maps that can both expose and hide data errors. The user-friendly interface of an integrated package may incorrectly make a complex model appear quite simple, which may lead to misapplication of the model by inexperienced users (Shamsi, 2005).

5.3 Asset Management

A definition of *asset management*, which fits well with the objectives of this manual for the prevention and control of sewer system overflows, is "…delivering a specified level of service to customers and regulators at an optimal life-cycle cost within a strategy that ensures long term sustainability of public assets" (Sklar, 2005). The asset

TABLE 4.5 Comparison of GIS integration methods.

Aspect	Interchange method	Interface method	Integration method
Automation	None; requires manual batch processing to copy data	Some; must frequently switch between model and GIS software	Full; all tasks can be performed from within one software
Ease of use/ user-friendliness	Cumbersome	Easy	Very user-friendly
Learning curve	Steep	Average	Short
Data entry error potential	High	Moderate	Low
Computer programming and scripting	Optional	Moderate	Extensive
User qualifications	GIS	GIS and engineering (or science)	GIS and engineering
User training	GIS	GIS and modeling	GIS and modeling
Suitability	Small projects, one time use	Medium projects, periodic use	Large projects, frequent use

management strategy can provide a framework for the collection system manager to work effectively with internal utility staff (planning, engineering, finance, etc.) and external stakeholders (regulators, elected and appointed officials, public boards, etc.) to make the best, most defensible decisions related to the maintenance, rehabilitation, and replacement of infrastructure and equipment assets. Key components of the asset management decision framework include service levels, asset inventory, condition assessment, criticality (consequence) evaluation, effective useful life, and asset valuation. While the modified approach to compliance with GASB 34 accounting standards makes it mandatory to establish an asset management program, utilities that have adopted the more popular conventional approach to GASB 34 compliance can still benefit from the asset management principles of risk management to help make the most of both capital investment and maintenance program spending.

5.3.1 Service Levels and Performance Measures

The asset management framework begins with defined commitments to deliver a specified level of service to utility customers. Service-level commitments typically cut across the following categories: reliability, customer service, quality, and regulatory. Service levels focus on high-level measures, which link back to the utility's mission and strategic plan and must align with, and respond to, external influences such as customer expectations and regulatory requirements.

Performance measures are internally focused on key activities, which provide the means to track and report progress against service-level commitments. Performance measures rely heavily on the utility's IMSs to support necessary work practices and to provide data management and reporting mechanisms.

The prevention and control of sewer system overflows represents a key service level commitment, which is typically referred to as the *sewer overflow rate* and can be reported as the number of sewer overflows per 100 km or miles of collection system piping per year (number of overflows/100 km or miles/year). A similar service level related to overflows could also be defined for WIB events and reported per 1000 customers per year (WIB/1000 customers/year). Both service levels are strategic and align with the utility's mission to protect human health and the environment. However, both measures are so broad that neither can provide specific guidance to the collection system manager for the following reasons: (1) they only relate to the result and not the underlying cause(s) for sewer overflows and (2) internal factors such as the type of system (separate vs combined) and external influences such as storm intensity (e.g., 5-year vs 10-year storm), which are outside the manager's control, greatly influence the result.

The collection system manager also requires more specific measures addressing the underlying cause(s) of potential problems in the collection system, which are within the manager's ability to remedy. An example of a more specific service level related to sewer system overflows and WIB events could be the number of collection system failures per 100 km or miles of collection system piping per year (failures/100 km or miles/year). A "failure" is then defined as a structural condition (collapse, root intrusion, joint displacement, etc.) or operational condition (deposition, blockage, etc.), which reduces carrying capacity and could, therefore, result in a sewer overflow or basement backup.

A comprehensive set of performance measures in support of a specific service level will track and report the following: (1) quantity, effort, and cost of activities to identify and define the problems; (2) quantity, effort, and cost of activities to prevent or remedy the problems; and (3) quantity of problems prevented and remedied. Table 4.6 shows examples of performance measures related to typical maintenance activities performed to avoid collection system failures (collapses, blockages, and overflows) that could be tracked and reported via the CMMS.

TABLE 4.6 Example collection system performance measures.

Type of activity	Performance measure	Example reporting
CCTV gravity main inspection	Length per month and % System per year	Length requiring cleaning vs. length inspected and total system length Number of structural defects requiring repair by type vs length inspected and total system length
Manhole inspections	Quantity per month and % System per year	Quantity requiring cleaning vs. quantity inspected and total system quantity Quantity requiring repair vs. quantity inspected and total system quantity
Gravity main maintenance	Activity (cleaning, root foaming, etc.) completed per month and % System per year	Percent length identified by inspection prior to overflow event (failure avoided) Percent length required by failure (e.g., overflow event)

The data to support reporting of the performance measures are tracked continuously via the work order and/or service request functions in the CMMS. Therefore, it is critical that the CMMS have work order types specific to the activities within the performance measure. For example, there is no way to track and report manhole inspections completed per month unless the CMMS can define a specific work order type called "manhole inspection". It is also critical that work orders be associated with specific collection system assets and not just street addresses such that the collection system manager can effectively report on SSOs and CSOs and provide system-wide analysis of overflow events.

The collection system manager can find guidance on service levels and performance measures, along with data from annual utility benchmarking studies, via industry sources such as the QualServe Benchmarking Program, which is a joint program of the American Water Works Association and Water Environment Federation.

The ability to track and analyze failures to determine the underlying, or "root cause", of the failure is referred to as *root-cause failure analysis* (RCFA). Many CMMS packages provide some level of RCFA capabilities, which should be carefully considered when making the software selection. Failure analysis combined with the use of performance measures provides the collection system manager a "problem-cause-remedy" approach to identify the true causes of sewer system overflow problems on a system-wide basis, implement the appropriate remedies, and report on the effectiveness of those remedies.

5.3.2 Assessing Condition, Consequence, and Risk

Condition within the asset management framework defines the "probability of failure", which, in the case of collection system gravity mains and manholes, can be assessed with standard techniques as those found in the *Pipeline Assessment Certification Program* and the *Manhole Assessment Certification Program* (see Chapter 6).

In addition to defining the various work order types for manhole and gravity main inspections, the collection system manager may want to store the actual results from those inspections, typically from CCTV work in *Pipeline Assessment Certification Program* format, in the CMMS. Having access to inspection data via the CMMS allows the collection system manager to perform reporting, which cross references both historical work history (cleaning, previous inspections, etc.) with the latest inspection results to better identify, document, and manage problem areas within the collection system.

Accessing current and historical inspection data via integration between the CMMS and GIS allows the collection system manager to quickly categorize large areas of the system for the most appropriate maintenance or capital approach.

Consequence, also referred to as *criticality,* defines the consequence of failure for collection system assets. Consequence of failure for the collection system is often evaluated using principles from the "triple bottom line", which assigns and ranks consequences related to financial (economic), environmental, and social impacts resulting from failure. Methodologies for evaluating consequence include a weighted criteria approach wherein specific criteria are developed for the various types of potential impacts and the collection system pipes are then scored against the criteria on a numerical scale. The final consequence rating is determined by summing all the products of the criteria scores multiplied by the weights.

An alternative approach involves assigning collection system pipes a consequence rating ("3" = high, "2" = moderate, and "1" = low) based on an evaluation using "automatic criteria" for each rating, which can also incorporate triple-bottom-line principles. Criteria such as pipe size and depth can be used to differentiate the direct financial cost to repair a collapse. Criteria related to the pipe's proximity to waterbodies or other environmentally sensitive areas can be used to differentiate potential environmental consequences of a collapse or overflow. Similarly, a pipe's location in a business or commercial district versus a rural area can be used to differentiate potential social impacts such as customer impact and traffic or business disruption.

Regardless of the approach, the criteria development and the scoring methodology selected should be specific to the utility and the characteristics of the collection system.

The collection system manager can use the results of a comprehensive condition assessment, structural potential for collapse combined with operational potential for blockage, and a consequence evaluation to develop an overall risk assessment of the collection system where risk potential equals condition times consequence. Therefore, the mains with the worst condition scores and highest consequence scores have the highest risk potential.

Assigning a risk to gravity mains allows the collection system manager to then have performance measures in terms of risk reduction and not just quantity of work performed. For example, giving preference to cleaning the gravity mains with the highest criticality and worst potential for blockage reduces the risk of an overflow event more effectively than simply cleaning a set percentage of the system each year. Similarly, preferential inspection of the highest-risk gravity mains provides more opportunity for intervention prior to a collapse or an overflow event than inspecting a set percentage of the system each year.

Table 4.7 shows a conceptual rehabilitation and replacement plan over a 20-year period using the structural condition (potential for collapse) and the estimated time to failure from the *Pipeline Assessment Certification Program* combined with the criticality ratings. The least critical pipes (criticality = "1") are scheduled at the high end of the estimated time to failure, the moderately critical pipes (criticality = "2") are scheduled before the high end of the estimated time to failure and the most critical pipes (criticality = "3") are scheduled before or at the low end of the estimated time to failure.

5.3.3 *Mobile Geographic Information Systems*

Using a GIS directly in the field for data collection into a GIS database is referred to as *mobile GIS* or *field GIS*. The goal of a mobile GIS is to link maintenance crews with the most recent GIS data to make their job easier and more efficient. Location-based services allow for the wireless transmission of a GPS location from the field (e.g., for a sewer cleaning truck); the location is displayed (even in real time) on a GIS

TABLE 4.7 Conceptual rehabilitation and replacement plan.

Potential for collapse condition grade	Estimated time to failure	Gravity main rehabilitation and replacement 20-year planning horizon		
		Criticality = 1 (low)	**Criticality = 2 (moderate)**	**Criticality = 3 (high)**
5	< 5 years	4 to 5 years	2 to 3 years	1 year
4	5 to 10 years	7 to 10 years	5 to 7 years	3 to 5 years
3	10 to 20 years	15 to 20 years	10 to 15 years	7 to 10 years
2	> 20 years	NA	20 years	15 to 20 years
1	>> 20 years	NA	NA	NA

map via wireless Internet. Mobile GIS technology is being used to inspect manholes, catch basins, inlets, and outfalls; document smoke and dye tests; and review television inspection of sewer pipes. Indeed, GISs are integrating field inspections, digital photos and videos, and GPS data in one manageable system. "Video mapping" is allowing digital photos and videos to automatically find their correct geographic location on maps and also allowing users to click on map features to review inspection results, photos, and videos. To comply with their CSO consent order, Pittsburgh Water & Sewer Authority (PWSA), Pittsburgh, Pennsylvania, used a GIS-based sewer system inspection program in 2007 and 2008 for 28 500 manholes and 39 000 catch basins. The authority's inspection application used ESRI ArcPad mobile GIS software on Panasonic Toughbook (Panasonic Computer Solutions, Secaucus, New Jersey) field computers. Created using ArcPad Application Builder software, the application records digital photos and videos using Envirosight LLC's (Randolph, New Jersey) QuickView Zoom Camera.

5.4 Work Order Management

Work management is one of the key business processes involved in the day-to-day operations and maintenance of the collection system. Important considerations for successful work management include (1) efficient internal practices for planning, scheduling, and tracking the progress and results of work performed; (2) effective business systems to allow the collection system manager and other stakeholders to know the status of all active, planned, and historical work; and (3) established performance measures the collection system manager can use to gauge the effectiveness of the overall work management process.

5.4.1 Work Order Assignment

Typically, two information systems are responsible for generating "work" within the utility, that is, the CIS and CMMS. Work generated via the CIS is externally driven by customer calls and is typically referred to as a *service request*. Work generated via the CMMS is internally driven by the utility's needs and is typically referred to as a "work order". The processes for how service requests and work orders are created, completed, and closed out are typically different and often occur in different parts of the organization, particularly in large utilities. However, service requests and work orders have one important aspect in common, that is, they both represent time and money spent investigating, inspecting, and/or maintaining the utility's assets. Therefore, for the most complete picture of the performance and condition of those assets, it is beneficial for the collection system manager to have access to current and historical information for both work orders and service requests.

The principal roles of the work management process include the following:

- Planning and scheduling work including defining the specific activities required, assigning resources (crew, equipment, and materials), and estimating time and cost;
- Recording the actual materials, time, and cost to perform work;
- Verifying quality and timeliness of completion;
- Capturing data to build a historical record of asset performance;
- Analyzing data (e.g., for root cause of overflow events); and
- Reporting results and progress toward efficiency and effectiveness goals.

Fulfillment of these roles requires that the supporting processes to create, complete, and close out work be fully defined within the utility and then properly supported by the CMMS. In addition to the staff performing the work, a key role is the "planner scheduler" responsible for prioritizing, assigning, and scheduling all the work to be completed. Best practices recommends a dedicated planner schedule role. However, depending on the size of the organization, this is not always possible and a shared role often exists where a maintenance supervisor or foreman is also responsible for planning work within the CMMS. The planner scheduler, maintenance supervisor, or foreman can also perform work order closeout, which involves reviewing the data for completeness and accuracy and determining the need for follow-up activity (inspection, etc.). Large utilities may audit a portion (e.g., 5%) of closed work orders on a quarterly basis for uniformity, completeness, and accuracy. Such an audit is recommended as part of post-implementation assessment for a new CMMS.

The CMMS application provides the ability to define various types of work and specific work activities within each type. This historic record of activities and costs associated with specific assets can be useful to the collection system manager in developing budgets and resource needs for maintenance activities and in planning for capital projects. Therefore, the types of work and associated activities defined within the CMMS need to reflect the maintenance program the utility is required to perform. The list of work types should be broad, but brief. Examples of common broad work types include emergency, corrective, preventive maintenance and, inspection. Additional types might include capital (for work not part of operations and maintenance), safety (for prioritization of work related to safety issues), contractor (for work performed by outside contractors), predictive (for work related to a reliability-centered maintenance program), and so on. Each work type may also be associated with one or multiple cost account numbers within the CMMS. For example, the cost account for "capital" work would typically be different than for "preventive maintenance" work. A list of 8 to 12 work types, which reflect the utility's business practices, is typically sufficient.

The list of specific work activities should be more detailed to describe all the work the utility performs. For example, if the utility performs gravity main inspection using different methods (e.g., CCTV and zoom camera), separate work activities for each should be defined in the CMMS under the inspection work type. Lastly, CMMS work management should also classify work as "planned" or "unplanned". For example, a gravity main inspection, which is performed as a result of an overflow event, would be an unplanned inspection. Conversely, a CCTV inspection performed for condition assessment purposes would be a planned inspection. The differentiation between planned and unplanned work is important to facilitate reporting against service-level measures, as discussed in Section 5.3, *Asset Management* (e.g., collection system failures). For example, a gravity main cleaning performed as a result of a CCTV inspection before an overflow occurs is a "failure avoided". Tracking and measuring the avoidance of failures (e.g., corrective work performed as a result of preventive maintenance or inspection) is the best method to demonstrate the return on investment from a proactive and performance-based collection system management approach.

Additional metrics associated with work order management, which could be useful to the collection system manager, typically include

- *Operation and maintenance* (O&M) *cost ratios*—measures business efficiency:
    - Operations and maintenance cost per account: O&M costs/total number of active accounts;
    - Operations and maintenance cost per million gallons: O&M costs/volume processed; and
    - Operations and maintenance cost ratios are reported within the annual QualServe Benchmarking Report.
- *Planned maintenance ratios*—measures maintenance planning efficiency:
    - Maintenance time planned/total time (planned and unplanned) and
    - Planned maintenance ratios are also reported within the annual QualServe Benchmarking Report.
- *Work order turnover*—measures maintenance work efficiency (Campbell, 2001):
    - Work order turnover = work orders completed last month/work orders generated last month; and
    - A number greater than 1 indicates more work was completed than new work was requested in the time period. Therefore, the work backlog shrank, which should result in a corresponding increase in the overall level of service. Work order turnover should be tracked as an overall measure and also by work activity. A continuous trend of shrinking or growing backlog is a key indicator of the need for additional resources or reassignment of resources between activities.
- *Work order backlog*—measures maintenance work efficiency (Campbell, 2001):
    - Work order backlog = work orders overdue/work orders completed this month;
    - The result is the decimal equivalent of 1 month. Example: 720 overdue work orders/3200 completed work orders in month = 0.225 of 1 month = 6.8 days; and
    - Similar to turnover, backlog should also be tracked as an overall measurement and by activity. Initial backlog targets can be established based on staffing levels and historical workload information.

5.4.2 *Work Order Tracking*

Work order and service request information is most useful to the collection system manager when it is directly associated with an actual asset (a specific gravity main, manhole, overflow structure, etc.). Most CMMS applications allow a work order to be created against an asset by referencing the asset identification (ID), against an internal cost account by referencing the account identification (ID), or both. Best practices always associate work performed with both an asset and an account. The CIS application typically associates a service request with a customer identification number, customer account number, a customer address, and/or a cost account number. If the service request then results in work performed (e.g., a gravity main inspection), that work often remains associated with only the customer account, identification, or address and not with the actual asset worked on. Best practices would provide for the association of work performed from a service request with the specific asset(s) involved. Most CMMS applications also provide the capability to generate a customer service request. For utilities planning the implementation of a new CMMS, combining the two functions into a single application could be a viable alternative. Integrating an existing CIS with the CMMS is also a viable and common solution. The preferred approach to integration would include the GIS such that staff in customer service can view open and completed work orders in the vicinity of the current customer call and the collection system manager can view and report on customer call activity throughout the service area.

Whatever measures the collection system manager selects to monitor the overall work management process, they must be clearly communicated to all staff such that clear expectations are set and the underlying data behind the measures are understood. Staff will be confused if they do not understand the purpose of these measures and how they are determined. For example, is a work order "overdue" based on the original requested date or the final completion date set by the planner scheduler? Should all work activities have the same targets? Should there be seasonally adjusted targets for some activities based on variations in flow conditions experienced in the collection system (e.g., spring wet weather periods vs summer dry weather)? Properly applied, work order management measures can be a useful tool for the collection system manager to focus resources on shifting priorities while providing both upward feedback to utility management and downward feedback to utility staff.

5.5 Regulatory Compliance Monitoring

Many cities in the United States are dealing with regulatory enforcement actions for sewer overflows, such as consent orders and consent decrees. These communities must comply with various mapping, monitoring, inspection, and rehabilitation requirements and develop and implement sewer overflow control plans. To comply with regulatory requirements, cities are collecting massive amounts of data on the inventory and condition of their sewer system infrastructure. A dilemma that all stakeholders are facing is how to cost-effectively manage this data and monitor what has been accomplished versus what still needs to be done. The GIS represents a cost-effective technology to manage and analyze these datasets. Above and beyond the conventional GIS mapping of inspection data, integration of field inspection data with GIS allows development of a sewer rehabilitation decision support system that can be used to plan the rehabilitation work required to control sewer overflows.

Now more than ever, U.S. EPA has strict regulations on water pollution and water quality under the Clean Water Act. Overflows and failing processes can be troubling for agencies and cities, especially when dealing with aging equipment and infrastructure. For example, 83 communities in Allegheny County, Pennsylvania, have administrative consent orders and consent order and agreements (COAs). Administrative consent orders are enforced by the Allegheny County Health Department for SSO communities. Consent order and agreements are enforced by the Pennsylvania Department of Environmental Protection for CSO communities. These communities must complete various mapping, inspection, monitoring, and modeling tasks to develop overflow control plans. Consent orders mandate one-fifth of sewer system deficiency correction each year until 2010. Field inspection of sewer manholes and pipes and GIS mapping with specific layers, attributes, and format is required by all consent orders. To meet their COA requirements, PWSA, which represents the largest consent order municipality, used a GIS-based sewer system inspection program for 28 500 manhole inspections, 39 000 catch basin inspections, and 10 200 dye-testing inspections.

5.5.1 *Remote Monitoring System Requirements*

Remote monitoring within the collection system includes the automated collection of operational data from remote locations (e.g., flow, rain, and pumping stations) and requires the following technical elements: data measurement, collection, communication, and storage. The critical data management work practices for the utility include data review, approval, and reporting.

Field devices for remote flow monitoring typically include primary devices (for measuring level over a weir), open channel meters (for gravity mains), or closed pipe meters (for pumping stations). The selection and application of a particular flow monitoring device is discussed in Chapter 5. The field devices typically generate a 4- to 20-ma output to a remote terminal unit (RTU), programmable logic controller (PLC), or similar device, which then converts the reading to the desired output (inches of flow depth, feet per second, gallons per minute, etc.) based on a defined range and set of physical parameters for the installation.

Pumping station monitoring is critically important to effective collection system management for the prevention of sewer system overflows. In addition to station flow, the collection system manager may need to monitor wet well level, pump run time, and pump failure. Depending on station size, additional equipment could be monitored for run and failure status including backup generators and odor control equipment. Remote monitoring of ingress/egress (in addition to general security alarms) and the use of live video monitoring are becoming common with the recent availability and low cost of high-bandwidth data transmission. The extent and complexity of overall pumping station remote monitoring should be developed on a case-by-case basis considering the station's consequence of failure and the utility's operational practices.

In addition to in-line and pumping station monitoring, many utilities also have rain gauges located at key points within the collection system service area to measure, record, and transmit rain data. Total rainfall and rainfall intensity can then be calculated throughout the service area. Access to quality rainfall data, along with the system flow and level data, is critical to meaningful analysis and reporting of collection system performance for the prevention and control of overflows.

Water quality data can also be continuously monitored via remotely installed probes for common constituents (dissolved oxygen, pH, temperature, and conductivity). As discussed in Chapter 5, collecting automated data for bacteriological sampling is very difficult. Once collected, the samples need to be analyzed in the laboratory. Therefore, the results cannot be transmitted remotely. The status of automated sampling equipment (running, failure, and full) can be remotely monitored.

5.5.1.1 Alternate System Approaches

A typical arrangement for monitoring and collection of remote data is shown in Figure 4.11. Multiple field devices are generally connected to the RTU or PLC, which is cabinet-mounted in the field. The communication link, "A", between the field

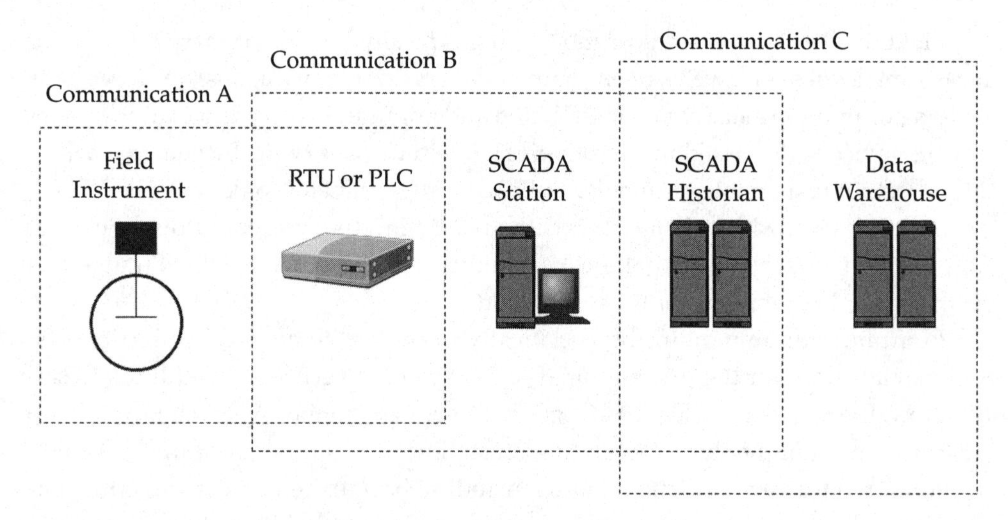

FIGURE 4.11 Typical remote data collection arrangement.

instrument and the RTU or PLC, can be hardwired or wireless using the protocols supported by the vendor.

The communication link, "B", between the field PLC and the SCADA system, can be via dedicated phone line, radio, wireless, or cellular-based communications. Options exist whereby wireless or cellular communication combined with a "data collector" or "hub" in place of the traditional RTU or PLC can transmit field data directly to the SCADA historian. Cellular phone communication is also used by some application service provider vendors to transmit data to their remotely hosted location, which is then made available to the utility for viewing and download via a Web site. A utility may also be able to take advantage of a city's fiber optic network or wireless mesh network, particularly in more urbanized locations, for a portion of the data communication. The many available and changing options for data communication and aggregation make it essential for the utility to do a comprehensive evaluation and cost comparison before selecting the final approach.

The communication link, "C", between the SCADA historians and data warehouse, is generally via the utilities local area or wide area network. The data warehouse may collect data from multiple historians and from other operational data sources such as the laboratory system.

5.5.1.2 *Need for System Redundancy*

The entire data collection system is vulnerable to a variety of failures including field sensor; communication hardware (e.g., PLC and RTU); communication transmission (loss or interruption of communications due to damage, weather, power outage, etc.); and SCADA hardware/software and data warehouse hardware/software failures. The consequence of each failure should be evaluated by the utility and the appropriate redundancy provided. Examples include (1) backing up overflow sensors with in-line level and flow monitors at upstream locations; (2) a redundant RTU or PLC for overflow locations critical to regulatory reporting; (3) backup power supplies (batteries, uninterruptible power supplies, etc.) at critical locations in the field and on the SCADA and data warehouse hardware; (4) redundant SCADA servers; and (5) redundant data warehouse servers.

Equally important to equipment and power supply redundancy are good practices via SOPs for inspecting, calibrating, and general preventive maintenance on all field instrumentation, hardware, and software components. Good practices are also required for computer network management, security, and data backup. The backups from the data warehouse done via the utility's network should be to a remote location. If done by tape, the tapes should also be stored in a remote location.

5.5.2 *Data Management and Reporting*

The magnitude of data collected from remote locations can be quite large and will vary with the number of locations and the frequency of the collection interval (1-minute, 5-minute data, etc.). Consider the number of data points collected from 10 remote collection system level monitoring sites using a 1-minute collection interval. The daily data quantity would be 1440 (the number of minutes in a day) multiplied by 10 points for a total of 14 400 data points per day. The monthly total would then be 432 000 data points for a 30-day month. The annual total is then 5,256,000 data points. Five years of historical data for reporting and trending is more than 25 million records. This magnitude of data requires well-defined and automated procedures for quality assurance, review, and approval.

Remotely monitored water quality data related to common constituents (dissolved oxygen, pH, temperature, and conductivity) can be managed via the SCADA system with the other remotely monitored flow and level data. Bacteriological sampling results, which must be done in the laboratory, should be maintained in the LIMS, if possible. Many LIMS applications provide for the creation of "field" or "mobile" sampling programs, which can be configured to hold results from remote water quality sampling locations.

5.5.2.1 Current and Historical Database Archive

The SCADA historian will typically hold raw data for a set period of time (e.g., 30 to 90 days), and will provide onscreen trends and tabular and graphical reports of system flows and levels. The data warehouse sits upstream from the SCADA historian and will typically collect data from the historian at predetermined intervals (e.g., hourly or daily) and provide storage for multiple years in support of long-term analysis and trending. Compliance reporting is also typically performed from the data warehouse. The data warehouse may apply summarization routines to reduce the volume of stored data. For example, 1-minute values from the SCADA historian could be summarized to 5- or 15-minute averages. Generally, if data are also collected from a LIMS database, they are not summarized within the data warehouse, but represent the final approved analysis results as stored in the LIMS. The use of summarization routines within the data warehouse must be carefully coordinated with the overall quality review and reporting needs of the utility such that all collection system events are properly captured for historic analysis, trending, and reporting.

5.5.2.2 Report Development

Reporting is essential to getting the benefit from an IMS performing collection system monitoring. A data warehouse collecting simultaneously from SCADA historians and a LIMS will be able to produce integrated reports on collection system performance showing rain events, flows, levels, and water quality parameters. If data are reported separately, the collection system manager will need to manually align system flow reports from SCADA with the water quality data from LIMS to produce a comprehensive system-wide report.

Attention should also be given to reporting for data quality and data integrity. Data quality reports provide an overall summary of any data points, which deviate from the specified business rules for data validation. For example, a report could show all the data values outside of the predetermined minimum or maximum ranges. A similar report could show all the erroneous data values reported as a result of an instrument failure. Data integrity reports provide the history of the review and approval process indicating data that were not approved for use in reporting. These system level reports can typically be configured within the data warehouse for automated notification or delivery to the collection system manager.

5.5.3 *South Bend, Indiana, Case Study*

The city of South Bend, Indiana, undertook implementation of a system-wide monitoring network for its CSS using a decentralized approach with embedded sensors, repeaters, and remote data acquisition points deployed throughout the municipality. Flow and level data are monitored at multiple locations throughout South Bend's CSS including at 36 outfalls, 5 retention basins, and 69 strategic locations in the 7-mile-long main interceptor and major connecting trunk lines (Ruggaber et al., 2008). Figure 4.12 shows the extent of the decentralized monitoring network deployed in South Bend.

The decentralized network uses various types of nodes to gather flow and level data and communicate the data to a secure Web site for viewing and retrieval by South Bend officials. The three types of nodes include (1) an instrument node, which can accept up to four distinct sensors; (2) a gateway node, which serves as a data acquisition and communication point; and (3) a repeater node, which can relay data from

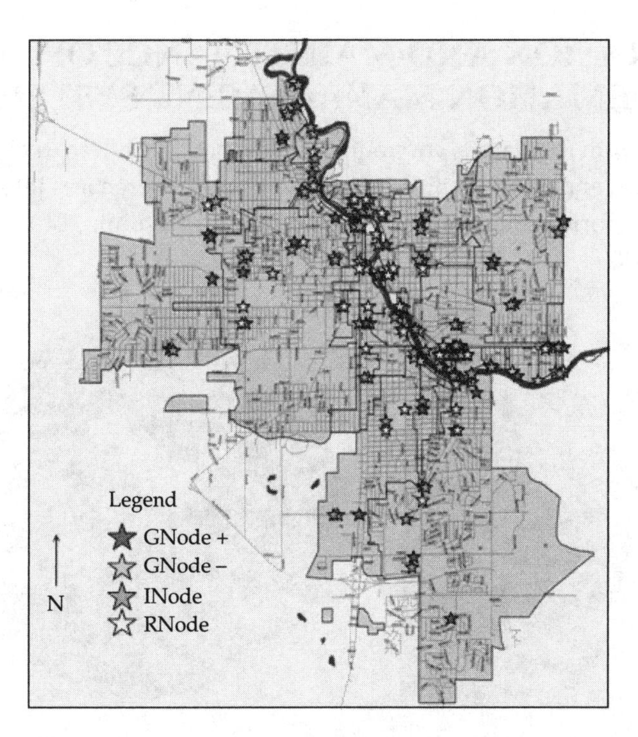

FIGURE 4.12 Decentralized monitoring network in South Bend, Indiana (Ruggaber et al., 2008).

an instrument node to a gateway node. The instrument nodes are battery powered and are typically installed with an integral noncomposite manhole lid, which uses the existing manhole frame and cover. Communication with the gateway or repeater nodes is wireless via an antennae built into the composite manhole. The instrument nodes use full power only for communication and very low power for data acquisition, thereby providing an expected battery life of 3 to 6 years. The repeater and gateway nodes are mounted on existing light or traffic signal poles. Power for the repeater nodes can be either battery or external 120-V alternating current power. The gateway node provides versatile communication options to upload data to a secure Web site including cellular connection and wireless fidelity (Wi-Fi). Figure 4.13 shows a pole-mounted gateway node and the underside of a composite manhole lid with the integral instrument node assembly.

The secure Web site provides South Bend continuous access to view and download flow and monitoring data in a variety of file formats for reporting and analysis.

6.0 OPERATION AND MAINTENANCE OF AN INFORMATION MANAGEMENT SYSTEM

With pressure from regulators, communities, and possible alternative providers, it is critical that cities and agencies incorporate industry best practices into their business processes and information management. A centralized group should have oversight

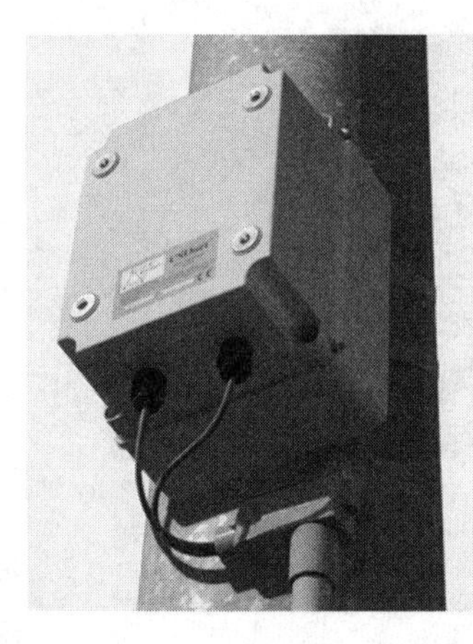
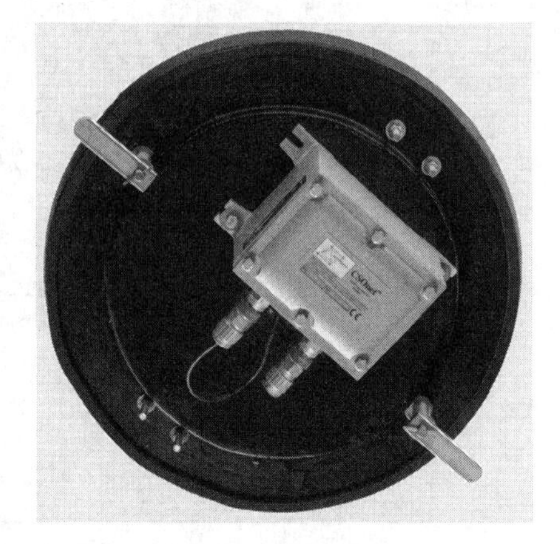

FIGURE 4.13 Gateway nodes: pole (left) and manhole (right) installations (Ruggaber et al., 2008).

of the acquisition and implementation of all information technology and IMS needs within an agency. For most organizations, the information technology department should take on this role. This oversight ensures that new and upgraded solutions remain compatible with the organization's technology vision and architecture, that information technology investments are more likely to meet the desired objectives, and that both short- and long-term IMS plans can be developed and maintained (Lanyon et al., 2009).

6.1 Quality Assurance and Quality Control

Quality assurances, by definition, are those processes and procedures that are incorporated to the collection and production phase of a project to ensure that high-quality products are produced. These include such items as detailed specifications, attention to workflow processes, automation, checklists, menu pick lists, and design standards. Each step in the production process should be designed with quality in mind. Possible errors should be anticipated and procedures instituted to prevent errors from occurring. Continuous review of procedures and corrective action to the processes will prevent future errors.

Quality control or validations are those measures implemented to identify errors after a production phase is completed. These include such items as automated geometry and attribute checks, design validation reports, and manual feature verification. Specific attribute values or a range of acceptable values can be verified. Relationships between different attributes can be checked.

Software for validation must be developed or acquired. Standardized screen displays can also be designed, and check plots can be customized so that errors are more easily identifiable. While manual or onscreen feature checking is time-consuming, these types of inspections may be the only way to check some types of graphic features against source documents.

Validation activities should be considered like any other production process. The standards and steps should be defined. Responsibilities for each validation step should be assigned. Acceptance criteria should be standardized and realistic and based on how the data will be used.

Production schedules should be established to account for specific quality control steps and time for any corrective actions to occur. After any corrections are made, additional quality control checks should be made to ensure the corrections were made properly.

Data developed as part of a data conversion project should be subjected to a series of comprehensive quality control checks. Although the goal for any production process is to produce 100% compliant data, most organizations recognize that they must allow for a small percentage of non-compliant data.

Often, a validation process will uncover types of errors that stem from conflicting source documents. These are sometimes called *discrepancies* and will require research (sometimes called *problem resolution*). It is important to establish early on who will perform the research to solve specific types of discrepancies. The solution may likely involve going to a secondary or, in some cases, a tertiary source to resolve the discrepancy.

Time limits for problem resolution should be established (e.g., 24 to 72 hours) so that production procedures are not adversely affected. Attributes can be used to flag features for which further investigation is needed to verify that it is correct.

6.1.1 *General Rules for Producing Quality*

Some general rules about QA/QC include the following:

- Quality is the result of careful planning.
- Define the specifications for measuring data quality and the specific criteria for final data acceptance.
- Define the procedures for meeting quality standards. Procedures should be as standardized as possible. Validation processes must be documented.
- Errors should be detected as early in the process as possible. The later a mistake is found, the more costly it is to correct.
- All errors should be logged.
- Staff should be encouraged to report errors, and especially to recommend improved procedures to prevent errors. Changes to procedures should be tested outside of the standard process until proven valid.
- Acceptance criteria should be achievable and realistic. A lot of quality criteria are subject to interpretation. Define the criteria by the way that the data will be used.
- Ideally, all errors should be corrected, even if they are within acceptance limits.
- Assign responsibility for performing validation tests, researching discrepancies, and fixing mistakes.

6.1.2 *Data Acceptance Criteria*

As part of the conversion program, each organization should establish the standards under which the data should be produced. Typical standards used to establish acceptance criteria are listed here. These standards may be modified depending on the specific requirements of any project. The following are facility network data acceptance standards and minimum percentage of data that should comply with these standards:

- Deliverable format/design adherence—100%,
- Positional accuracy (based on American Society of Photogrammetry or National Map Accuracy Standards)—90%,
- Graphic quality—98%,
- Annotation placement—90%,
- Completeness—99%,
- Priority attribute accuracy—98%, and
- Non-priority attribute accuracy—96%.

As part of establishing the standards, the unit area to be measured and the rules for counting features should also be defined.

When QA/QC procedures are established, a database should be designed to enable quality control personnel to not only record the types of errors, but also quantify the number of errors per type. Through implementing this database, reports can be generated that aid quality control personnel in verifying that acceptance percentages are being met. If the percentages are not being met, quality control personnel can communicate the discrepancies to the production department in a timely and precise manner.

6.1.3 *Data Quality Assurance and Quality Control and Approval*

The large volume of data and multiple potential failure modes in the data collection and transmission processes make it difficult to rely only on a manual review of the data. The data warehouse and the SCADA historian must both work together to support overall data management and reporting. Configuring the alarm functions of the SCADA system to alert operations staff to potential high flow or level situations, which could result in overflow events, is the critical first step in the data management process.

The next step is to configure the data warehouse to execute quality checks and apply business rules during the data collection process. For example, if a data warehouse is using routines to summarize the 1-minute historian data to 15-minute averages, the data warehouse could be configured to also store the highest value and the lowest value during that period, which would capture any short duration changes within the collection system. A warehouse that is collecting raw historian data with no summarization could also be configured to identify minimum and maximum values over a set time period or series of time periods (15 minutes, hourly, daily, etc.) to facilitate reporting. The configuration of the data warehouse to automatically identify data outside of predetermined range can also greatly reduce the volume of data requiring a manual review. The data warehouse could also collect the alarm events generated by the SCADA system for equipment, power, and communication failure.

Business rules within the data warehouse should be developed for how to assist the collection system manager to better identify data anomalies associated with these types of failures and exclude this data from reporting. Monitoring locations within the collection system could have no flow until a wet weather event occurs, which results in the continuous collection of the value zero by the SCADA system. The data warehouse could be configured to only collect this data point from the historian if the value were other than zero. A similar strategy could be applied for rain gauges, which will also read zero until a rain event occurs.

6.2 Data and System Security

Operating and maintaining the IMS is vendor-specific and should be addressed by its supplier. However, there is one aspect of the IMS that only the collection system manager can supply, that is, security of data against accidental loss because of fire, employee mistakes, or vandalism.

Every labor hour that is put into building and using the IMS increases the economic worth of the system in replacement-value terms; the IMS could eventually become the utility's single most valuable asset.

Therefore, the collection system manager must use his or her authority to provide off-site storage for duplicate copies of the database(s). Merely keeping duplicates in a file cabinet next to the computer will be useless in the event of a catastrophic loss; even if a recovery were attempted from such a loss, anecdotal attributes could never be recovered.

Another approach is to complete initial conversion of all data based on documents that are "frozen" as of a certain data. After the conversion and all QA/QC and acceptance checks have been completed, the backlog of projects that were completed since the original data was converted is updated.

Many organizations elect to have all of their updates processed on-site using the same tools used during the conversion effort. This approach has the following advantages:

- It will serve to fully test the update and maintenance tools in an operational environment,
- Staff are able to get fully trained on the update process and tools,
- It is often easier for in-house staff to perform QA/QC on the data that has been converted, and
- It is often a more time-effective approach rather than sending the data back to production offices for conversion.

Some large utilities are also now outsourcing some of their maintenance operations to private companies. This approach is becoming more common for private investor-owned utilities (e.g., electric and gas). Outsourcing may involve periodically sending updated source documents to a maintenance contractor or it may involve that maintenance contractors provide on-site staff augmentation services to assist with all or a portion of the updating work. It is also important to note that, irrespective of the approach, effectively maintaining data in a GIS environment requires a change in the work routines of staff, which often involves retraining and new responsibilities.

6.3 Data Maintenance

Organizations have two choices regarding maintenance of data throughout the conversion process. The first option is to maintain the facilities' data during the conversion process. If a vendor is performing the conversion, this means that those copies of the new as-builts and/or scanned as-builts that show the work has been completed must be provided to the vendor.

The other approach (as stated in the previous section) is to complete initial conversion of all data based on documents that are "frozen" as of a certain data. After the conversion and all QA/QC and acceptance checks have been completed, the backlog of projects that were completed since the original data was converted is updated.

Many organizations elect to have all of their updates processed on-site using the same tools as those used during the conversion effort. This approach has the following advantages:

- It will serve to fully test the update and maintenance tools in an operational environment,
- Staff are able to get fully trained on the update process and tools,
- It is often easier for in-house staff to perform QA/QC on the data that has been converted, and
- It is often a more time-effective approach rather than sending the data back to production offices for conversion.

Some large utilities are also now outsourcing some of their utility maintenance operations to private companies. This approach is becoming more common for private investor-owned utilities (e.g., electric and gas). Outsourcing may involve periodically sending updated source documents to a maintenance contractor or it may involve that maintenance contractors provide on-site staff augmentation services to assist with all or a portion of the updating work. It is also important to note that, irrespective of the approach, effectively maintaining data in a GIS environment requires a change in the work routines of staff, which often involves retraining and new responsibilities.

7.0 REFERENCES

Azteca Systems, Inc. (2010) Azteca Systems Home Page. www.azteca.com (accessed May 2010).

Campbell, J. D.; Jardine, A. K. S. (2001) *Maintenance Excellence—Optimizing Equipment Life-Cycle Decisions;* Marcel Dekker: New York.

Environmental System Research Institute (1992) *ArcView User's Guide,* 2nd ed.; Environmental System Research Institute; Environmental System Research Institute: Redlands, California.

GBA Master Series, Inc. (2010) GBA Master Series Home Page. www.gbamasterseries.com (accessed May 2010).

Heaney, J. P.; Sample, D.; Wright, L. (1999) *Geographical Information Systems, Decisions Support Systems, and Urban Stormwater Management;* U.S. Environmental Protection Agency: Edison, New Jersey.

Infor, Inc. (2010) Infor Hansen Asset Management, formerly Infrastructure Management System from Hansen Information Technologies, Web Site. http://www.infor.com/hansen (accessed May 2010).

Lanyon, R.; Harp, D.; Buskirk, R.; Smith, K. (2009) It All Starts with IT. *Water Environ. Technol.*, 21 (2), 24–31.

Lovely, R. (2008) Risky Business. *Water Wastewater Digest*, 48 (11).

Maidment, D. R. (2002) *Arc Hydro GIS for Water Resources;* ESRI Press: Redlands, California.

Michael Baker Corporation (2008) *NYC GIS Sewer Mapping Project;* Project Flysheet: Virginia Beach, Virginia.

MWH Soft (2010) MWH Soft Home Page. www.mwhsoft.com (accessed May 2010).

National Association of Sewer Service Companies (2001) *Pipeline Assessment and Certification Program*; Reference Manual Version 4.3.1; National Association of Sewer Service Companies: Owings Mills, Maryland.

National Association of Sewer Service Companies (2006) *Manhole Assessment and Certification Program*; Reference Manual Version 4.3; National Association of Sewer Service Companies: Owings Mills, Maryland.

O'Brien, J. (1999) *Management Information Systems—Managing Information Technology in the Internetworked Enterprise;* Irwin Professional Publishing: Boston, Massachusetts.

RJN Group, Inc. (2010) RJN Group Home Page. www.rjn.com/CW (accessed May 2010).

Ruggaber, T.; Garnes, G.; Osthues, G.; Radcliff, R. (2008) High Density CSO Monitoring Using an Embedded Sensor Network in South Bend, Indiana. *Proceedings of the Urban Water Management Conference;* Louisville, Kentucky, Mar 3–April 2. http://www.heliosware.com/news/papers/High_Density.pdf (accessed March 2011).

Shamsi, U. M. (2002a) *GIS Tools for Water, Wastewater, and Stormwater Systems;* American Society of Civil Engineers: Reston, Virginia.

Shamsi, U. M. (2002b) Chapter 14: GIS Applications in Inspection and Maintenance of Collection Systems; In *Best Modeling Practices for Urban Water Systems Monograph 10. Proceedings of the 2001 Conference on Stormwater and Urban Water Systems Modeling;* Toronto, Ontario, Canada, Feb 22–23; Computational Hydraulics International: Guelph, Ontario, Canada.

Shamsi, U. M. (2005) *GIS Applications for Water, Wastewater, and Stormwater Systems;* CRC Press: Boca Raton, Florida.

Shamsi, U. M. (2009) *GIS Applications in Water, Wastewater, and Stormwater Systems,* Continuing Education Seminar; ASCE Press: Reston, Virginia.

Shamsi, U. M.; Fletcher, B. A. (2000) Chapter 7: AM/FM/GIS Applications for Stormwater Systems; In *Applied Modeling of Urban Water Systems Monograph 8. Proceedings of the 1999 Conference on Stormwater and Urban Water Systems Modeling;* Toronto, Ontario, Canada, Feb 18–19; Consolidated Hydraulics International: Guelph, Ontario, Canada.

Sklar, D.; Pomerance, H.; Westerhoff, G. (2005) Envisioning the Future Water Utility. *J.—Am. Water Works Assoc.,* 97 (11), 67–74.

U.S. Environmental Protection Agency (1994) *Combined Sewer Overflow (CSO) Control Policy;* EPA-830/94-001; U.S. Environmental Protection Agency: Washington, D.C.

U.S. Environmental Protection Agency (1999) *Combined Sewer Overflows: Guidance for Monitoring and Modeling;* EPA-832/B-99-002; U.S. Environmental Protection Agency, Office of Water: Washington, D.C.

U.S. Environmental Protection Agency (2000) *Environmental Planning for Communities—A Guide to the Environmental Visioning Process Utilizing a Geographic Information System (GIS);* EPA-625/R-98/003; U.S. Environmental Protection Agency, Technology Transfer and Support Division, Office of Research and Development: Cincinnati, Ohio.

U.S. Environmental Protection Agency (2002) *SSO Fact Sheet—Asset Management for Sewer Collection Systems;* U.S. Environmental Protection Agency, Office of Wastewater Management: Washington, D.C.

U.S. Environmental Protection Agency (2005) *Guide for Evaluating Capacity, Management, Operation, and Maintenance (CMOM) Programs at Sanitary Sewer Collection Systems;* EPA- 305/B-05-002; U.S. Environmental Protection Agency, Office of Enforcement and Compliance Assurance: Washington, D.C. http://www.epa.gov/npdes/pubs/cmom_guide_for_collection_systems.pdf (accessed June 2010).

U.S. Environmental Protection Agency (2010) *Storm Water Management Model User's Manual, Version 5.0;* EPA-600/R-05-040; U.S. Environmental Protection

Agency, Water Supply and Water Resources Division, National Risk Management Research Laboratory: Cincinnati, Ohio. http://www.epa.gov/ednnrmrl/models/swmm/epaswmm5_user_manual.pdf (accessed Oct 2010).

Water Environment Federation (2005) *GIS Implementation for Water and Wastewater Treatment Facilities;* Manual of Practice No. FD-26; McGraw-Hill: New York.

Wikipedia (2009) *Information Management* Web Site. http://en.wikipedia.org/wiki/Information_management (accessed June 2010).

8.0 SUGGESTED READINGS

Bolotsky, S. (2008) The Asset Management Bandwagon. *Water Wastewater Digest,* 48 (11).

Shamsi, U. M. (2007) *Navigating a Smart User Strategy,* Water and Wastewater News Online, Environmental Protection Web Site. http://eponline.com/articles/2007/10/01/navigating-a-smart-user-strategy.aspx?sc_lang=en (accessed June 2010).

Shamsi, U. M. (2008) Chapter 14: Arc Hydro: A Framework for Integrating GIS and Hydrology; In *Intelligent Modeling of Urban Water Systems, Monograph 16;* James, W., Ed.; Computational Hydraulics International: Guelph, Ontario, Canada.

Chapter 5

System Characterization

(continued)

1.0 INTRODUCTION

To develop a plan for controlling combined sewer overflows (CSOs) or sanitary sewer overflows (SSOs), the existing system must first be understood. The better the understanding of what causes overflows, the more targeted the selected plan for control of those overflows can be. *System characterization* is the assessment of a system's existing configuration, condition, capacity, flows, and effects on receiving water quality. System characterization relies on use of existing information, collection of new information, and use of tools to assess the behavior of the system in a variety of conditions that are not directly observed. The culmination of a system characterization effort is a quantitative and qualitative assessment of the system's configuration, capacity, generated flows, and effects on receiving water quality. The system characterization evaluations and tools can be used to evaluate overflow mitigation technologies, assist with mitigation and implementation plan creation, and be updated as the system changes to evaluate predicted performance.

1.1 General Considerations for System Characterization

The effort associated with characterizing an existing system can be costly and time-consuming. Generally, the detail associated with system characterization increases as the stage in the implementation effort advances. More specific data and analysis are required as programs move from the feasibility phase to the planning, design, and assessment phases. The investment in collecting data and using modeling leads to an improved accuracy of system understanding that helps ensure that capital facilities are adequately sized to accomplish their objectives, yet not oversized beyond what is required. System characterization should be viewed as a continuing process through the life of plan implementation. For example, large facilities that are being implemented in the collection system and the treatment plant may change the behavior of the system. Predicting changes in the service area tributary to the collection system over the course of 15 to 20 years is difficult, leading to a degree of uncertainty. Some items to consider include

- Considerations relative to data that describe the system:
    - System control plans typically project the performance of the system over all seasonal conditions. Data should support year-round evaluations (e.g., flow monitoring programs should reflect all types of seasonal conditions);
    - Historical data used in analyses may have been collected during different operational and maintenance practices. These conditions need to be understood (e.g., a pumping station where discharge has been limited to protect the wastewater treatment plant [WWTP] downstream or where pumps were out of service during data collection); and
    - Changes in the collection system during periods of historic flow records (e.g., changes in population, major industrial flow, or construction of local sewer projects may affect system behavior).

- Changes in system performance during program implementation:
  - Changes in the physical configuration of the sewer system during program implementation are likely. As such, characterization of the system needs to define not only overflow from the system, but the flows generated in the system;
  - Performance of constructed facilities, that is, information that can be gathered from constructed facilities, such as the quality of effluent, should be considered for updating the projected post-construction program performance as facilities are implemented;
  - Each project that provides flow reduction, hydrologic modification, conveyance, storage, or treatment is likely to have an effect on the performance of the system that requires some level of re-evaluation based on post-project implementation data collection; and
  - Changes in population, flow generation (development, redevelopment, and plumbing codes) or climate may have an effect on the proposed plan that requires adjustment during implementation periods.

Data collection and assessment can be performed by both internal staff and hired consulting or contracting firms. Understanding the time scales for data needs may help to determine whether a utility would like to establish internal staff to manage or maintain the programs. The need for long-term information at a location may support the installation of permanent metering equipment versus short-term installations. In all cases, the quality of the data collected should be documented and understood.

1.2 Overflow Quantification

In programs that evaluate CSOs and separate sanitary overflows, several quantifiable parameters are used to describe the system and the overflows. While the goal of CSO control programs is to meet water quality standards (WQS), these parameters help to characterize the system behavior. The following metrics are used to describe discharges from combined systems and to characterize CSOs:

- *Frequency of overflow*—number of events in a representative year or longer period, where frequency assumes some inter-event period without overflow. Typical inter-event periods include 6, 12, 24, or 48 hours (measured at the outfall).

- *Number of system events*—number of times in the course of the representative year that the system overflows either to a specific receiving water or as a whole. This definition is consistent with the U.S. Environmental Protection Agency (U.S. EPA) CSO policy.
- *Volume of overflow*—total volume of overflow generated in a representative year or during a specific design event.
- *Duration of discharge*—number of hours in a representative year during which the system is overflowing to receiving waters.
- *System capture*—total percent of wet weather volume in combined areas that is captured by the system and conveyed to treatment. It is typically evaluated on a representative year basis.
- *Pollutant discharge and pollutant capture*—total load of a pollutant of concern that is discharged during the representative year or during a particular design storm of interest; also, the amount of specified pollutant captured.
- *Treated discharge*—amount of system discharge (CSO) that is treated in wet weather facilities should be distinguished from untreated overflow.
- *Threshold event*—the size of precipitation event that results in overflow, either from an individual outfall or a system as a whole.
- *Other items specifically related to state regulations.*

Wet weather flows in separate sanitary sewer systems (SSSs) are more typically characterized by the response of the system. Some metrics used in wet sanitary systems to reflect the system behavior include the following:

- Total percent capture of rainfall (volume of excess flow that is captured by the sanitary system, expressed as a percent of the rainfall volume that generated the flow), also expressed as "R" value.
- Peaking factor from dry to wet weather (ratio of peak wet weather flowrate under specified condition to the average dry weather flow). This may be misleading dependent on base infiltration or limitations in system capacity.
- Unit flowrate per unit of main line sewer (expressed as gallons/day/inch diameter mile or similar value).
- Unit flowrate per population (or population equivalent), such as gallons/capita/day or gallons/day/residential equivalency unit.

- Unit flowrate per area, such as gallons/acre/day.
- Unit flowrate per parcel (used primarily in residential areas), such as gallons per m/house.

These quantitative metrics support both hydraulic modeling of sanitary systems and help to rank the relative response of different portions of the system.

Sanitary sewer overflows are more typically evaluated for theoretical storms of various return frequencies versus the use of a representative year. This is a result of the expectation that SSOs should not occur, or at least be rare, at a minimum less than once per year. The Clean Water Act (CWA) does not define an acceptable frequency of SSOs as they are considered illegal. However, planning studies and system characterization do evaluate potential for SSOs under various theoretical storm conditions to assist in evaluation of alternatives. For those theoretical storms, the following characteristics may be identified:

- *Volume of overflow*—total volume of overflow generated in the specific event;
- *Duration of discharge*—number of hours overflow occurs during the specific event;
- *Peak flow generated*—peak flowrate that would need to be conveyed to avoid overflow during the specified storm event; and
- *System surcharge*—extent and severity of system surcharging during the specified theoretical storm.

Overflows may be caused by issues other than wet weather, including physical or operational problems and power failure. Those types of overflows are not fully discussed in this chapter as these investigative and analytical techniques are primarily focused on flow quantification and response. A comprehensive overflow control program should identify these other causes of overflows so that they can be remedied.

1.3 Characterization Approach

1.3.1 Physical System Characterization

The physical characteristics of the collection system must be defined to evaluate its behavior. This includes such elements of the system as sewers, regulator structures, pumping stations, inverted siphons, and other key system components. Identification of the existing system generally relies on system maps (either geographic information systems [GISs] or hand-drawn maps) supported by construction records and field investigations. Collection system physical characterization includes a determination of which sewers are combined, separate sanitary, or storm. Physical characteristics of the collection system include information on the condition of various elements and their age and construction history (including sequencing of sewer construction and other items such as plumbing codes), which can shed light on the design intent of certain system components. Data such as pipe size, elevation, slope, regulator chamber hydraulic characteristics, pumping station capacities, and other system elements are important in determining how flow will behave once it enters the system.

Collection system physical characterization includes an understanding of the tributary areas to different components of the system. This also includes such characteristics as sewershed delineations to various portions of the system, recognizing that, in combined areas, the sanitary and storm tributary areas are not necessarily the same. In addition, internal flow diversions may result in variable tributary areas from event to event. Each sewershed can also be characterized by land use, population, time frame of development, and plumbing characteristics prevalent on private property. Hydrologic characteristics such as soil type, imperviousness, and slope are examples of relevant information for system flow generation in combined areas.

Collection system capacity at various locations is based on the aforementioned physical characteristics, supplemented by operational data. Most collection systems have key hydraulic control points represented by the WWTP, major pumping stations, and other hydraulic controls (such as a river crossing on the interceptor system). In combined systems, regulators, which split flow between the interceptor and the receiving stream, serve as critical hydraulic control points.

Not all overflow locations are associated with intentional regulators. Some overflows may occur at low points on the system. System capacity limitations may also be realized when basement flooding occurs.

Care should be taken in developing an understanding of system capacity. Hydraulic constraints that occur at system pinch points (such as a pumping station or siphon) may result in system surcharging, which causes flow to travel in the opposite direction as that anticipated. Overflow locations or cross connections with storm sewers may not only allow flow into the system, but also act as relief locations. Deteriorated physical condition, including loss of diameter (crushing) or obstructions (sediment, roots, grease, and debris), may result in a loss of capacity. Collection system capacity should be validated through monitoring programs, interviews with operational staff, and flow records at key locations. Hydrologic and hydraulic (H&H) modeling can also support system capacity assessment.

1.3.2 *Collection System Flow Response*

A key aspect of collection system characterization is the determination of flows in the system and how they are transported through the system. This work is supported by flow monitoring programs and hydraulic modeling, which is the subject of Section 4.0 of this chapter. Flow monitoring quantifies the flows produced in actual conditions during dry and wet weather. Flow monitoring can also be valuable in identifying hydraulic constraints in the collection system. Hydrologic and hydraulic modeling is used to project system response to a variety of changed conditions (either physical system changes or different precipitation conditions) based on the measured performance obtained from flow monitoring programs. Hydrologic and hydraulic modeling is also the principal tool for evaluating how the system will behave following changes in its configuration.

1.3.3 *Receiving Water Effects*

Evaluation of receiving water effects from CSO and SSO discharges requires an understanding of the overall receiving water H&H conditions and the variety of sources that contribute pollutant load to the waterbody. Measuring the effect on the receiving water requires an understanding of the pollutant load discharged from the outfalls and how that pollutant mass affects the receiving water. Water quality evaluation may rely primarily on data evaluation or may include computation models to represent the conditions in the receiving waters following discharge. Data collection to support the characterization of the receiving water includes outfall sampling, receiving water sampling, and monitoring and flow characterization of both the outfalls and the receiving water. Receiving water characterization can be highly complex, as water quality models consider the effects of multiple pollutants (and how they change over time) and also the travel in the receiving water (potentially in three dimensions, reflecting complex hydrodynamics effects such as tidal influences). Receiving water effects are discussed in Section 5.0 of this chapter.

2.0 COLLECTION SYSTEM MAPPING AND ATTRIBUTE DATA

An understanding of the collection system needs to be developed to evaluate the behavior of the system. This work should begin with the development of overview figures that provide a summary of the system's flow diagram, hydraulic profile, and tributary area map.

2.1 Review of Existing Records

A review of existing information should include not only the physical description of the system, but also documents that shed light on the purpose of system elements and the sequence of system construction. Geographic information system records or system maps provide a good overview of the physical elements of the system, showing overall pipe configurations and locations of regulators and pumping stations and other major elements. Details on regulator structures, pumping stations (including capacity), and other complex system elements will require data gathering of a greater level of detail than is typically included in a GIS system. Construction or inventory drawings of these components, including data on pump curves or capacities, should be obtained, with care given to ensuring the most current drawing for the facility is obtained. Prior reports, including basis-of-design documents and facilities' plans that summarize the key elements of the system, should be consulted to gain an overview and understanding of the previous system and the intent behind various system elements. Understanding the timing of system improvements is also helpful, and timelines should be developed to clarify such items as what portions of the system were installed as relief, where sewers were added for drainage subsequent to sanitary sewer construction, or where the system was extended to consolidate treatment facilities from prior outlying systems. Additional information on data management systems is included in Chapter 4.

A review of existing records should include interviews with operations and maintenance (O&M) staff, long-term employees, and others with knowledge of the history of the system to capture the unique perspective of such individuals and to describe the maintenance history of the system. Information should be summarized onto a flow schematic that includes such elements as

- Treatment plants (water and wastewater),
- Intercepting sewer networks (combined, separate, and storm),
- Primary trunk sewer networks,
- Key relief sewers,
- Force mains,
- Pumping stations (stormwater, sanitary, and flood pumping stations),
- Combined sewer overflow regulators and diversion structures,
- Hydraulic control points (flow divides, gates, separate/combined system junctions, siphons, gates, etc.),

- In-line storage and treatment facilities, and
- Off-line storage/treatment facilities.

The older and larger the existing system, the more complex the search will be. Ownership of the pipe and facility network that comprises the system may be distributed amongst multiple parties, including a regional sewer authority, local municipalities, departments of transportation, and county-level road or drainage departments. A listing of potential agencies that may be needed for data acquisition should be identified and systematically contacted. Typical places for locating representative drawings and records are

- Wastewater collection or water utility records,
- Municipal department archives,
- State or county regulatory agencies,
- City planning commissions,
- Repositories in town halls or borough halls,
- Public libraries,
- Local public works departments,
- Local contracting and consulting firms, and
- State and county transportation archives.

Once the available drawings are obtained and key facilities are inventoried, historical operational data must then be compiled to characterize operation of key facilities identified. Useful operational data sources include

- Existing temporary flow survey and permanent monitoring data,
- Historic rainfall data,
- Water billing records (dry weather flow, infiltration and inflow estimates),
- Treatment plant headworks data,
- Monitoring reports from pretreatment programs,
- Pumping station data,
- Maintenance crew inspection reports,
- Public complaint files,

- In-line storage and treatment facility data, and
- Off-line storage and treatment facility data.

The following sources can provide additional information that can greatly improve an understanding of the development and operational status of the collection system:

- Bases of design documentation for treatment facilities,
- Previously completed in-house and consultant reports,
- Facilities' plans, and
- Regional water quality commission reports.

2.2 Drainage Area Mapping

As the performance of collection systems (whether combined or separate sanitary) is defined in terms of hydrologic response, an understanding of the tributary area to points of study on the collection system is critical. Drainage area mapping is based on the sewer system configuration and the topography of the area. Drainage areas may be overlapping, with multiple outlets serving the same area as a result of relief sewers or modifications to a system over time.

2.2.1 *Drainage Basins and Subbasins*

A definition of *tributary areas* should include clear identification of the sanitary and stormwater tributary areas to an outlet and an identification of an unserved area. It is rare, even for combined areas, to have exactly the same tributary area for both sanitary flow and stormwater, and this distinction needs to be quantified for modeling purposes. Unserved areas are those which, by topographic characteristics, would be tributary to a sewer system, but there are no sewers in the specific area. Examples of these areas include undeveloped areas, parks, and cemeteries.

The tributary area should be subdivided into appropriately sized drainage basins and subbasins. These should be defined, as a minimum, to a regulated overflow location, noting that the tributary area to an overflow that results from a surcharged system is often different than the area upstream of the overflow. Tributary areas subdelineation should also consider the original development timeframe of the tributary area, different land use characteristics, previously separated areas, and other key characteristics. Drainage basins are often subdivided as level of detail of study increases, and the delineation of the tributary area should be determined based on the current study objective.

2.2.2 *Land Use*

Land use characteristics for various drainage basins and subbasins should be categorized by property type (e.g., residential, industrial, and commercial); significant plumbing characteristics (e.g., constructed before/following foundation drain connection ban); presence of basements; age of construction (potential pipe type/construction method effects); and other characteristics relevant to system performance. In many studies, only those characteristics related to zoning code are included, but the other categories of information identified will support system understanding and improve the ability to characterize the system.

2.3 Collection of Missing Information

Missing information should be collected in a hierarchy of needs based on study requirements. Critical hydraulic controls are typically the most significant data to be collected, including regulator and overflow physical details, pumping station capacities and controls, and overflow physical controls. Missing information should be gathered using paper and document records, as previously noted, and field investigations. Accuracy of information should be comparable, with careful attention paid to datum used (including datum verification) and reliability of construction records (recognizing that many aged construction documents were not updated following completion of construction). Care to confirm physical configuration of older sewers constructed along waterbodies is recommended as older sewers are more likely to be modified subsequent to original construction because of hydraulic modifications for navigation, flood protection, or land use changes. A minimal level of field verification should be included to make sure that these critical elements of the system are properly understood.

2.3.1 *Balancing Information Needs, Costs, and Benefits*

As previously discussed, the phase of a project will affect the level of detail required. Additional detail is required as projects move from a planning to a design stage. The degree of precision of the data should be appropriate to the needs of the project stage.

2.3.1.1 Data Needs for Combined and Sanitary Sewer Modeling

Combined and sanitary sewer modeling requires data to define the H&H characteristics of the system that extend beyond the basic data previously described. Of critical importance for modeling purposes are data related to boundary conditions. These data include physical characteristics that control the hydraulic elevation, such as river stage, presence or absence of backwater protection, and wet well levels in pumping stations or at treatment works and other critical locations. Critical hydraulic structures such as siphons and junction chambers need to be defined in additional detail for proper modeling to occur. An understanding of flow routing changes between dry and wet weather conditions needs to be understood.

2.3.1.2 Information Management Needs

As data requirements and collected data increase, more attention to management of the data is required. In particular, the quality and source of data need to be understood. While GIS provides an excellent tool to store data, conflicts between data sources can lead to confusion. Each layer should be properly annotated with source (meta-data) information. A library and cross-reference of data sources should be provided to ensure ready access to relevant information. An approach to resolving data conflicts should also be developed.

Extensive data can be overwhelming and make it difficult to orient to the system. Multiple reports that address similar issues can be confusing if an up-to-date "current version" is not maintained. The effort to keep an executive summary of the current understanding of the system will be beneficial to the program team.

2.3.2 *Collection Methods*

Methods of collecting data are varied and provide various levels of detail. Physical data collection can involve survey of manholes and chambers, including either global positioning satellite data collection techniques or traditional surveying. System investigations that are intended to be comprehensive need to be prepared to follow system networks and confirm connectivity of the system through advanced investigation techniques such as dye testing, pole cameras, or closed-circuit television (CCTV).

Large hydraulic chambers often require confined space entry to confirm dimensions and critical components. This should be provided for in investigation planning. Pumping stations may require wet well drawdown tests to confirm capacity. Interceptors along creeks or rivers may need to be walked to determine potential for river inflow. Such investigations may include surveying structure elevations, if they are not previously available, to compare against flood elevations. Investigations should provide a budget to confirm system limits if questions exist on the tributary area limits.

2.4 Typical Collection System Map Products

The following map products are recommended for system planning:

- *System map*—a map depicting tributary areas, locations of overflow, waterbodies, land use, combined and sanitary areas, critical hydraulic structures (junction chambers, pumping stations, and treatment facilities), and major sewers;
- *System flow schematic*—a schematic representation clarifying flow splits, overflows, pumping stations, hydraulic controls, and capacities. Flow schematics should also include clarification of metering locations;
- *Interceptor (and other major sewer) hydraulic profile*—an overall hydraulic profile of interceptor system showing hydraulic controls, elevation of regulator dams/weir crests, and other critical elements. Receiving water hydraulic gradient under various return frequencies should also be provided; and
- *Regulator flow diagram and schematics*—detailed drawings of critical hydraulic splitter elements showing the flow control under dry and wet weather conditions.

3.0 COLLECTION SYSTEM CONDITION

Collection system condition refers to the structural integrity, operational functionality, and the potential for infiltration and inflow of various elements of the collection system. As collection systems age, the ability of the system to perform as originally intended deteriorates. In many instances, previous studies have attempted to characterize the system condition.

3.1 Review of Existing Records

Reviewing existing records, as described in Section 2.1 of this chapter, will provide background on the condition of the system and its original history. The various data sets should be evaluated for items specifically related to sewer system condition.

3.2 Infiltration and Inflow Considerations

Overloaded sewer systems are evidenced by surcharging conditions that can result in basement backups and the discharge of untreated wastewater to streams, lakes, or other waterbodies. Although constructed to collect and convey wastewater, separate

SSSs also inevitably convey a certain quantity of extraneous clear water. This clear water, or infiltration and inflow, can originate as groundwater, precipitation, surface water infiltration and inflow, or piped water flow (such as leaking water mains). The quantity of infiltration and inflow present depends on the physical condition of sanitary sewers and the number of connections that can contribute additional flow to the sewer system. The terms, *infiltration and inflow, infiltration*, and *inflow,* are defined by U.S. EPA as follows:

- *Infiltration and inflow*—the total quantity of water from both infiltration and inflow without distinguishing the source;
- *Infiltration*—the water entering a sewer system and service connections from the ground through such means as, but not limited to, defective pipes, pipe joints, connections, or manhole walls. *Defective* is used to describe both poorly constructed and deteriorating structures. Water distribution system leakage can contribute to excessive infiltration levels, and its contribution to the sewer system should also be understood; and
- *Inflow*—water discharged to a sewer system including service connections from such sources as, but not limited to, roof leaders; cellar, yard, and area drains; crushed laterals; foundation drains; cooling water discharge; drains from springs and swampy areas; manhole covers; summit manhole plugs; cross-connections from storm and combined sewers; tide gate leakage; catch basin laterals; stormwater; surface runoff; street wash water; or drainage.

In collecting data to support an initial characterization of the sewer system, it is necessary to distinguish between infiltration and inflow that occurs on an ongoing basis as a result of groundwater or stream inflow conditions and rainfall-dependent infiltration and inflow. Infiltration and inflow that occurs on an ongoing basis can be a potential source contributing to dry weather overflows in separate and combined sewer systems, and the addition of rainfall-induced infiltration and inflow can introduce additional capacity problems. Infiltration and inflow can be influenced by flooding and effects associated with tidal variations. When SSSs and wastewater treatment facilities are overloaded, either infiltration or inflow must be reduced or the sewer system and treatment facility capacities must be revised to accommodate the excess flow. Thus, it is important to account for excessive quantities of infiltration and inflow that can cause hydraulic overloading of both the sewer system and wastewater treatment facilities in evaluation of CSO and SSO abatement technologies.

Many programs begin with a broad evaluation of the quantity of infiltration and inflow in sewer systems to rank areas for further study. Rankings include such unit measures as "percent capture", flow rate per unit of sewer, and peaking factors. However, all measures should be used with care as hydraulic limitations and upstream overflows may limit the capacity of the system to convey flows.

3.3 Collection of Additional Condition Information in a Field Inspection Program

Field inspection programs are structured to identify sources of infiltration and inflow that occur from public and private sources. Inspection programs include such aspects as system mapping, manhole inspections, smoke and dye testing, internal inspection of sewers, and private property investigations. These work activities can be performed as part of a comprehensive sewer system evaluation survey (SSES) intended to comprehensively identify the condition of the system and the sources of infiltration and inflow or as individual targeted activities.

3.3.1 *System Mapping*

System mapping is used to identify the connectivity of the system and is particularly important if existing mapping is unclear or outdated. It is important to have clear definition of the storm drainage system in the vicinity of the sanitary system (where systems are separate) to confirm absence of cross connections and to validate the limits of the system. System mapping can help to identify anomalies when performed by an experienced field crew.

3.3.2 *Manhole Inspections*

Manhole inspections are intended to confirm the condition, pipe connections, and placement of manholes. The placement of manholes in areas susceptible to flooding or where evidence of overflow has occurred should be noted in such an investigation. The condition assessment should be performed in conjunction with a standard coding system for consistency of measurement. One such industry coding system is the National Association of Sewer Service Companies' (NASSCO) *Manhole Assessment and Certification Program* (NASSCO, 2006).

3.3.3 *Smoke and Dye Testing*

Smoke and dye testing are investigative techniques that help identify potential inflow sources to the SSS. Smoke testing entails introducing smoke to the collection system to identify the location of physical defects without entering the system.

For instance, introducing smoke to the sewer can identify connected roof drains, as smoke would be emitted from the rain gutters of connected properties. Similarly, smoke could rise from pervious surfaces overlying a defective lateral connection that may be contributing infiltration to the sewer. The most effective use of smoke testing includes visual observations of all potential sources. To the maximum extent practicable, all areas tributary to a sewer, including rear yards and building rooftops, should be observed during smoke testing. Smoke testing effectiveness is limited when existing sewers are large in diameter as smoke travel is limited. Plumbing codes and practices in the jurisdiction should be understood to clarify results that might be affected by practices such as building traps that prevent the passage of smoke.

Dye testing entails placing a concentrated dye in the conduit, appurtenance, or other potential source in question and checking the downstream sewer system for traces of the dye. This type of testing has proved useful in identification of illicitly connected laterals in storm and sanitary sewers. The techniques (i.e., smoke and dye testing) are often used in tandem for identification and verification or dye testing may be used alone where systems connections are trapped and smoke cannot pass.

3.3.4 Internal Inspection of Sewers

A variety of techniques are available for internal inspection of sewers, and this technology is rapidly advancing. Technologies include traditional CCTV inspection, sonar and radar inspection, and electrical and electromagnetic technologies and other innovative methods (U.S. EPA, 2009). Zoom cameras provide options for system inspection from manholes. Internal inspection provides information on the sewers' structural condition, connections, grade, solids deposition, and sources of infiltration.

Closed-circuit television inspection systems offer a cost-effective means of conducting detailed physical inspections in sewers that are not flowing full at the time of inspection. It can readily document system physical condition, as previously noted, and can support identification of sources of infiltration and inflow if conducted at times when it is occurring. Data are typically captured in a video format on either DVD or computer hard drive, with integrated software to quickly locate visual images of defects. Coding systems such as NASSCO's *Pipeline Assessment and Certification Program* (NASSCO, 2001) have become industry standard. A *Lateral Assessment and Certification Program* has also been developed by NASSCO.

3.4 Quality Assurance and Quality Control

Quality collection of information is important in planning follow-up actions. Detailed quality review processes should be considered. In all efforts of this type, the reliability of the data requires that it be as comprehensive as possible. Otherwise, misleading information could be developed. Examples include manhole inspections that do not clearly identify the source of water from every pipe that enters the manhole or smoke testing that does not position staff on the roof of large commercial properties to determine whether smoke is emitted from roof sumps.

Confirmation of the connection of all potential stormwater sources is difficult and requires a commitment to use of multiple types of data collection (such as smoke and dye testing in tandem with CCTV work). Adequate time should be provided for the necessary effort. For example, dye testing may require extensive periods for dye travel if the sewer system has minimal flow and significant debris. Documentation of field investigation should identify what is confirmed and also clearly identify those items that could not be confirmed for follow-up investigation.

4.0 CHARACTERIZING SYSTEM FLOW AND RAINFALL RESPONSES

The fundamental objective of any flow characterization program is to determine the system's hydrologic (i.e., flow generation) and hydraulic (i.e., flow transport) characteristics. Flow characterization includes flow monitoring (to measure the actual flow conditions, specifically as it relates to volumes, rates, and system hydraulics); precipitation monitoring (to understand the rainfall that results in wet weather flow); monitoring of key hydraulic controls (such as critical wet well levels, the WWTP capacity at the outlet of the collection system, and the elevation of the receiving water); and H&H (computational tools that mimic system flow generation and behavior). As noted in the introduction to this chapter, the purpose and stage of the study influence the required level of detail and precision of calculations. A program can be driven by regulatory requirements (CSO and SSO control, infiltration and inflow, or SSES studies); system maintenance (prevention of basement backups and monitoring major users' contributions); or service needs (extension of interceptors or additional connections to the collection system). In any case, the underlying questions are

- How much flow is being generated?
- How quickly is it generated? and
- How is the flow transported through the system?

The characterization effort may include objectives for both understanding current system behavior and the anticipated behavior as changes in the system occur in the future, either in terms of flow contribution or system components.

4.1 Planning a Monitoring Program

A successful monitoring program is dependent on proper planning. The monitoring program must begin with study objectives in mind. Planning of the program should clearly define the questions that will be answered, how accurately they need to be answered, and at what level of detail. The program should be clear in whether it is only intended to evaluate current conditions or whether it will be used to support modeling that will, in turn, project future conditions. The following discussion includes suggestions to ensure major objectives are defined and addressed. Other references such as *Guidance for Quality Assurance Project Plans for Modeling* (U.S. EPA, 2002) should also be consulted for additional detail. Some of the key issues that need to be considered are

- What existing flow data are available (e.g., pumping records, WWTP records, overflow records, and prior studies)? What does it tell us about performance of the system, source of flow origination, and critical wet weather conditions?
- What amount of flow data are required to answer the study objectives? This must be considered both in terms of spatial density and varying conditions (rainfall and seasonal changes in response).
- What level of accuracy is required for the data? This will influence the type of meters selected.
- What level of investment is appropriate? This should consider the scale of the project that will be defined by the data.

Metering logistics—are suitable locations for metering to answer the study objectives available? If not, what adjustments can be made to collect quality data that accomplishes a majority of the objectives?

- How will the integrity of the data be checked?
- How will data management be accomplished?
- How will data be analyzed and where will it be stored?
- How will flow data that is collected be projected to design conditions?

4.1.1 *Identifying Spatial and Time-Frame Data Requirements*

Flow monitoring programs are designed based on study objectives and the project phase. Programs may be intended to provide information to support initial assessment, planning, design, project performance, or ongoing system performance. Each of these phases requires a different level of information to meet the objective.

Density and placement of metering equipment is a primary consideration. Metering density should ideally be determined based on relatively consistent tributary area characteristics. Metering density may be based on sewer system configuration (such as metering each major input to an interceptor), sewershed size (such as a pre-determined acreage or lineal foot of pipe), jurisdictional boundaries (passing from a system under one owner to another), or other similar factors. Density will generally increase as the level of detail required increases. Placement of metering equipment should be performed to ensure that direct measurement of generated flow is accomplished. For example, measuring overflow discharge does not identify the amount or rate of flow generation, but reflects a combination of the flow generation characteristics and the hydraulic function of the system. It is recommended that flow measurement be performed upstream of overflow and flow-splitting locations. Placement should minimize the need to have one meter subtracted from another. This may lead to significant flow calculation error (Czachorski et al., 2008; McCulloch et al., 2007).

Duration of metering is also important. In locations with significantly varied seasonal response, the data collection period should span a variety of hydrologic conditions. The duration of metering should be sufficient to capture the number of events required for model calibration and validation. This requires a minimum of five to six representative events (U.S. EPA, 1999). More than the minimum are required if the system response varies depending on antecedent moisture conditions. If long-term (seasonal) changes in antecedent moisture conditions significantly affect the response of the system to rainfall, 2 years of monitoring may be required to calibrate and validate a hydrologic model. Planning of the overall program schedule to allow for appropriate seasonal monitoring may require additional planning time for the deployment of flow metering equipment.

4.1.2 *Temporary and Permanent Monitoring Concepts*

4.1.2.1 *Temporary Programs*

A temporary metering program is typically one in which a system operator is seeking to determine flow characteristics for a specific, narrowly defined purpose. It is often the basis for sewer system modeling and making decisions on capital improvements projects. A temporary metering program is intended to capture the conditions most relevant to the modeling efforts or capital improvements being considered. This could be the volume of overflow needed for CSO control, quantifying the amount of inflow, or estimating the rate of flow that must be transported to service a new area.

Site installation complexity associated with temporary metering programs is typically less involved than permanent metering programs and relies on batteries for power supply. Data downloads are typically not integrated into the system owner's supervisory control and data acquisition (SCADA) system, and often require visits to the monitoring site. The program almost always includes wet weather infiltration and inflow evaluation so that they are scheduled to coincide with weather conditions that generate the highest infiltration and inflow rates.

4.1.2.2 Permanent Programs

A permanent metering program is designed for the long-term management of a sewer system. Data collection supports ongoing operation and maintenance, regulatory reporting, and future projects planning. As long-term control plans incorporate adaptive management objectives, permanent metering networks should be considered to support those objectives. Another example of a permanent network consists of flow meters placed to enable a metropolitan collection utility to bill major users or communities for sewer use based on the wastewater flowrates generated by that user. Permanent metering networks have more intensive capital requirements that will result in less labor-intensive operation over the life of the network, although adequate resources must be provided to ensure accuracy of metering equipment.

Permanent metering programs have to deal with a significantly different set of installation and operation needs than do temporary programs. Permanent metering programs will have more extensive siting, equipment considerations, power, communications, and data management considerations. Long-term data collection should be performed with a higher level of precision than a temporary metering program. To accomplish this, construction of high-quality hydraulic monitoring locations that enhance the reliability of metering data may be appropriate. In addition, higher-end metering equipment that offers a greater degree of precision may also be appropriate. Permanent meters tend to have a permanent power source so convenient access to the electrical utility is a necessity. Permanent meters can be battery-powered, but the replacement of new batteries and their installation will need to be factored into O&M costs. Data are almost always downloaded from permanent meters by means of telephone, cellular, or radio devices. These save labor costs because the metering sites do not need to be physically accessed to acquire flow data, but they do require an analysis of which type of communication is the most accessible, cost-effective, and easy to use. Data from long-term metering programs will require management. Often, this is incorporated into a SCADA system's data management software. The data need to be stored and permanently archived such that they are retrievable in a usable format.

4.1.3 *Selecting Meter Locations*

The success of a flow monitoring program is highly dependent on the proper site selection of a meter location. The difficulty of a site selection lies in the paradox that the logical subdivision of a sewer system may guide the program to use a structure that does not provide good hydraulic flow characteristics for the meter technology used. For example, the junction of two major interceptors or sewers conveniently divides a major area into two equally sized subdistricts, but a hydraulic interference between the two sewers or similar problem may prevent the collection of accurate data. This is a common problem encountered in flow monitoring programs that often results in a compromise between the subdistrict delineation and meter operation.

Good hydraulic characteristics are critical for obtaining quality data. The following should be considered:

- Free-flowing conditions and reasonable velocities away from turbulent or disturbed flow are preferred. Flow meters work best when the hydraulic conditions are sub-critical flow under the widest possible range. Very high velocities (often supercritical flow), turbulent conditions, or very low velocities can contribute to inaccuracies. For primary devices, a high "velocity of approach" can skew the calculated flowrates significantly. Area-velocity meters can still compute a flowrate under these conditions, but the meters are considered somewhat less accurate. Pang (2008) suggests consideration of surface water profiles in site selection, with installation of metering equipment in a subcritical section followed by a supercritical section preferred and indicating generally better performance in a section of pipe that has a downstream steeper pipe section. The metering equipment should be located far enough into the pipe to avoid flow transitional areas. Carlson et al. (2007) report best results when velocities are in the 61- to 122-cm (2- to 4-ft)/sec range.
- Backwater conditions should be avoided. This is the phenomenon whereby a restriction, blockage, or excess flow downstream causes the flow at the meter location to back up and the level in the monitored pipe to rise above what

would be expected for the flowrate observed at that time. This may occur constantly or intermittently depending on the cause and location of the cause. The advantage of area velocity meters is that they can still calculate flow under these conditions, but the accuracy is typically less than under more ideal conditions. Some primary devices can be subject to moderate backwater, but the accuracy is diminished. Look for debris that collects on the walls or steps of the structure for an indication of regular backwater or surcharging conditions when inspecting a structure.

- Evidence of silt or debris accumulation should be minimal at selected sites. The existence of silt or debris reduces the area of a pipe being measured and can impair the operation of sensors in area-velocity meters. While silt may be removed through cleaning, it is quite common for silt to re-accumulate due to slope of a particular pipe section. Many metering technologies have routines for calculating the effect of silt; however, these are approximations at best and often vary during the course of a monitoring period, making those assumptions inaccurate much of the time.

Selection of metering locations also should consider needs associated with installing and maintaining equipment. By its nature, flow monitoring requires staff to enter confined locations on the sewer system; as such, the following effects need to be considered:

- *Accessibility*—can the structure be accessed by personnel under all conditions? Structures located in floodplains, traffic areas, security zones, or in buildings may present difficulties to staff attempting to access the meter for data collection or maintenance.
- *Physical characteristics*—does the metering structure size or physical condition present any special problems? A manhole that is extremely deep, in poor structural condition, or subject to grease, gas, or other noxious and dangerous substances due to upstream industrial discharges may make it difficult, if not impossible, to work safely in the structure.
- *Data collection requirements*—if the device is a permanent device, are communication systems available? If not, is it practical to collect the data manually?

Another element unique to selecting sites for CSO monitoring programs is related to estimating overflows. Monitoring to directly measure overflow is challenging because of the intermittent nature of the flow and the complex hydraulics at the overflow location. Weir equations have often been used to estimate CSO discharges, and these are often misleading as the regulator dam may have significant approach velocity, may be affected by downstream control, or may function as an orifice if the ceiling height forms a constriction under certain conditions. An alternative is to install meters located upstream of a regulator and downstream in the connection to the WWTP. The overflow rate would be obtained by subtracting the flow in the connection to the WWTP from the flow upstream of the regulator. This may be a better option, although the connection to the WWTP may be significantly turbulent, resulting in poor flow data; additionally, use of meter subtraction is inherently likely to induce error.

4.1.4 *Selecting Meter Technology*

The type of meter best suited for an application is dependent on a number of site-specific and project-related factors, not the least of which is cost. Consequently, the collection system manager should ask the following questions:

- Will the device be installed in a newly constructed structure and conduit or retrofitted into an existing one?
- Is the meter installed in an open-channel or a closed-pipe?
- Is the flow monitoring program going to be installed on a permanent or temporary basis?
- How will data be downloaded from the device?
- How much money does it cost to install and operate the device?
- Is the device subject to surcharging or backwater conditions?

4.1.4.1 *Primary Devices*

In the sewer flow monitoring field, the term, *primary device,* generally refers to something that requires a simple level measurement to calculate a flowrate. These are all open-channel monitoring devices. Primary devices are typically limited to weirs (e.g., broad-crested, sharp-crested, and V-notch) and flumes (e.g., Parshall and Palmer-Bowles).

Primary devices typically require more head loss to operate effectively than other types of meters. For that reason, they are generally more suited for installation in newly constructed facilities. Placing them in an existing sewer section may be impractical and cost-prohibitive unless the head requirements can be attained relatively easily.

While primary devices theoretically result in clear discharge versus depth relationships, in practice they are affected by local hydraulics and installation. Primary device equations typically assume a stilling pool upstream of the device, which may not be available. Additionally, backwater or high-flow conditions may submerge a flume. Testing of flumes using dye dilution has shown that correction values may be required (Pistilli et al., 2006).

4.1.4.2 In-Pipe Flow Meters

The technology of in-pipe flow meters has greatly improved over the last 25 years and provides analysts with opportunities to measure flowrates in far more situations than were previously available. In-pipe flow meters can be either closed-pipe or open-channel. Closed-pipe meters are discussed in the following section.

Open-channel, in-pipe meters are also called *area-velocity meters* due to the method by which they compute flowrate. A physical area of flow is calculated by measuring the depth in a pipe, and the velocity is measured by a type of ultrasonic, radar, magnetic, or Doppler device. The continuity equation yields the flowrate from the product of these two computations.

There are many types of area-velocity meters available on the market today that use a wide variety of level and velocity measurement technologies. One area of difference in area-velocity meters is the number of individual velocity sensors used. Single-path meters are generally smaller and less expensive and can be adapted to temporary metering applications. Multipath meters are generally used for larger conduits in permanent installations; they are more expensive and can be more accurate. Transit time meters may provide high-accuracy flow measurement, although their ability to operate in situations with entrained air is limited.

The means for measuring velocity varies widely and is the subject of much debate, especially among manufacturers. Velocity can be measured at a single point (either the top or the bottom or the flow regime) and converted to an average velocity based on a proprietary methodology or multiple readings can be used to determine the average velocity.

Single-point velocity measurements depend on a consistent relationship between the average velocity in a conduit and the measured point velocity. A profile of the average velocity in the conduit must be determined under varying conditions, if possible.

The level measurement is typically performed by either a pressure transducer or an ultrasonic or radar sensor above the flow. Some programs use redundant-level sensing devices. Limitations associated with each type of device need to be understood (e.g., "dead bands" for ultrasonic devices and blocked or damaged pressure sensors). An understanding of the flow calculation method that correlates the depth measurement location with the velocity measurement location may be critical.

All types of flow metering equipment have an inherent best application, and no equipment can overcome poor site hydraulics, so a combination of appropriate equipment selection with good site selection is critical to a satisfactory program.

4.1.4.3 Pumping Stations

Flow metering at pumping stations most often use some type of "closed-pipe" metering device. While closed-pipe meters have been available in drinking water systems for many decades, many of these technologies are not feasible in wastewater applications. The development of new technologies has enabled the installation of metering devices in pumping stations. By far the most common and longest used are the electromagnetic meter and the Venturi meter. Most recently developed is the transit-time meter, which measures velocity in a manner similar to the transit-time open-channel meters.

4.1.4.4 Other Techniques

In some instances where budget constraints do not allow the use of expensive flow meters or where the need for quantitative data is not a high priority, a number of qualitative tests can be performed to provide an estimate of high flow conditions. A common technique is the use of elapsed time meters in pumping stations. If the pump flow characteristics are reasonably well defined, then the time a pump runs is multiplied by an average pumping rate to yield a flowrate. Sometimes, a wet well test is conducted to verify the pump characteristics when no other reliable information is available. This is performed by measuring the time it takes to draw down a wet well of known diameter to calculate the pump capacity. An estimate of the influent flowrate to the wet well should be accounted for in this test to improve accuracy. The influent flowrate can be estimated by timing the rise in the wet well with all of the pumps turned off immediately before the draw down test.

Dye dilution is a methodology that is used to test the accuracy of flow metering devices. A sample of fluorescent dye is injected into the sewer and the concentration of dye is measured at the meter location or downstream point. The amount of dilution observed is inversely proportional to the flowrate. This principle can be applied in the same manner to estimate flowrate at entry points to the sewer downstream from the dye injections. This can yield highly accurate results, but is limited in the time in which it can be conducted because of cost and labor requirements (Kumpula and Minor, 2006).

Other techniques have been tried successfully when all that is required is an indication of surcharging in a sewer. This information can be used as a crude basis for estimating flowrate or overflow from a combined sewer structure. The inside of a manhole structure can be marked with chalk and examined after an event to determine the high water reached at that location of the sewer system. Although this provides good evidence of the extent of surcharging, it does not indicate how long the high water level was reached. There have even been instances of ping pong balls being placed in a regulator or similar structure to indicate the level of surcharging. While these can be useful tools, they should not be interpreted as more than gross estimates of the flow conditions being observed.

4.2 Measuring Rainfall

Collection of rainfall data and their integration with flow monitoring data is essential for understanding the process of flow generation and the development of collection system hydraulic models. Rainfall data, when coupled with system-wide flow monitoring data, close the loop and establish the cause and effect relationship of the collection system response to rainfall events. Rainfall duration and intensity are so variable that parts of the collection system may be dry while others are deluged. Knowing that rain occurred only in a part of the collection system can have significant implications on model calibration, which, in turn, can have significant implications on the most environmentally sound storage and overflow options. Additionally, rainfall data can be helpful in determining spill thresholds for CSOs and in defending the municipality or agency against flooded basement claims.

Long-term airport rainfall records are typically adequate for establishing historic rainfall patterns useful for planning level models used to size facilities that serve a relatively small portion of the system. Airport data, or other National Weather Service gauges, generally have excellent quality of data due to strict standards for data collection and gauge configuration that are used in the program. Airport records are not sufficient for understanding flow monitoring data being collected in a flow monitoring program or for projecting the effect on a large system of a storm event. These activities require more detailed representation of the magnitude and directional path of the storm event.

Rainfall data can be collected through a series of gauges, or through radar–rainfall methods. Rainfall data can be used to understand system behavior as documented through monitoring programs or they can be used in real-time CSO control facilities. Real-time rainfall data access and predictive software packages support real-time control practices in CSO facilities. Real-time control systems have also used other rainfall predictions and operator experience. Predictive capabilities for projecting rainfall intensity can be used to actuate collection system pumps and hydraulic controls to minimize system-wide overflows and to support decisions for the use of storage facilities.

4.2.1 *Rain Gauge Density Considerations*

Historically, there have been several schools of thought regarding the density of a rain gauge network. In Canada, rainfall and collection system modelers recommend one gauge every 1 or 2 km (0.6 or 1.2 miles). In Britain, the Water Research Centre has recommended only one-half that density, or one gauge every 2 or 5 km (1.2 or 3.1 miles). In the United States, current spacing recommendations are related to thunderstorm size. The average thunderstorm is 6 to 8 km (3.7 to 5 miles) in diameter and round in "footprint." Therefore, rain gauges are frequently spaced every 6 to 8 km on center.

In general, and to maintain quality control, the typical minimum rain gauge network density for a study basin of any size is two gauges (Lei and Shilling, 1993). Sites should not be sheltered and should be relatively secure from vandals. Public buildings often offer a good combination of optimal site characteristics, including security, freedom from rain shadow, localized wind patterns, and abruptly varied topography. Additionally, locally collected rainfall data can be compared with regional airport gauges as a quality assurance/quality control (QA/QC) check on the rain gauge siting using cumulative frequency distribution curves. In the case where significant deviation in rainfall volumes occurs, the gauge site should be calibrated and investigated for any sources of interference (Lei and Schilling, 1993).

4.2.2 *Ground Gauges and Radar–Rainfall Data*

Radar–rainfall data are provided by a combination of radar weather information with local ground-located rain gauge networks. The process entails the use of local rain gauge information to calibrate radar reflectivity images, provided either in real time or as historical data files, to yield rainfall data with a pixel-based geographic resolution of either 1 or 4 km^2. Radar-based rainfall is provided by service vendors that specialize in this type of data processing. Radar–rainfall data do not replace the need for local ground gauges, as these are required both for calibration and for rainfall monitoring during less-intense wet weather periods. Most radar-based rainfall data will be coupled with gauge data to compile a complete data history.

4.2.3 *Snowmelt*

The presence of snowmelt is a factor that can greatly influence an inflow or CSO runoff event and can skew analytical results if not properly accounted for. To account for the effect of snowmelt during winter monitoring conditions, different approaches may be used including local weather services records of water equivalency in snow pack or heated gauges to identify the water volume equivalency of snowfall. Both methods have limitations, such as limited recording locations for snowpack and timing issues, when considering the data from heated rain gauges. At a minimum, data should be available to characterize whether a significant change in the snowpack occurred during wet weather events during the winter.

Rain gauges used in locations and at times of the year that have subfreezing temperatures need to be heated and have wind protection. These gauges will record precipitation in a rainfall equivalent, whether it occurs as rain or snow.

4.2.4 *Evaluating Rainfall Data*

Evaluating rainfall data for both temporary and permanent meter applications is critical to the analysis of flow meter data. In both instances, rainfall data are compiled and related to historic rainfall conditions at a variety of timescales, including short (intensity), event (hours or days), seasonal (months), or long term (year or more). Because monitoring programs are intended to relate system flow response to normal or extreme conditions, the rainfall during the monitoring period needs to be understood in that context. Rainfall data analyses will often identify and rank events in terms of intensity, total volume, and duration, particularly in comparison to a theoretical design storm condition. Monthly, seasonal, and yearly data are typically evaluated for total rainfall and distribution and the number of events of each magnitude. Historic rainfall data (such as a National Weather Service gauge with an extended period of record) may be evaluated to determine characteristic rainfall volumes and distributions and the size of large events for the area under study.

4.2.5 *Groundwater Monitoring*

Groundwater levels play an important role in the amount of wet weather flow generated in collection systems. Groundwater conditions in various locations can affect the amount of runoff or rainfall-derived infiltration and inflow (RDII) that is generated. Widespread fluctuations in groundwater table and changes in localized groundwater (such as in sewer trenches or basement backfill areas) can each result in changes in flow response. Monitoring of groundwater levels is discussed in *Reducing Peak Rainfall-Derived Infiltration/Inflow Rates—Case Studies and Protocol* (WERF, 2004).

4.3 Collecting and Analyzing Data

Flow data collected from a monitoring device needs to be collected in a manner that safeguards its integrity and assures the analyst that it represents an accurate account of the flow conditions observed for the monitoring period. Flow data collection may be through direct field or telemetered downloads or through a SCADA system.

4.3.1 Field Data Collection

Field data collection can be through a data collection unit or laptop. The field visit where the download occurs provides an opportunity to check the site and perform independent measurements that can be used to verify the data (such as independent depth and velocity measurements). The field technician should note whether the equipment is operating properly, whether any debris has become attached to the sensor or in the flow channel, and whether there is any indication of system operational issues. A field review of the data may support immediate troubleshooting. The following considerations should be noted when collecting data:

- Condition of the metering location; note the presence of silt, debris, oil and grease, or anything else that may possibly affect the reading of the sensors. Record any actions taken to correct those conditions.
- Condition of the site; items such as surcharged condition, high water marks, debris on steps, or other visual observations of extreme wet weather response or other unusual operation should be noted.
- Remaining battery life.
- Does the data appear consistent with previous data collected or is there any drift in level or velocity sensor readings?
- Reference measurements (e.g., depth and velocity) to provide assurance that meter data are representative of actual conditions.

4.3.2 Supervisory Control and Data Acquisition

Supervisory control and data acquisition is a software tool whereby data from remote sensing units are stored in a central processing device such as a personal computer.

Data are transmitted to the central device through direct serial link, modem, radio, Ethernet, or similar communication technology. The software stores information from external devices such as flow meters, pumps, level sensors, gauges, and switches, and often displays the data in an illustrative and animated diagram that represents the process being monitored and controlled. Properly designed, SCADA has the ability to take a fairly complex system and give the user a logical and intuitive tool to make decisions on system control and management.

Permanent flow monitoring systems are best suited for taking advantage of the capabilities of SCADA due to the costs of developing the software. Real-time monitoring of flow data in CSO systems can provide valuable control of storage and treatment facilities to minimize overflows. Flow monitoring and rainfall gauge data can be combined through SCADA to support samplers, level sensors, and even alarms. More sophisticated programs can even perform analysis of data received and provide graphical displays such as pie charts, bar graphs, and tables.

Supervisory control and data acquisition systems do not replace the need to perform maintenance visits to sites to ensure that sensors and monitoring equipment are working properly. During site visits, information similar to that listed in the previous section (Section 4.3.1, *Field Data Collection*) should be noted. Additionally, the format of data storage in SCADA systems should be evaluated relative to how well it will support future data analysis. In some instances, SCADA systems are not amenable to retrieval of data in a manner that benefits the analyst. Quality reviews of data should be performed regularly to ensure the future value of the data. In essence, SCADA systems do not replace the need for ongoing efforts to ensure the quality of the collected data and its storage and retrieval in a useful format.

4.3.3 *Initial Data Review*

It is essential to review data promptly after it has been downloaded. This evaluation should occur weekly and biweekly at a minimum during the data collection period so that any problems can be addressed. Techniques used both routinely during the monitoring period and at its conclusion include (ADS Environmental Services, 2007)

- Level versus velocity comparisons;
- Flow versus depth scatter graphs; and
- Flow per unit (acre, person, etc.).

A comparison of the measured flowrate to flow depth provides information on the hydraulic conditions present during the metering period. Presented in a scattergraph of the data, backwater conditions can be identified and the characteristics of the flow can be compared with theoretical values. A sample scattergraph is shown in Figure 5.1.

4.3.4 *Unitized Flowrates*

Statistics for various flow components may be presented based on several physical characteristics of the tributary area. Flowrates may be "unitized" to different descriptive factors. The dry weather flow or groundwater infiltration may be compared to some descriptive value to unitize data to enable the analyst to compare sub-sewersheds of differing size. Descriptive factors identified in literature include (Bennett et al., 1999; Lukas, 2007)

- Collection system dimensions
    - Pipe length;
    - Quantity of pipe expressed as the diameter of the pipe times its length.; and
    - Pipe trench footprint (pipe length multiplied by standard trench width).

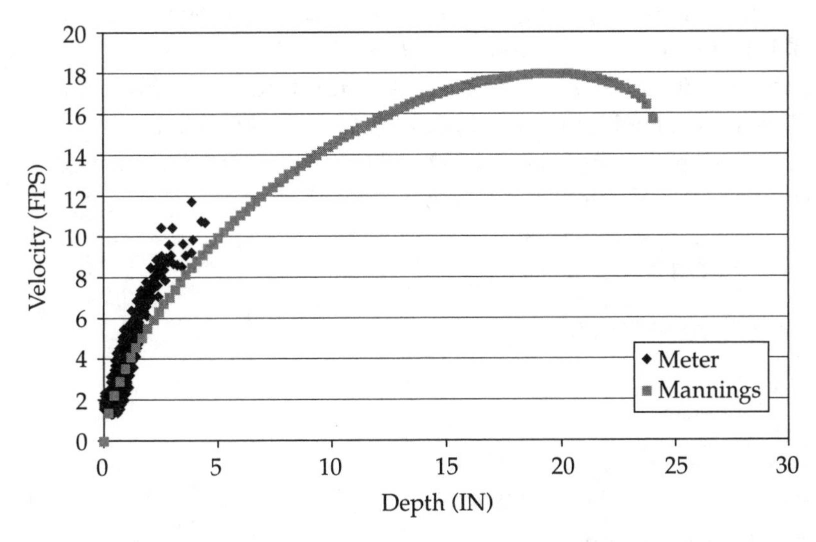

FIGURE 5.1 Sample scattergraph (Nelson et al., 2007).

- Base wastewater
  - Amount of base sanitary flow;
  - Number of lateral connections;
  - Equivalent residential units or equivalent dwelling units, includes both residential units and non-residential properties with flow expressed as an equivalent residential unit; and
  - Equivalent population.
- Hydrologic factors
  - Total sewershed area upstream of the meter and
  - Sewered area tributary to the meter.

4.3.5 *Dry Weather Flow Evaluation*

Dry weather flow is composed of base wastewater flow (the actual sanitary component) and dry weather infiltration. The dry weather infiltration is seasonally influenced and can change significantly from one time of the year to another. The dry weather flow pattern exhibits a typical diurnal pattern in the absence of any unusual outside influences such as rainfall.

Base wastewater flow can be estimated from an evaluation of water consumption in the tributary area, tributary population (most appropriate for predominately residential areas), or from an evaluation of the flow data. A number of methods can be used to estimate the base wastewater and infiltration components of the flow (Mitchell et al., 2007), with the minimum flow factor method being perhaps the easiest to apply for sewersheds between 750 and 7500 m^3/d (0.2 and 2.0 mgd). The infiltration component of the flow can be approximated as

$$\text{WWP} = (\text{ADF} - \text{MDF})/X \tag{5.1}$$

$$\text{BI} = \text{ADF} - \text{WWP} \tag{5.2}$$

where

BI = base infiltration;
WWP = daily average total wastewater production;
ADF = average daily flowrate;

MDF = minimum daily flowrate; and

X = fraction of WWP that accounts for nonzero nighttime wastewater production (0.7 to 0.75).

Each of the above flowrates (BI, WWP, ADF, and MDF) would be expressed in consistent flow units.

4.3.6 *Surface Runoff*

Surface runoff is modeled using hydrologic parameters that include the effective drainage area size and the physical imperviousness and soils characteristics of the tributary area. Other parameters are also considered, including directly and indirectly connected impervious area, drainage area shape, ground slope, and depression storage. An evaluation of small and large storms of varying intensity from the flow monitoring data set can be used to quantify the runoff generated from impervious and pervious areas and to support quantification of such hydrologically significant items as effective tributary area and depression storage. The model is then calibrated to match flowrates and volumes that have been recorded by flow metering results. Physical and hydrologic parameters such as initial abstraction, slope, and drainage area shape, among others, are primarily used for calibration.

Increasing consideration of low-impact development results in the need for additional evaluation of system hydrology. Low-impact development affects the initial abstraction and effective impervious area, and may act as distributed storage nodes.

4.3.7 *Rainfall-Derived Infiltration and Inflow*

Rainfall-derived infiltration and inflow is the term used to identify the hydrograph component that is the wet weather response in a separate sanitary system. The sources of flow present in separate systems are also present in combined systems, and inclusion of an RDII component supports evaluation of alternatives such as separation.

Rainfall-derived infiltration and inflow is calculated from flow data analysis as the additional flow that occurs during and following a wet weather event. As with other flow components, this may be expressed in unitized values, related to tributary area characteristics. Rainfall-derived infiltration and inflow sources include catch basins, inlets, leaking manholes, yard drains, roof downspouts, and similar defects. The volume of inflow generated during an event and the rate at which it is transported through the sewer system are both necessary to characterize. The volume of inflow will be important when sizing storage facilities or when estimating the total amount of flow to be treated. The rate of inflow is more important when determining what capacity is required for a new pipe to service an area.

Rainfall-derived infiltration and inflow quantification techniques used in the industry were included in a Water Environment Research Foundation study (Bennett et al., 1999) that evaluated eight broad categories of RDII prediction methods. The described methodologies are

- Constant unit rate method,
- Percentage of rainfall volume (*R* value) method,
- Percentage of stream flow method,
- Synthetic unit hydrograph method,
- Probabilistic method,
- Rainfall and inflow regression method,
- Synthetic stream flow regression method, and
- Methods embedded in hydraulic software.

One such approach embedded in hydraulic software recognized the influence of antecedent moisture on this flow response. This approach is described by Van Pelt et al. (2006). It is important to understand that the RDII response of a system changes with different conditions, such as antecedent moisture condition of the soil, prolonged rainfall, or the season of the year. As such, monitoring programs need to include data analysis of the various conditions to be able to predict the system response in subsequent modeling analyses.

Flows present in a separate sewer system include base wastewater flow, groundwater infiltration, and RDII. These are present in varying proportions culminating in the total flow. Typically, base wastewater flow and groundwater infiltration are the most constant, although they are affected by water use variation in the tributary area, seasonal changes in groundwater, and long-term rainfall effects on groundwater. RDII is typically the most variable, changing with immediate response to rainfall and subsiding after wet weather. Figure 5.2 depicts the various components of flow in a separate sewer system affected by RDII.

Figure 5.3 depicts the hydrograph de-convolution process that is used in the "RTK method" of RDII flow quantification. This method is one application of the unit hydrograph approach, where the total volume (*R*) is developed from flow monitoring data and hydrograph shape constants related to time and recession limb (*T*, *K*) describe the distribution of the total wet weather volume. It is one of a number of approaches to describing wet weather flows in a modeling application.

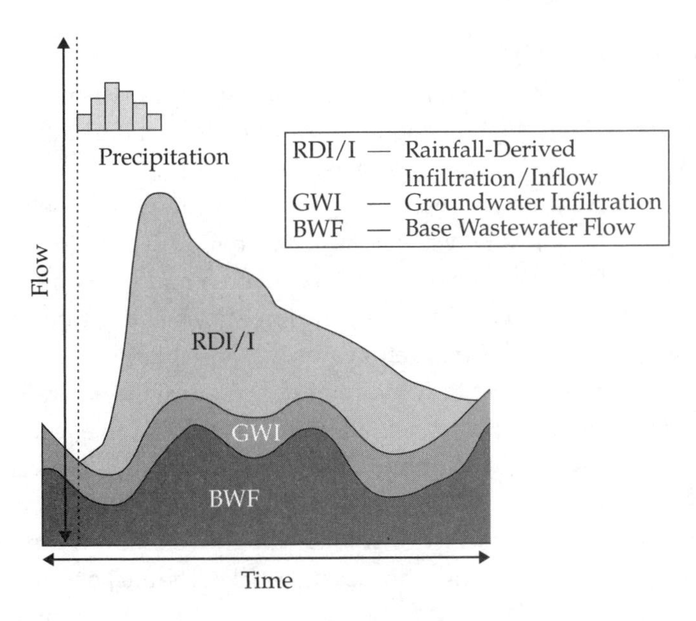

FIGURE 5.2 Three components of wet weather wastewater flow (Lai et al., 2007).

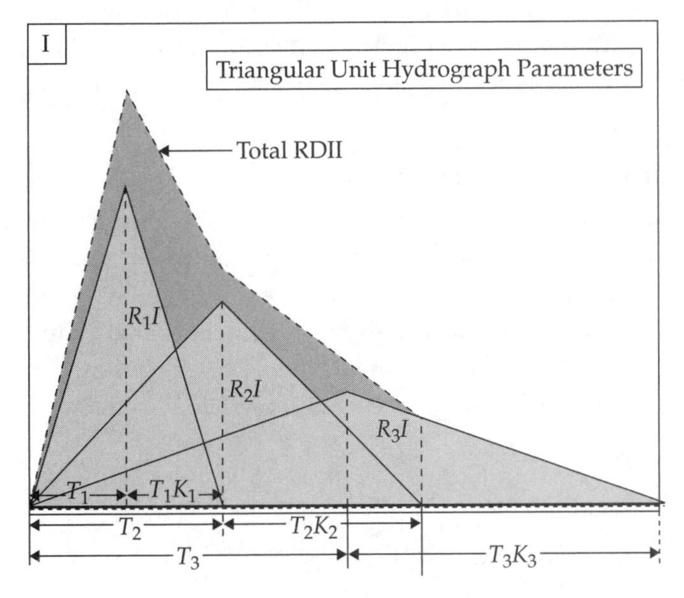

FIGURE 5.3 Rainfall-derived infiltration and inflow RTK hydrograph (Lai et al., 2007).

4.3.8 Hydrograph Analysis

Analysis of hydrographs to determine the quantity of infiltration and inflow requires a certain amount of interpretation and judgment on the part of the analyst. The first step is to separate the dry weather, or "base flow", from the inflow portion of the hydrograph. The base flow is intended to represent the flowrate at a monitoring point that does not include any wet weather contribution. The inflow component of flow would be the difference between the wet weather and base flow hydrographs. This may be performed during seasonally low times (such as dry summer months) or throughout the course of the year to reflect seasonal variability.

Ideally, the source for a base flow hydrograph would be the flowrate immediately before the wet weather event. Sometimes, the period preceding an event does not accurately represent the base flow for the time of the wet weather event. This can occur when there are significant differences between weekend and weekday flows or if there are significant users that discharge at non-regular intervals. The base flow component changes seasonally and with rainfall over the prior several weeks. The analyst may need to select a different period such as the week before the event. In permanent monitoring locations, there may be enough data to develop typical dry weather hydrographs that can be scaled up or down depending on the conditions observed for the event being analyzed.

Excess flow generated is separated from the base wastewater to assess the effect of the wet weather response. The volume and distribution of this flow component are quantified and used in prediction techniques described in Sections 4.3.6 and 4.3.7 of this chapter.

4.3.9 Tabulating Wet Weather Events

The desired outcome of a flow monitoring program and hydrograph analysis is the collection of multiple events from which one can draw conclusions and make projections on flow characteristics of a study area. The methodology for compiling events and performing a statistical analysis of the data can be highly influenced by local practices and regulatory requirements. Typically, the following information is required when tabulating wet weather events at each flow meter:

- Date of event,
- Total amount of rainfall,
- Average and peak rainfall intensities,
- The volume of inflow and/or rainfall-induced infiltration generated by the event,

- Peak flowrate, and
- Other specific information that might have a bearing on the event analysis, such as weather conditions at the time of the event, occurrence of an overflow event, amount of overflow volume, and whether the overflow was accounted for in the values derived from the inflow hydrograph analysis.

4.4 Hydraulic and Hydrologic Modeling

Hydraulic and hydrologic modeling is a critical component in planning and evaluating sewer system response to wet weather events. This chapter presents only a broad overview of H&H modeling. More detail is available in Chapter 4 of *Existing Sewer Evaluation and Rehabilitation* (WEF et al., 2009). A number of public domain, commercially available, and proprietary models have been used to analyze sewer systems. Hydraulic and hydrologic models need to be able to mimic existing system behavior and project future system performance following changes to the tributary area, routing, or under varying storm conditions. While H&H models are important tools for sewer system understanding, the application of the models must consider the intended use of the output.

Use of H&H models to size facilities needs to consider the acceptable range of precision and accuracy. An evaluation of the effect of the H&H model, either over-predicting or under-predicting the resultant behavior of the system, should be considered in a sensitivity analysis.

Hydraulic and hydrologic modeling, as with flow monitoring, needs to be performed to a degree of accuracy and precision appropriate for program goals. Hydraulic and hydrologic models are often applied to extreme conditions, and can be better applied when the data used in calibration reflect conditions close to the design condition. Care needs to be taken with hydraulic modeling to ensure that system performance affected by surcharging and overflows is understood in both the data set used for analysis and the hydraulic model representation of the system.

4.4.1 *Selection of a Model*

A number of simplified and sophisticated H&H models are commercially available (WEF et al., 2009). There are additional models that are available that are proprietary (i.e., the property of a consulting firm) and can only be used when the model owner is hired to perform the analysis. Considerations for model selection are based on issues both of hydraulic modeling capabilities and ease and cost of use. Models vary in complexity, general simulation methodology (dynamic vs static), hydrologic calculation methodology, and input and output data format. Generally, proper simulation of wet weather flows requires a fully dynamic model. Modeling will never be completely precise due to the probabilistic nature of hydrology, and this needs to be well understood by those evaluating modeling results.

Considerations for collection system model selection need to be based on site-specific needs. Models represent large investments for agencies, and software, development, and maintenance costs are significant. Requirements of the model will vary and the selected model should meet the objectives. In the event the existing model is well-established, the capabilities and limitations should be understood by those that work with and use the model for management purposes so that the use of the modeling tool is appropriate and optimized. Model selection should include an owner and modeler evaluation of the capabilities required in the model. The following characteristics of models should be considered in model selection:

- Hydrologic computation options; available methods of flow development for dry weather flows, groundwater, RDII and runoff, suitability for representing stormwater best management practices (BMPs) or low-impact development, and availability of user options;
- Hydraulic options such as available pipe configurations, complex hydraulic structure options, and time scales (note that most overflow control programs require use of dynamic models and the ability to represent relatively complex hydraulics);
- Number of nodes or links that can be supported by the model;
- Model capabilities for importing and exporting of data, integration with GIS, graphical displays, and general user interface tools;
- User input options by time scale (such as seasonally influenced flow computations);
- Model stability and model run time;
- Modeling time scales and means of interface with rainfall records;
- Transparency of calculations and methodology used by the model;
- Overall cost of the model for owner and consultants;
- Ease of use, availability of skilled modelers, and approach to long-term maintenance of the model by the owner;

- Regulatory acceptance; and
- Vendor support availability and long-term track record.

In specific circumstances, the following additional items may be considered in model selection:

- Ability for the user to edit model code to allow for site-specific concerns and
- Sophisticated hydraulic analysis such as real-time control or surge analysis.

Selection of a particular model is based on an assessment of the model's ability to meet project objectives. For significant overflow control programs, the most commonly used modeling packages include the public domain and commercial versions of U.S. EPA's *Storm Water Management Model User's Manual, Version 5.0* (U.S. EPA, 2010) (including InfoSWMM [MWH Soft, Broomfield, Colorado], DHI's MIKE Urban [DHI, Portland, Oregon], PCSWMM [Computational Hydraulics International, Guelph, Ontario.], and XPSWMM [XP-Software, Portland, Oregon]) and non-*Storm Water Management Model*-based proprietary packages, including InfoWorks (MWH Soft, Broomfield, Colorado) and Mike Urban with the MOUSE extension (DHI, Portland, Oregon).

Specific information on a particular model is best located through a "users group" that may be found through an Internet search of the model name.

4.4.2 *Model Scale*

As previously discussed with flow monitoring programs, model scale will vary with the purpose of the model. The extent of the sewer system included will vary with the planning or design levels. Simplification of collection systems is inherent in all models and limitations associated with the simplification need to be understood. In skeletal models, the hydrology represented is often reflective of not only the flow generation in tributary areas, but also the hydraulic limitations of the delivery systems. Changing the scale of the physical model may also change the results of the model simulation. Physical scale is most appropriately matched to the level of detail of the system that was used in flow monitoring, with a more detailed physical model requiring more detailed flow data to support it. Examples of model scale are presented in Table 5.1.

Use of low-impact development or "green infrastructure" is increasingly being used to support the control of CSOs. The scale of modeling required to accurately reflect green infrastructure is much more complex than that required for end-of-pipe

Table 5.1 Data collection levels (Wastewater Planning Users Group, 2002).

Data	Level A	Level B	Level C	Level D
Pipe data				
Prime source	A complete survey or resurvey of the sewers should be carried out.	Ground and pipe levels should be taken from existing records as far as possible.	Existing records should be used for the majority of ground levels, invert levels, and pipe sizes.	Existing records should be used.
Missing data		Surveys should be carried out to provide the missing data	Occasional missing items of data levels are estimated or interpolated from the data at neighbouring manholes or pipes, subject to a maximum of two consecutive manholes and a maximum of 5% of the total data.	Where data is missing: Ground levels may be estimated or interpolated from other levels (which could only be OS spot heights); Invert levels may be estimated from known levels or depths, or if no other information is available the depths can be assumed; Missing pipe sizes may be estimated from the upstream and downstream sizes.

Checks on sewer record data				
	A complete consistency check should be carried out on the input data and a sample manhole survey carried out to check the accuracy of the sewer record data supplied. Resurveys should be organized in any areas where significant errors are found.	A complete consistency check should be carried out on the input data before modeling work commences.	A representative sample of the data used in the model should be checked for internal consistency, either visually or using suitable software and any obvious discrepancies should be checked on site.	No routine checks on the supplied data need to be carried out unless problems are highlighted by the model software.
Ancillary (CSO and Pumping Station data)				
	Surveys should be carried out at all major ancillaries.	Plans of the ancillaries should be obtained to provide detailed information to the modeler. Site surveys should be organized where there is any uncertainty as to how the structure will operate.	Data for the ancillary structures should be obtained from sketch plans or from record sheets or surveys should be organized.	Data for major ancillary structures should be obtained from sketch plans or record sheets or surveys should be organized. Other ancillary data may be estimated.

(continued)

TABLE 5.1 (Continued)

Data	Level A	Level B	Level C	Level D
Contributing area data survey				
Identifying impermeable areas	The determination of which areas are impermeable should be carried out by detailed survey.	The determination of which areas are impermeable should be carried out from sewer record plans and a sample survey.	The determination of which areas are impermeable should be based on experience and examination of record plans.	Impermeable areas may be estimated using flow survey data.
Determining contributing area boundaries	Detailed surveys should be carried out to determine connectivity.	For foul and combined systems connectivity should be determined from a sample survey. For partially separate systems detailed surveys should be carried out to determine connectivity.	For foul and combined systems connectivity should be determined from plans by judgement. For partially separate systems sample surveys should be carried out to determine connectivity.	Connectivity should be determined from plans by judgement.
Calculation of contributing area data	Contributing areas calculated from the GIS or taken from paper plans using a planimeter.	Contributing areas calculated from the GIS or taken from paper plans using a planimeter	Contributing areas calculated from the GIS or taken from paper plans using a planimeter or by counting grid squares.	Contributing areas calculated from the GIS or taken from paper plans by counting grid squares.
Calculation of impermeable area data	The percentages of paved and roofed areas are calculated from measurements of the areas.	The percentages of paved and roofed areas are calculated from measurements of the areas.	The percentage paved and roofed determined from measurements of sample areas.	

Operational data				
Temporary changes to the system	Obtain from operations staff	Obtain from operations staff	Obtain from operations staff	Obtain from operations staff
Flooding and surcharge data	Detailed data on flooding and surcharge should be obtained from records and surcharge surveys should be carried out.	Detailed data on flooding and surcharge should be obtained from records.	Detailed data on flooding and surcharge should be obtained from records.	A basic knowledge of major flooding points should be established from records.
Other incident data	Obtain from operations staff	Obtain from operations staff	Obtain from operations staff	Obtain from operations staff
Baseflow data				
Dry weather flow data				
Daily per capita values	Estimated from flow measurements within the catchment.	Use standard values of water usage.	Use standard values of water usage.	Use standard values of water usage.
Geographic distribution	Distribute according to detailed population estimates based on house counts, and using metered water supply and trade effluent figures.	Distribute according to population estimates based on house counts and using information from major trade effluent increases.	Distribute global population according to length of pipe or connected areas.	Distribute global population according to length of pipe or connected areas.

(*continued*)

TABLE 5.1 (Continued)

Data	Level A	Level B	Level C	Level D
Diurnal variation	Estimate from detailed flow measurements.	Estimate from detailed flow measurements.	Use standard values.	Assume no variation.
Infiltration data				
Fixed element	Data from detailed infiltration survey and from long-term records at outfall	Data from detailed analysis of long-term records at outfall	Standard values	Use standard values of infiltration.
Seasonal variation	Data from detailed infiltration survey and from long-term records at outfall	Data from detailed analysis of long-term records at outfall	Estimated from long-term records at outfall	Not included
Rainfall-induced variation	Data from detailed infiltration survey and from long-term records at outfall	Data from detailed analysis of long-term records at outfall	Not included	Not included
Geographic distribution	Detailed infiltration survey across the system	Estimated from available CCTV data and from short-term sewer flow survey used for verification.	Estimated from available CCTV data and from short-term sewer flow survey used for verification.	Distributed uniformly

Boundary conditions				
River levels	Use continuous level monitor	Use continuous level monitor	Periodic (e.g. daily) level measurements	Exceptional levels recorded otherwise normal levels assumed
Tide levels	Use continuous level monitor	Values inferred from tide tables – adjusted from peak level measurements	Values inferred from tide tables – adjusted from level measurement elsewhere.	Values inferred from tide tables
STW inlet water levels	Use continuous level monitor	Use continuous level monitor	Peak levels recorded.	
Pipe roughness data				
	Information on roughness, and hydraulic problems should be obtained from available CCTV records.	Where sewer condition is known to be poor, available CCTV records should be inspected and the result used to assess roughness	Global roughness values should be assumed.	Global roughness values should be assumed.
Sediment level data				
	Information on sediment depths should be obtained from available CCTV records.	Information on sediment depths should be obtained from available CCTV records where there are known sediment problems.	Assumed sediment depths should be included where there are known sediment problems	Sediment depths should not be included

facilities. Best management practices affect the hydrology at a localized scale. The effect of these practices is localized. From a modeling perspective, they may affect the initial abstraction value, the flowrate at which runoff occurs, or a combination of effects on volume and rate. Success in reflecting the effects has been accomplished with highly detailed modeling (Liebe and Collins, 2007). U.S. EPA released the SUSTAIN model in 2009, which is a public domain software for BMP hydrologic assessment and optimization (Shoemaker et al., 2009).

4.4.3 *Model Calibration and Validation*

Model calibration and validation is an important process of model use. Care should be taken in ensuring that critical conditions that will drive future facilities are well matched between model results and monitoring data. Volume and peak flowrates for these conditions, in particular, should be well matched. Modeling guidance (WaPUG, 2002) provides guidance for standards of model calibration that provide a benchmark for the industry. Wastewater Planning Users Group (2002) criteria are presented in Table 5.2. However, it is up to the user to ensure that model results are satisfactory for the specific application for which the model is being developed. As an example, excellent calibration to large events, high-volume events, and high-intensity events on an individual event basis should be evaluated if these are the conditions from which facility sizes will be determined. Such presentations of data such as "average net error" may result in cancelation of high and low error and mask problems with critical event representation.

Validation is an important aspect of model development and involves the process of testing the model against an independent data set. Validation is needed to ensure that the model retains its performance and has not been significantly skewed by any one event (U.S. EPA, 1999).

4.4.4 *Modeling Applications to Project Design Conditions*

Design conditions vary between control of CSOs and separate sewer overflows. Combined sewer overflow controls are typically based on the performance of the system over the course of a year. As such, volumetric sizing is generally accomplished satisfactorily with a defined typical year of rainfall (or, in some instances, a longer period of record). The results of this type of modeling will typically be sufficiently accurate for determining the volume and frequency of overflow from combined systems. Larger storm conditions need to be evaluated for the hydraulic conveyance sizing portions of the system, and should reflect a more critical conveyance condition,

TABLE 5.2 WaPUG model calibration criteria (Nelson et al., 2007).

Parameter	Dry weather calibration criteria	Wet weather calibration criteria
Peak flow rate	–10 to +10% of measured, or ±0.1 mgd	–15 to +25% of measured, or ±0.1 mgd
Flow volume	–10 to +10% of measured, or ±0.1 mgd	–10 to +20% of measured, or ±0.1 mgd
Maximum, average, and minimum depth	–0.33 to +1.67 ft at surcharged locations ± 33 ft at non-surcharged locations	–0.33 to +1.67 ft at surcharged locations ± 33 ft at non-surcharged locations
Shape	The shape of modeled and metered curves should be similar for flow and depth	The shape of modeled and metered curves should be similar for flow and depth
Timing	The timing of the peaks, troughs, and recessions of modeled and metered curves should be similar for flow and depth	The timing of the peaks, troughs, and recessions of modeled and metered curves should be similar for flow and depth
Flooding		Predicted flooding will be corroborated using customer complaints, flooding/overflow records and other historical records

such as a 10-year high-intensity storm. Wherever possible, the results of typical year analysis should be reviewed against performance monitoring of the sewer system, with an understanding of how "typical" the monitoring period was.

Separate sanitary overflows caused by excessive RDII typically require an approach to limit the frequency of discharge to a design storm (typical ranges of value are on the order of one per two years to one per 25 years, depending on locale). In many locations, the wet sanitary system response is most significant in times of the year when rainfall intensities are lower, but the volume produced by the tributary area is more significant. Because of the non-uniform response to rainfall in different seasons, selection of a condition that represents the critical flow condition for the return frequency is needed. Selection of design conditions may require the evaluation of multiple years of record, varying season theoretical design storms, or other another approach that effectively reflects the probability of a critical volume or flowrate being produced rather than the probability of a certain rainfall condition occurring.

To evaluate the various conditions for performance of a system, both design storms and continuous simulations of varying durations are needed. Continuous simulations result in longer modeling computation times that can affect time scales of getting results if a large number of scenarios are required. Generally, screening model runs may be performed with shorter durations or design storms and refined analyses may be performed with the continuous time frame identified for the model.

4.4.5 Ongoing Model Enhancement

The usefulness of models as tools is only as good as their ability to represent the collection system. Even the most aggressive monitoring programs rarely exceed 1-year duration for initial model development. Results of flow monitoring may be affected by the year being either wetter or dryer than normal or by the particular distribution of events. Ongoing monitoring programs should be used to continually enhance and "ground-truth" the model.

Long-term CSO control programs and SSO control and abatement programs result in changes to the system. Facilities are added, new cross connections are discovered, and relief capacity may be added. The capacity of the downstream WWTP may be modified to accept additional wet weather flow. Locations of river inflow to the system through an outfall may be discovered. In short, the understanding of the collection system and the physical components of the collection system may change significantly.

5.0 CHARACTERIZING EFFECTS OF OVERFLOWS ON WATER QUALITY

Water quality monitoring, when combined with flow monitoring or modeled flow outputs, allows for the pollutant load generated by the sewer system to be quantified as the product of the discharged volume and pollutant concentration. This information can then be used in a receiving water model to establish the effect of incremental sewer pollution abatement on water quality. Additionally, receiving water quality must also be assessed to both close the loop and to fully understand the effect of sewer-related pollution sources on receiving water. Similar to flow monitoring programs, water quality monitoring programs should have objectives and an approach documented in a quality assurance project plan (QAPP) or work plan. This section provides guidelines for defining the relative effect of sewer-related pollution sources.

5.1 Water Quality Objectives

The stated goal of CSO control is the attainment of WQS in the receiving water to the extent that they can be achieved by overflow control. Combined sewer overflows should not "cause or contribute to the violation of [WQS]" (U.S. EPA, 1994). In most programs, the effect of CSO discharges is most readily apparent in the effects on bacteria levels and dissolved oxygen levels in the surface water. Nutrients and toxics are other pollutants of concern, although the effects from CSO discharges are less apparent.

Sanitary sewer overflows are prohibited by CWA. Water quality objectives or standards do not drive the elimination of these sources of pollution. Despite this regulatory limitation, some municipalities or collection authorities have invested in water quality assessment associated with SSO discharges. In the context of this chapter, receiving water monitoring and modeling is intended for the support of CSO programs.

5.2 Assessing Effects of Overflows on Water Quality

Receiving water characterization needs to assess attainment of WQS to support public health protection and health of aquatic species. To evaluate the receiving waters from this perspective, characterization should include

- Attainment of WQS (chemical and bacteriological numeric standards),
- Hydrologic and hydraulic characteristics of the receiving water necessary to perform modeling, and
- Biological community characterization and limitations to healthy aquatic life.

5.2.1 *Receiving Water Characterization Parameters*

The following receiving water characterization parameters should be assessed at either a qualitative or quantitative level.

5.2.1.1 Hydraulic

An understanding of hydraulic conditions in the receiving stream is necessary to support characterization or modeling of the receiving water and the effect of overflows on receiving water. Hydraulic characteristics include such aspects as distribution of flows, depth of flow, tidal influences, or dispersion in lakes. Hydraulic characterization in rivers is aided by river monitoring data (gauging stations) maintained by the U.S. Geological Survey, Army Corps of Engineers, and the Bureau of

Reclamation. These stations provide historic and current estimates of flowrates that can be correlated to times of overflow from the system.

Where possible, the data should be able to answer the questions of the total volume of overflow as a fraction of the receiving water flow on a daily, seasonal or annual basis. The source of flow in the receiving water should be quantified into volume from areas upstream of all overflows (or the overflows under study) from the overflows in the study area or other discharges in the study area (such as WWTP effluent, other NPDES discharge effluent, stormwater, or other tributaries). A flow balance should be performed at varying time scales.

An understanding of the hydraulic characteristics of the receiving stream will help determine the most appropriate type of receiving water hydraulic model to use, if one is needed. River models may be one-dimensional (linear, such as along a stream), two-dimensional (uniform depth, non-stratified bay, lake, or estuary), or three-dimensional (varying depth, stratified bay, estuary, or lake). The amount of data collection and complexity associated with these additions of dimension makes the modeling increasingly complex.

The level of hydraulic characterization required will depend on whether a receiving water quality model is required and the complexity of the selected model.

5.2.1.2 Water Quality Standards

Combined sewer overflow control programs need to demonstrate compliance with WQS. Clarification of how WQS will be measured is important to determine at the onset of a program to avoid problems with reporting later on. Primary pollutants of concern and some specific considerations are

- *Bacteria*—bacterial WQS are typically written based on a set of samples (either as a numerical limit on the sample sets' geometric mean or as a limit to the percentage of samples that exceed a certain value). In some states, a single sample maximum has been used. Water quality standards for pathogens are expressed as concentrations of some indicator species of bacteria. As of 2010, fecal coliform and *Escherichia coli* were predominant indicator bacteria; however, other indicator organisms that more closely reflect risk to human health may be used in certain circumstances or in the future. Some northern states have a defined recreational season (typically May through October) during which standards apply.
- *Dissolved oxygen*—standards may vary by the designated use of the waterbody. Dissolved oxygen standards will often express a minimum value or an average value.

- *Total suspended solids*—although not a water quality standard, reporting of this value is often expected by regulators.
- *Biochemical oxygen demand (BOD)*—although not directly a water quality standard, but influences the levels of dissolved oxygen in-stream.
- *Other parameters*—in some locales, nutrients or metals may be a concern due to total maximum daily load listings.
- *"Free froms"*—most standards list a variety of "free froms", such as "nuisance growth", that may result in regulatory limitations not directly related to a numerical standard.

Generally, state reports or 303(d) listings will provide information on which standards are believed to be violated due to the influence of wet weather discharges.

5.2.1.3 Aquatic Life

A biological assessment of aquatic life helps to define the abundance and diversity of aquatic life. Surveys may focus on macroinvertebrates or fish. Streams with significant overflow and degradation may be characterized as dominated by pollution-tolerant species. A healthy environment would typically have many species that are relatively small in number intermixed with a few abundant species. Surveys of this aquatic diversity are often performed on a semi-regular basis by state environmental or regulatory agencies. Fish consumption warnings should also be cataloged for the receiving water of interest.

5.2.1.4 Physical Habitat

Physical habitat assessment reflects the physical characteristics of the waterway to support aquatic life. This may include such items as pools and riffles, impoundments, and low-flow periods. Stream modifications such as channelization for flood control or modifications for navigation may have significantly altered the natural environment.

5.2.1.5 Recreational Use

Actual recreational uses of receiving waters should be sought out and documented in conjunction with an overflow control program. This should include interviews with stakeholders and observations at various times of day or surveys of the public. Potential existing uses may include boating, swimming, water skiing, jet skiing, wading, fishing, or passive (shore-based) recreation.

5.2.2 *Monitoring Program Design*

Monitoring program design should provide a clear discussion of the study objectives and a statement of the questions to be answered. A water quality monitoring work plan and/or a QAPP should be developed before data collection.

Monitoring programs typically characterize the water quality characteristics upstream, downstream, and in the vicinity of overflows. Monitoring programs need to sufficiently quantify spatial and temporal variations in water quality under a variety of conditions. Many aspects of water quality monitoring programs are driven by various input data required for water quality models. However, as with flow monitoring programs, water quality data collected should be able to stand on its own for the benefit of understanding effects on water quality.

5.2.2.1 *Water Quality Parameters*

Monitoring programs should focus on those variables likely to directly or indirectly indicate either a violation of WQS or assist in defining plumes or mixing zones. Any variables that are measured either in overflow or receiving waters downstream of outfalls should also be measured in upstream and background sources of wet weather flow. A preliminary evaluation of water quality relative to bacteriological concentration and levels of dissolved oxygen should be undertaken to guide the selection of monitoring parameters. If dissolved oxygen is impaired in the waterbody, a full suite of parameters relevant to dissolved oxygen modeling should be evaluated. While other pollutants from CSO discharges may affect receiving waters, the level of control required to address bacteriological and dissolved oxygen issues is generally sufficient to meet other goals.

Sampling for indicator bacteria requires crews that are deployed to collect grab samples and available laboratories to perform analysis within short holding times. As indicator bacteria exist in other sources to the receiving water including stormwater runoff from both urban and agricultural lands, failing septic systems in upstream areas, pet waste, and geese, sampling programs may include quantification of some of these other sources. Consideration should be given to whether bacteriological source tracking or other pathogens and viruses are worthy of study, although the costs associated with these efforts can be significant. In instances where both background and wet weather sources contribute to in-stream bacteria concentrations, defining compliance sampling requirements should be done during the development of any post-construction monitoring program.

Dissolved oxygen provides a measure of the ability to support aquatic life. Dissolved oxygen is influenced by the concentration of oxygen-demanding pollutants in the water column and sediment, and the activities associated with photosynthesis and respiration. Dissolved oxygen is readily and inexpensively monitored with various probes either continuously or through a series of discrete samples. Often, the relative cost difference of deploying continuous probes versus collecting multiple measurements with handheld probes will also favor use of continuously recording equipment. A continuous recording meter will also provide diurnal readings that are important in assessing both compliance with dissolved oxygen standards and understanding the large diurnal dissolved oxygen swings, which are a potential indication of eutrophication in receiving streams. Water chemistry parameters (e.g., BOD, particularly 20-day or ultimate BOD) are needed for modeling. An evaluation of sediment oxygen demand (SOD) can be important in predicting dissolved oxygen, particularly in smaller streams where dissolved oxygen levels in dry weather remain depressed. Sampling may also be needed to understand photosynthesis and respiration, which influences diurnal dissolved oxygen levels.

Other water quality parameters (pH, temperature, and conductivity) can often be measured with the same probes as dissolved oxygen. Temperature is critical to understanding dissolved oxygen levels. The other parameters are generally not necessary, although they may be convenient in defining plumes or other changes related to the physical properties of wet weather discharge.

Additional parameters should be selected based on pollutants of interest to the particular waterbody, with sampling of the water column and sediment contemplated for inclusion in the program. If preliminary monitoring or agency historic data indicate that standards related to nutrients or toxic substances are violated in receiving water, then those specific variables should be measured in receiving waters. In some instances, it may be acceptable to choose the most common pollutant causing a violation and use that to indicate relative concentrations of other pollutants.

5.2.2.2 In-Stream

In-stream monitoring should adequately address spatial coverage through selection of monitoring locations and temporal scale. Spatial distribution should be selected to capture boundary conditions upstream and downstream of discharge locations, locations in proximity to significant discharges, and locations where dissolved oxygen effects may be exerted (based on physical flow characteristics or travel time downstream of major discharge locations). Particular locations that may be susceptible to dissolved oxygen sags (such as a backwater-affected mouth of a stream as it discharges into a hydraulically modified river) should receive focus in monitoring programs. Areas of recreation should be well characterized for bacteria.

Wet weather event time scales should include dry weather background characterization and event response. Wet weather sampling should include multiple samples over the course of the event, round-the-clock sampling during overflow, and several days of post-overflow sampling to track recovery of the waterway. Sample time spacing should be relatively short during the overflow portion of the event, and then increase during the recovery phase of the event. Size of the stream, duration of overflow, and other hydraulic considerations (including tides, where applicable) are all considerations in sample timing. Dry weather samples should be collected in receiving waters during days with at least 48 and, preferably, 72 hours of antecedent dry weather (defined as no rainfall sufficient to cause runoff).

Temporal duration of sampling needs to cover the period when annual effects are most significant. In locations with four seasons, these effects are typically most significant in the warmer months, when dissolved oxygen is more significantly affected by water temperature and when bacteriological data are more relevant for water recreation. For maximum effectiveness, the duration of the monitoring program should extend over the full season and capture a minimum of three significant wet weather events. Contingencies about different criteria for qualifying storms should be made and discussed with regulators.

Dissolved oxygen monitoring is most descriptive when it includes deployment of recording water quality monitors that provide a clear description of daily activity and changes over the course of an event.

In-stream monitoring programs need to include provisions for quantifying flowrate and volume during the monitoring period. In many instances, local gauging stations can provide needed data.

Logistics of in-stream water quality monitoring programs can be very difficult and, therefore, should not be underestimated. The logistics are complicated by the difficulty in forecasting wet weather events and reaching desirable sampling locations. To the maximum extent possible, recording probes and automated sampling equipment should be considered to more effectively collect desired data.

5.2.2.3 Overflow Characterization

Overflow characterization should be synchronized with in-stream monitoring when possible. Overflow monitoring should include volumetric, timing, and pollutant information. In many programs, physical monitoring is performed at the largest discharge locations and hydraulic models are used to quantify discharges from other locations. Overflow characterization of representative outfalls should be performed to assess local conditions. The observed values can be compared to data from other studies for application to the system as a whole.

Sampling of overflows should include discrete sampling or flow-based composites to determine event loadings. Time-based composite sampling may not accurately determine pollutant loading. Discrete sampling is helpful in defining the benefit of certain control measures, particularly storage, which provides preference to the first flush component of the storm. Overflow sampling should ideally include some measurement of bacterial and dissolved oxygen concentrations, although these measurements are difficult due to the need for grab sampling and access issues, particularly at times of discharge.

Probe-monitored data are typically not collected in combined wastewater overflow (or in other wet weather discharges). Problems with interferences and decontamination limit the usefulness of collecting dissolved oxygen and pH in sewer overflow, in addition to the need to keep probes continually submerged. Temperature and conductivity probes are more durable and can provide some useful data. Often, WWTPs routinely measure influent dissolved oxygen and pH with dedicated equipment, and those measures can be used to estimate what might be discharged at overflow points. Regardless, dissolved oxygen in effluent is less critical to resultant water quality than carbonaceous biochemical oxygen demand. In some studies in which long-term deposition of overflow sediments is a problem, as in ship channels or impoundments ultimate, BOD might be measured on some subset of effluent samples.

5.2.2.4 Summarizing Receiving Water Data

Summaries of receiving water data should provide clear information relative to the attainment of WQS and the contributions of various sources to the water quality pollutant loadings based on the monitoring program. The following are examples of data that may be presented:

- Summary statistics reflecting the frequency, duration, and severity of WQS exceedances. These should be reported based on location, location characterization (i.e., upstream of study area, downstream of CSO discharges, or downstream of study area), and sampling condition (e.g., dry weather or wet weather).
- Summary of loading sources for events, months, or seasons potentially presented as pie charts. Such pie charts could include a summary of distribution

of the source of flow (upstream, CSO, stormwater, or WWTP discharge) and the pollutant loading.

- Time series data showing variability of pollutant concentrations over time.
- Spatial plots showing concentration of selected parameters of interest during wet weather and dry weather conditions.

5.3 Receiving Water Modeling

Receiving water models are used to project the water quality for existing conditions at greater temporal and spatial detail than what can be captured with monitoring and to assess the effects of future changes. Receiving water modeling choices will be based on the type of waterbody (whether the receiving water is a river, lake, coastal area, or estuary will affect modeling choice); the water quality parameters to be modeled (bacteria, dissolved oxygen, or other parameters); and the sources of pollutants to the waterway (U.S. EPA, 1999).

5.3.1 *Available Models*

Available models differ in their ability to simulate various complexities of hydrodynamic conditions and time-scale effects of event or continuous simulation. Models selected for various parameters require different capabilities, as follows:

- *Bacteria modeling*—bacterial models are based on short time scales and need to simulate decay, sedimentation, and re-suspension and as mixing appropriate to the waterbody. Clarity of whether a mixing zone applies for application of WQS should be considered as part of model selection.
- *Dissolved oxygen modeling*—models need to simulate decay, SOD, and photosynthesis and respiration. For river models, lateral mixing is less significant for dissolved oxygen than bacteria as the time scale of the effect of dissolved oxygen loading is longer.

U.S. EPA has established several significant resources both for developing and supporting water quality modeling tools. One trend in modeling application that is becoming more viable to wet weather modeling is the use of integrated modeling tools to capture all sources of pollutants of concern. As discussed earlier in this chapter, background sources of bacteria can make determination of causes of impairment difficult and controversial. A U.S. EPA white paper (U.S. EPA, 2008) defines the need for further development and application of such tools. The more traditional tools are also documented and accessible through U.S. EPA Web sites. (Currently, http://www.epa.gov/athens/wwqtsc/html/water_quality_models.html provides access to all of the combinations of static and dynamic simulation models commonly in use in the United States). Each model is typically built around a hydraulic simulation program that incorporates both internal and external physics. Hydraulic models divided into reaches or other segments of the waterbody, and within each segment of the model various biologic or chemical processes, are simulated to illustrate fate processes of contaminants. One of the more recent hydrodynamic engines is Environmental Fluid Dynamics Code (U.S. EPA, http://www.epa.gov/athens/wwqtsc/html/efdc.html), which provides more detailed and realistic hydrodynamics of complex systems and can be used to provide the hydrodynamic input to Water Quality Analysis Simulation Program (WASP).

5.3.2 *Water Quality Goals*

A significant challenge in water quality modeling for CSO control is the assessment of the effects of "upstream" sources. This becomes significant in the application of language regarding whether CSO discharges "cause or contribute" to violations of WQS. It stands to reason that if the upstream waterbody does not meet WQS, any additional discharge of pollutants will further "contribute" to that lack of attainment. However, this presents an unachievable goal for control of CSOs and some reasonable evaluation process must be determined. Use of an upstream boundary condition with water quality slightly better than attainment of WQS is an approach that has been used in some studies.

Another aspect of water quality evaluation is the need for attaining water quality standards at times of overflow. Discharges may result in water quality effects that are manifested at times other than during the overflow. For example, dissolved oxygen levels can be depleted by SOD and bacteria in sediment can result in delayed violations of WQS. Monitoring and modeling should consider such delayed effects.

The evaluation of water quality should evaluate both whether an exceedance has occurred and the magnitude of that exceedance. Combined sewer overflow monitoring and modeling reports often treat attainment of WQS as "pass or fail", rather than acknowledging the significance of the exceedance. These measurements will be important in evaluating controls and the subsequent monitoring of progress following implementation of controls.

5.3.3 *Design Storms Versus Continuous Simulation Modeling*

A majority of systems that merit development of water quality models should be set up to simulate continuous periods of time, such as a recreation season or "typical" year. Those continuous simulations will often be used to define the critical events that will drive the level of control required for wet weather facilities. The continuous simulation better integrates the various factors associated with water quality effects than a single design storm condition.

5.3.4 *Hydraulic and Water Quality Calibration*

As with hydraulic modeling, water quality model calibration should be performed to demonstrate the ability of the model to replicate monitored events. Calibration would typically follow a two-step process of hydrodynamic calibration and water quality calibration. Sensitivity analysis should be used to assess the significance of parameters with limited measured data or the effect of highly variable inputs. It is important to note that the amount of data required to calibrate a model to a higher degree of certainty will increase dramatically when moving from planning level prediction to design quality. Careful planning and pre-defining the uses of the model are essential to ensuring that adequate flow and load data are collected to provide the desired level of accuracy. Following calibration of the integrated hydraulic and water quality model, it will typically be verified by comparison with a set of data from a time period or series of storms not used during calibration. Depending on logistics, a model can also be verified against continuous metering or other new monitoring data if a network of stations is set up independently.

6.0 REFERENCES

ADS Environmental Services (2007) *Scattergraph Principles and Practice.* http://www.adsenv.com/default.aspx?id=73 (accessed June 2010).

Bennett, B.; Rowe, R.; Strum, M.; Wood, D.; Schultz, N.; Roach, K.; Spence, M.; Adderley, V. (1999) *Using Flow Prediction Technologies to Control Sanitary Sewer Overflows;* 97-CTS-8; Water Environment Research Foundation: Alexandria, Virginia.

Carlson, T., Guerrero, G., Lipinski, S., Mass, S. (2007) Using the Operations Perspective to Maximize the Effectiveness of Flow Monitoring. *Proceedings of the Water Environment Federation Collection Systems Specialty Conference:*

Pioneering Trails to Collection Systems Excellence; Portland, Oregon, May 13–16; Water Environment Federation: Alexandria, Virginia.

Czachorski, R.; Prince, T.; Bennett, B.; Kaunelis, V.; Humphriss, C. (2008) Tools for Accurate Sewer Metering and Billing. *Proceedings of the Water Environment Federation Collection Systems Specialty Conference;* Pittsburgh, Pennsylvania, May 18–21; Water Environment Federation: Alexandria, Virginia.

Kumpula, G.; Minor, T. (2006) An Improved Dye-Dilution Flow Measurement Procedure for Sewers Using an In-Situ Apparatus. *Proceedings of the Water Environment Federation Collection Systems Specialty Conference—Infrastructure Stewardship: Partnering for a Sustainable Future;* Detroit, Michigan, Aug 6–9; Water Environment Federation: Alexandria, Virginia.

Lai, F.; Vallabhaneni, S.; Chan, C.; Burgess, E.; Field, R. (2007) A Toolbox for Sanitary Sewer Overflow Analysis and Planning (SSOAP) and Applications. *Proceedings of the Water Environment Federation Collection Systems Specialty Conference: Pioneering Trails to Collection Systems Excellence;* Portland, Oregon, May 13–16; Water Environment Federation: Alexandria, Virginia.

Lei, J.; Schilling, W. (1993) Requirements of Spatial Raindata Resolution in Urban Rainfall Runoff Simulations. *Proceedings of the Sixth International Conference on Urban Storm Drainage, Volume 1;* Marsalek, J., and Torno, H. C., Eds.; Niagara Falls, Ontario, Canada, Sept 12–17.

Liebe, M.; Collins, D. (2007) Modeling of Stormwater Removal and Peak/Volume Reduction Effects of Green Solutions (Inflow Controls) Using an Explicit Combined/Sanitary Sewer Model. *Proceedings of the Water Environment Federation Collection Systems Specialty Conference: Pioneering Trails to Collection Systems Excellence;* Portland, Oregon, May 13–16; Water Environment Federation: Alexandria, Virginia.

Lukas, A. (2007) Update on a Nationwide I/I Reduction Project Database. *Proceedings of the Water Environment Federation Collection Systems Specialty Conference: Pioneering Trails to Collection Systems Excellence;* Portland, Oregon, May 13–16; Water Environment Federation: Alexandria, Virginia.

McCulloch, J.; McCormack, K.; Ridgway, K.; Kalinowski, S. (2007) Identifying Peak Flow Rates From Differential Flow Metering Using an Integrated Flow Modeling Program. *Proceedings of the Water Environment Federation Collection*

Systems Specialty Conference: Pioneering Trails to Collection Systems Excellence; Portland, Oregon, May 13–16; Water Environment Federation: Alexandria, Virginia.

Mitchell, P.; Stevens, P.; Nazaroff, A. (2007) Quantifying Base Infiltration in Sewers. *Proceedings of the Water Environment Federation Collection Systems Specialty Conference: Pioneering Trails to Collection Systems Excellence;* Portland, Oregon, May 13–16; Water Environment Federation: Alexandria, Virginia.

National Association of Sewer Service Companies (2001) *Pipeline Assessment and Certification Program*; Reference Manual Version 4.3.1; National Association of Sewer Service Companies: Owings Mills, Maryland.

National Association of Sewer Service Companies (2006) *Manhole Assessment and Certification Program*; Reference Manual Version 4.3; National Association of Sewer Service Companies: Owings Mills, Maryland.

Nelson, D.; Onderak, E.; Cantrell, C. (2007) Getting Your Money's Worth—Collecting and Utilizing Accurate Flow Monitoring Data for Collection System Analysis and Modeling. *Proceedings of the Water Environment Federation 80th Annual Technical Exposition and Conference* [CD-ROM]; San Diego, California, Oct 13–17; Water Environment Federation: Alexandria, Virginia.

Pang, J. (2008) Application of Mean Velocity Profile, Depth and Velocity Data Time Series, and Foude Number to Flow Monitoring Location Selection and Flow Monitoring Data Analysis. *Proceedings of the Water Environment Federation Collection Systems Specialty Conference;* Pittsburgh, Pennsylvania, May 18–21; Water Environment Federation: Alexandria, Virginia.

Pistilli, J.; Lieberman, L.; Aichler, J.; Stonehouse, M.; TenBroek, M. (2006) Tuning the Flume Equation Fine Tunes Cash Register: Improved Flume Installed Accuracy by Equation Calibration. *Proceedings of the Water Environment Federation Collection Systems Specialty Conference—Infrastructure Stewardship: Partnering for a Sustainable Future;* Detroit, Michigan, Aug 6–9; Water Environment Federation: Alexandria, Virginia.

Shoemaker, L.; Riverson, J.; Alvi, K.; Zhen, J.; Paul, S.; Rafi, T. (2009) *SUSTAIN—A Framework for Placement of Best Management Practices in Urban Watersheds to Protect Water Quality;* EPA-600/R-09-095; U.S Environmental Protection Agency: Washington, D.C.

U.S. Environmental Protection Agency (1994) *Combined Sewer Overflow (CSO) Control Policy;* EPA-830/94-001; U.S. Environmental Protection Agency: Washington, D.C.

U.S. Environmental Protection Agency (1999) *Combined Sewer Overflows: Guidance for Monitoring and Modeling;* EPA-832-B-99-002; U.S. Environmental Protection Agency, Office of Water: Washington, D.C.

U.S. Environmental Protection Agency (2002) *Guidance for Quality Assurance Project Plans for Modeling;* EPA-QA/G-5M; U.S. Environmental Protection Agency: Washington, D.C.

U.S. Environmental Protection Agency (2008) *Integrated Modeling for Integrated Environmental Decision Making;* EPA-100/R-08-010; U.S. Environmental Protection Agency: Washington, D.C.

U.S. Environmental Protection Agency (2009) *Condition Assessment of Wastewater Collection Systems, State of Technology Review Report;* EPA-600/R-09-049; U.S. Environmental Protection Agency, Office of Research and Development, National Risk Management Research Laboratory–Water Supply and Water Resources Division: Washington, D.C.

U.S. Environmental Protection Agency (2010) http://www.epa.gov/athens/wwqtsc/html/efdc.html (accessed Oct 2010).

U.S. Environmental Protection Agency (2010) *Storm Water Management Model User's Manual, Version 5.0;* EPA-600/R-05-040; U.S. Environmental Protection Agency, Water Supply and Water Resources Division, National Risk Management Research Laboratory: Cincinnati, Ohio. http://www.epa.gov/ednnrmrl/models/swmm/epaswmm5_user_manual.pdf (accessed Oct 2010).

Van Pelt, T. H.; Czachorski, R. S. (2006) The Application of System Identification to Inflow and Infiltration Modeling and Design Storm Event Simulation for Sanitary Collection Systems. *Proceedings of the Water Environment Federation Collection Systems Specialty Conference—Infrastructure Stewardship: Partnering for a Sustainable Future;* Detroit, Michigan, Aug 6–9; Water Environment Federation: Alexandria, Virginia.

Wastewater Planning Users Group (2002) *Code of Practice for the Hydraulic Modelling of Sewer Systems Version 3.001;* http://www.ciwem.org/media/

44426/Modelling_COP_Ver_03.pdf. Chartered Institution of Water and Environmental Management (CIWEM): London, United Kingdom (accessed March 2011).

Water Environment Federation; American Society of Civil Engineers; Environmental and Water Resources Institute (2009) *Existing Sewer Evaluation and Rehabilitation*, 3rd ed.; WEF Manual of Practice No. FD-6; ASCE Manual and Report on Engineering Practice No. 62; McGraw-Hill: New York.

Water Environment Research Foundation (2004) *Reducing Peak Rainfall-Derived Infiltration/Inflow Rates—Case Studies and Protocol;* Water Environment Research Foundation: Alexandria, Virginia.

7.0 SUGGESTED READINGS

American Society of Civil Engineers; Environmental and Water Resources Institute; Water Environment Federation (2007) *Gravity Sanitary Sewer Design and Construction;* ASCE Manuals and Reports on Engineering Practice No. 60; WEF Manual of Practice No. FD-5; American Society of Civil Engineers: Reston, Virginia.

Bealer, A. (2008) Threading Multiple Vendor Flow Meters through a Single Telemetry Site. *Proceedings of the Water Environment Federation Collection Systems Specialty Conference;* Pittsburgh, Pennsylvania, May 18–21; Water Environment Federation: Alexandria, Virginia.

Enginger, K.; Stevens, P. (2006) Sewer Sociology—The Days of Our (Sewer) Lives. *Proceedings of the Water Environment Federation 79th Annual Technical Exposition and Conference* [CD-ROM]; Dallas, Texas, Oct 21–25; Water Environment Federation: Alexandria, Virginia.

Geldreich, E. E. (1978) Bacterial Populations and Indicator Concepts in Feces, Sewage, Stormwater and Solid Wastes. In *Indicators of Viruses in Water and Food;* Berg, G., Ed.; Ann Arbor Science Publishers: Ann Arbor, Michigan.

Goulding, G. (2008) Developing a Focused and Efficient Calibration Approach for Collection System Models. *Proceedings of the Water Environment Federation Collection Systems Specialty Conference;* Pittsburgh, Pennsylvania, May 18–21; Water Environment Federation: Alexandria, Virginia.

Kepner, C.; Maritime, M.; Goulding, G. (2007) Application of Continuous Simulations to Separate Sanitary Sewer System Planning. *Proceedings of the Water Environment Federation Collection Systems Specialty Conference: Pioneering Trails to Collection Systems Excellence;* Portland, Oregon, May 13–16; Water Environment Federation: Alexandria, Virginia.

Kurz, G.; Ballard, G.; Burgett, M.; Smith, J. (2003) A Proposal for Industry-Wide Standardization of I/I Calculations. *Proceedings of the Water Environment Federation 76th Annual Technical Exposition and Conference* [CD-ROM]; Los Angeles, California, Oct 11–15; Water Environment Federation: Alexandria, Virginia.

Lee, R. (2007) Interpreting Storm Flow Data to Determine Types of I/I. *Proceedings of the Water Environment Federation Collection Systems Specialty Conference: Pioneering Trails to Collection Systems Excellence;* Portland, Oregon, May 13–16; Water Environment Federation: Alexandria, Virginia.

LimnoTech (2008) Lateral Mixing Calibration of the EFDC Model of the Ohio River, Licking River and Banklick Creek. LimnoTech External Memorandum to Sanitation District No. 1.

LimnoTech (2009) Water Quality Calibration and Verification to Fecal Coliform in the EFDC Model of the Ohio River, Licking River and Banklick Creek. LimnoTech External Memorandum to Sanitation District No. 1.

Mays, L. (Ed.) (2001) *Stormwater Collection Systems Design Handbook;* McGraw-Hill: New York.

Mays, L. (Ed.) (2003) *Urban Stormwater Management Tools;* McGraw-Hill: New York.

Rowe, R.; Kathula, V.; Nelson, R. (2006) New Manhole Condition Prioritization Tool Produces Efficient and Robust Analysis Options. *Proceedings of the Water Environment Federation Collection Systems Specialty Conference—Infrastructure Stewardship: Partnering for a Sustainable Future;* Detroit, Michigan, Aug 6–9; Water Environment Federation: Alexandria, Virginia.

U.S. Environmental Protection Agency (2010) *Water Quality Models* Web Site. www.epa.gov/athens/wwqtsc/html/water_quality_models.html (accessed Sept 2010).

Vatter, B.; Nelson, D.; Cantrell, C.; FitzGerald, S. (2007) Accurate Representation of Inflow and Infiltration in Separate Sewer System Models. *Proceedings of the Water Environment Federation Collection Systems Specialty Conference: Pioneering Trails to Collection Systems Excellence;* Portland, Oregon, May 13–16; Water Environment Federation: Alexandria, Virginia.

Walski, T.; Barnard, T.; Harold, E.; Merritt, L.; Walker, N.; Whitman, B. (2007) *Wastewater Collection System Modeling and Design;* Bentley Institute Press: Exton, Pennsylvania.

Water Environment Federation; American Society of Civil Engineers (1992) *Design and Construction of Urban Stormwater Management Systems;* WEF Manual of Practice No. FD-20; American Society of Civil Engineers: New York.

Water Environment Federation (2006) *Guide to Managing Peak Wet Weather Flows in Municipal Wastewater Collection and Treatment Systems;* Water Environment Federation: Alexandria, Virginia.

Chapter 6

System Maintenance and Management

(*continued*)

1.0 INTRODUCTION

This chapter discusses how system maintenance and management are crucial to prevention and control of sewer system overflows. If a new system is designed and constructed properly, one would think overflows would not be possible. However, as a system gets older, many factors can contribute to the cause of overflows. If a system is well maintained, many of those problems can be minimized, although never completely eliminated. Improving system maintenance and management in a sewer system is a primary way to reduce overflows going forward. Many who work to maintain sewer systems have the "keeping-it-in-the-pipe" mentality when dealing with events that can cause overflows. These workers are on the front line in the fight to reduce or eliminate wastewater overflows that can damage the environment. Support of management in the effort to properly maintain sewer systems is of foremost importance. The goal should be to eliminate sanitary sewer overflows (SSOs).

2.0 TYPICAL SOURCES OF OVERFLOWS THAT CAN RESULT FROM MAINTENANCE ISSUES

There are many causes of sewer overflows that can be attributed to maintenance. Blockages are the main cause of overflows (U.S. EPA, 2004b) (see Figures 6.1 and 6.2). In a gravity system, the cave-in causes a blockage that results in the sewer system backing up and overflowing out of manholes, in basements, or at pumping stations. In a force main, the result is a geyser of wastewater. Many things can cause blockages

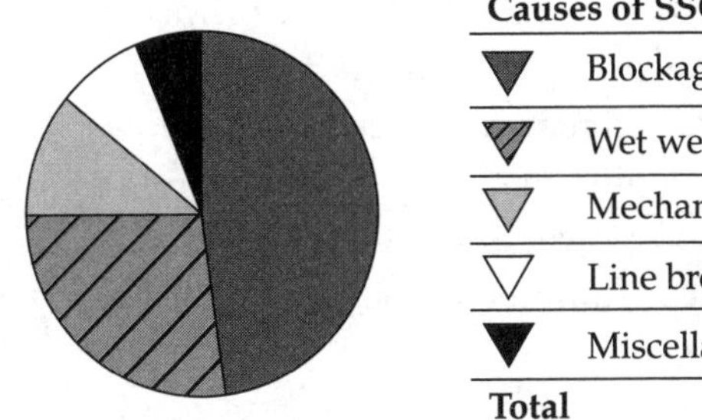

Causes of SSO Events	Percent
Blockages	48%
Wet weather and I/I	26%
Mechanical or power failures	11%
Line breaks	10%
Miscellaneous	5%
Total	**100%**

FIGURE 6.1 U.S. EPA wheel showing causes of SSOs (U.S. EPA, 2004b).

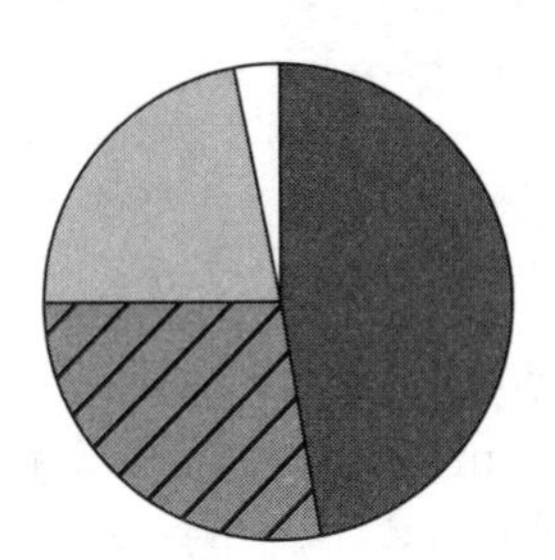

Causes of Blockage Events	Percent
Grease	47%
Grit, rock, and other debris	27%
Roots	22%
Roots and grease	4%
Total	**100%**

FIGURE 6.2 U.S. EPA wheel showing causes of blockages (U.S. EPA, 2004b).

in a sewer system besides cave-ins. Excessive fats, oils, and grease (FOG) in the sewer system can accumulate and cause backups resulting in overflows. Roots can get into the sewers along with FOG (referred to as *fats, roots, oils, and grease* [FROG]) and cause blockages. Sags in gravity lines caused by settlement can allow debris to accumulate, resulting in blockages. Excessive air or gas in force mains can actually cause a blockage and result in overflows.

Wastewater is corrosive by nature and is, at best, a terrible environment in which to try to work and maintain anything. Sewer lines can be attacked by corrosion from both inside and outside. Depending on the pipe material, slowing down or monitoring the corrosion process and attacking it preemptively before a failure occurs is the job of operations and maintenance staff. If corrosion is allowed to go too far, pipe failure resulting in a cave-in and the associated overflow of wastewater are to be expected. Corrosion in sewers attacks everything, including valves and vents. If unattended, these will eventually fail, resulting in a wastewater overflow.

Improperly maintained or malfunctioning mechanical and electronic equipment associated with pumping and emergency power, if unfit to perform when required, can result in overflows. Maintaining these equipment is critical to the overall goal of reducing overflows.

3.0 GETTING STARTED DEALING WITH OVERFLOWS

Whether developing a long-term program or refreshing an existing program, there will typically be immediate problems to solve while completing the long-range program. Minimizing overflows, whether system-wide or in targeted areas, requires skill sets that may not be present in the organization.

3.1 Setting up Teams

Each person brings unique skills, experience, and knowledge to a job. Creating effective teams to begin a long-term program will require a mix of experienced analytical and field personnel who understand customer service and the importance of the work to public health, and are open to learning new things. Information sharing in a neutral manner respectful of all participants is critical to long-term success.

Typical sewer operation and maintenance work teams perform physical inspections, preventive maintenance including cleaning, and small- and large-pipe and appurtenances repair. Liaison teams provide information such as trouble reports and detailed system data and support-contracted work such as physical inspection and system data validation. To fully use closed-circuit television (CCTV) or other electronic inspection and cleaning equipment, organizations commonly create separate work groups for those activities, scheduling the work in follow-up to a blockage or overflow or in planned maintenance and system assessment. Data gathered in television or other electronic inspection and cleaning is commonly reviewed by a single team using common standards to ensure consistent condition evaluation throughout the sewer system. Related teams include industrial pretreatment permitting and inspection, capital planning, work scheduling, and engineering support. Figure 6.3 shows some typical positions used in the SSO process along with typical assignments.

Sanitary Sewer Overflow

SOP No: EPD-1

Page

Issued: 11-9-07

Revised: 12-3-08

Administrative Staff

1. Upon receipt of call, staff shall document applicable information on Work Order form.
2. Staff will dispatch applicable utility crew for response.

Responding Staff

1. Staff responding to the incident shall promptly investigate the cause and stop the sanitary sewer overflow. **Appropriate safety procedures shall be observed.**
2. Staff responding should make every attempt to protect storm drain inlets and entrances to waterways from the overflow material.
3. If extra help is necessary, staff should immediately notify their Superintendentor the utility worker in-charge in their absence, and call for extra help (Mechanics, Utility Workers or Environmental Program Staff). The number of extra help depends on the severity of the incident.
4. Staff responding to the incident must complete the **Sanitary Sewer Overflow Report (attached).** This report will be submitted to the Environmental Program Division **no later than 24 hours after the incident.**
   a. In the event that the sewer overflow occurs on a weekend or holiday, the Sanitary Sewer Overflow Report will be submitted to the Environmental Program Division on the next scheduled workday.

Environmental Program Division

1. Once the Sanitary Sewer Overflow (SSO) Report has been received it must be electronically entered into the State Water Resources Control Board's Sanitary Sewer Overflow database, which is accessed through California Integrated Water Quality System (CIWQS) http://ciwqs.waterboards.ca.gov, based on category of spill.
   a. Category 1 spills/overflowsmust go through "Draft Submittal" (SSO database) within 3 days after overflow occurred and must be certified (finalized) by the Legally Responsible Official within 15 days of the conclusion of the response and remediation. Upon conclusion, ensure that "Ongoing Investigation" item iscorrectly answered in the SSO database.
   b. Category 2 spills/overflows must be reported within 30 days after the month in which the spill occurred.
   c. If no spills occurred in the month, a "No Spill Certification" must be completed within 30 days of the end of that month.
2. Environmental Program Division (EPD) staff will submit a copy of completed reports to the Public Works Manager.
3. EPD staff will file original report in the Sanitary Sewer Overflow binder.
4. EPD staff will prepare a confidential council newsletter at the beginning of the month, summarizing all reportable SSO's for the previous month.
5. After determining the responsible party, (i.e., contractor, engineering, etc.), EPD staff will prepare a letter/invoice to the appropriate party requestingreimbursement of costs incurred by the city.

Utility Superintendent During SSO event, Superintendent or designee will direct notification to outside agencies when necessary.

Figure 6.3 Sanitary sewer overflow team (City of West Sacramento, 2008).

Public Works Department

Sanitary Sewer Overflow

SOP No: EPD-1

Page

Issued: 11-9-07

Revised: 12-3-08

Completion of Sanitary Sewer Overflow Report

Definitions/Instructions as they appear on Report

SSO – Sanitary Sewer Overflow, any overflow, spill, release, discharge or diversion of untreated or partially treated wastewater from a sanitary sewer system.

Estimate Volume of discharge – Total volume of SSO.

Location – Physical address of SSO.

Description – Type of area where SSO is located.

Source – Where is the SSO coming from?

Cause – What was the cause of the blockage?

Route of Flow – Describes the direction of flow and the area(s) that the wastewater flowed over and/or through.

Destination – Final destination, describes the area that the wastewater ultimately reached.

Clean Up /Recovery – Describes procedures, equipment used and estimated volume of recovered material.

Remediation – What was done to correct the SSO? Who did repair/cleared blockage? Who inspected job?

Ways to Prevent Recurrence - What can be done to prevent the event from happening again?

Was the blockage/problem in the City owned collection system – (e.g., Forced Mains, Lift Stations, etc.)

Category 1 SSO – Meets at least one (yes) of the questions listed SSO Category 1 Determination table.

Category 2 SSO – Sewer overflow that does not meet any of the requirements of a Category 1 SSO, but caused by problems in the City owned collection system.

SSO Volume – Estimated volume, reference item 4 of report. The sum of the three lines should equal the total volume.

Response – What was done to SSO and the cause of SSO?

Samples Taken – If samples are necessary, contact the Environmental Program Division.

Notification – Notify the following individuals/agencies according to the following:

OES – Category 1 overflow, must be notified **within 2 hours** of becoming aware of discharge.

Yolo County Env. Health – Category 1 overflow, must be notified **within 2 hours** of becoming aware of discharge.

Regional Board – Category 1 overflow, must be notified **within 2 hours** of becoming aware of discharge.

Additionally, certification must be submitted within 24 hours of discharge stating that OES and Env. Health were notified.

Environmental Program Division - As necessary to aid in remediation.

Utility Supervisor – As necessary to aid in remediation and cleanup.

Mechanics – If necessary to shut down sanitary sewer lift station.

RD 900 – If overflowed material reaches canal/ditch/channel that they are responsible for.

Business/Contractor – As necessary to aid in remediation and cleanup.

FIGURE 6.3 (Continued)

Description of Responsibilities

Position	Responsibilities
Public Works Clerk / Secretary	Receive initial call and dispatch staff for response Staff communication center
On-Call Utility Maintenance Worker	Provide initial response for after hour calls. Investigate and assess SSO Contact additional staff for assistance Correct cause Contain SSO Post signs Clean-up Update communication center
Utility Maintenance Worker	Investigate and assess SSO Contact additional staff for assistance Correct cause Contain SSO Post signs Clean-up Update communication center
Chief Utility Maintenance Worker	Dispatch response staff Investigate and assess SSO Direct response staff Communicate with City Divisions SSO documentation Correct cause Contain SSO Initiate SSO report requirements Clean-up Update communication center Submit SSO field report
Engineering Construction Management Staff	Work with contractor on construction related SSOs Update communication center
Utility Maintenance Superintendent	Confirm Category 1 SSO Direct staff Update communication center
Assistant Utility Maintenance Superintendent	Staff communication center Direct staff Update communication center
On-Call Treatment Plant Mechanic	Operation of sanitary/storm pump stations SCADA Update communication center

FIGURE 6.3 (Continued)

Public Information Officer (PIO)	Issues news release Media notification
Public Works Manager	Determine whether drinking waters are impacted from SSO Provide information to PIO/City Manager/Public Works and Community Development Director
West Sacramento Fire Department	Provide boat Emergency response
West Sacramento Police Department	Traffic and crowd control
Environmental Program Manager	Review SSO field reports Complete certify electronic SSO reports Preparation of additional requirements involving SSO Coordinate with Yolo County Environmental Health Direct sampling operations as necessary Finalize Report
Environmental Program Specialist	Provide requested information on SSO Collect / analyze samples Route samples for analyses Initiate SSO electronic report requirements Finalize Report

FIGURE 6.3 (Continued)

3.2 Capacity, Management, Operation, and Maintenance Checklist

The U.S. Environmental Protection Agency's (U.S. EPA's) *Guide for Evaluating Capacity, Management, Operation, and Maintenance (CMOM) Programs at Sanitary Sewer Collection Systems* (U.S. EPA, 2005) provides a capacity, management, operation, and maintenance (CMOM) checklist that gives utilities a comprehensive list of actions and responsibilities. (The list is available at http://www.epa.gov/npdes/pubs/cmom_guide_for_collection_systems.pdf.) Elements include general information, collection system management, collection system operation, equipment and collection system maintenance, sewer system evaluation survey (SSES), and rehabilitation. The list is helpful in quickly determining what is lacking in an existing program. Many of the program elements, such as inspection and maintenance planning, will directly help staff reduce SSOs.

While beginning or revising a comprehensive sewer overflow control program, short-term actions commonly called "quick wins" may help improve conditions for some customers and build staff morale. Van Buren and Safferstone (2009) suggest evaluating possible quick wins based on the following:

- *Value*—cost reduction and urgent need;
- *Cost and feasibility*—no substantial distraction from other responsibilities and no new resources required;
- *Collective effect*—a win recognizable to the team as a win;
- *Opportunity to learn*—learn strengths, weaknesses, and motivations of the team; and
- *Opportunity to engage*—communication up and down the organization.

Short-term actions tend to be based on experiential information and should not be confused with sustainable program development. Quick wins can create real improvement in a system where actions have not already been taken to find and correct large sources of infiltration and inflow, such as missing manhole covers and damage along waterways; schedule recurring cleaning in areas with chronic blockages until physical inspection can identify the primary cause; or repair chronic problems correctable with in-house forces or with contracted support.

Input from field forces about recurring problems and pipeline conditions provides knowledge of how the system performs while building employee commitment to the program. Although very helpful, the information lacks the ranking process obtained with organized system evaluation. Experiential information about problem areas should be combined with physical inspection follow-up. Creating a feedback loop to inform those who provide input of what is done with the information can enhance morale and encourage the flow of information from operations critical when new programs or approaches are needed.

Whether modifying or creating a program, a key step is reviewing procedures for overflow reporting to the responsible regulatory agency. In the absence of other system information, comprehensive overflow reports are a meaningful database of system shortcomings, particularly recurring overflows.

3.3 Use of Contractors

Contractors are available to assist, whether they work in a specific area, perform a specific type of work, augment in-house forces, or provide special services.

Engineering contractors are available to help organize the program and bring expertise to the organization. Skills available include physical inspection and evaluation of system conditions, flow monitoring and modeling, long-range planning, preparation of plans and specifications, and knowledge transfer to staff.

Construction contractors can make specialty rehabilitation repairs, make area-wide improvements, or provide work crews to augment in-house forces. A typical construction contract for sewer rehabilitation involves the following multiple types of repairs in a defined area: replacing lengths of pipe, lining pipe, repairing service connections, and rehabilitating or replacing manholes. Alternatively, specialty contractors may be hired to make a specific type of improvement, such as manhole rehabilitation or pipe lining, in a defined area or at defined locations.

"On-call" contracts can provide repair crews to augment in-house forces and specialty equipment and skills not regularly required in the organization. One common use of on-call contracting by utilities is pavement repair. Developing a scope of services that minimizes contractor downtime and mobilization expenses is important to minimizing the cost per unit performed for this service.

A catastrophic problem may require use of an emergency contract. Contracting requirements vary depending on the bidding laws of the state and governing body. Standard methods and approval processes developed in advance for emergency contracting shortens the time between problem and resolution and should be incorporated into the utility's emergency preparedness plan. The procedure should identify the level of detail required to describe the work to the contractor, how advertisement will be made and for what duration bids will be taken, who has authority to mobilize the contractor and how that mobilization will be issued, and who will represent the owner in review and direction of the contractor's work.

More recently, water and sewer utilities have begun creating mutual aid or inter-local agreements with other utilities for emergency assistance. Including neighboring utilities' contact information and equipment in the utility's emergency response plan, along with information about rental contractors available to provide pumps and generators on short notice, makes the information accessible in an emergency.

3.4 Assessing Problems

Overflows occur when pipe flow is blocked or damaged, or the pipe receives more wastewater than the pipe can carry. To end the overflow condition, the primary cause must be identified.

Water Environment Research Foundation's *Effective Practices for Sanitary Sewer and Collection System Operations and Maintenance* (WERF, 2003) describes the Line

Blockage Assessment Program process steps of emergency relief, customer contact, internal inspection, evaluation, and response.

Common causes of pipe blockages include pipe structural failure, root intrusion, grease buildup, or buildup of other materials. Finding the actual cause of the problem can be complex. For example, the cause of heavy root intrusion may be a broken joint or pipe structural failure. Grease builds up where there are changes in pipe grade or the smooth pipe wall is interrupted by heavy corrosion or a protruding-building sewer connection. For example, the root cause of a grease blockage may be a sag or belly in the pipe for which the solution is relaying a few joints of pipe in new bedding material, a poorly maintained grease interceptor requiring assistance of an industrial pretreatment inspector, or a protruding piece of pipe or roots catching the grease and requiring repair.

Maintenance and operations activities related to flow overloads focus on identification and repair of excess flow sources. They include general condition deterioration, private property contributions, and system damage from storms. Fully defining a system- or basin-wide problem will, in most circumstances, require completion of an SSES by a firm specializing in such work.

Whenever possible, sewers are built following natural drainage patterns to minimize depth and maximize slope. As such, they are vulnerable to stormwater flow impacts along waterways including erosion of pipe supporting material and loss of manhole covers. Sewers in streets are not immune to the problem where flooding streets are relieved by sewer manhole rings and covers. Visual inspection of pipes near waterways and at stream crossings should be made following storm events.

Geographic information systems (GISs) can provide information about sewer locations, customer connections, land use, flood plains and drainage systems, past and planned capital improvements, and customer classes and industrial waste permittees, all depending on the time and effort the organization puts in to building and maintaining the system. Combined with a maintenance management or work order system, especially when the SSO cause is tied to an asset number rather than to merely an address, can help visually identify trouble areas by type. Analysis of maintenance management and GIS data can help determine whether factors such as pipe age, pipe material, proximity to flood zones, or soil type are unique sources of problems.

Field personnel usually appreciate sharing information about system conditions and grow more open as relevant improvement actions are taken. They know where there is gravel in a concrete pipe, where a 25.4-cm (10-in.) line is actually 15.24 cm

(6 in.), and whether the equipment provided is the most efficient to do the job. A comfortable discussion focused on system problems will produce new insight, but should only be held if follow-up to help solve those problems will occur.

In reliability centered maintenance, work is prioritized based on the severity of impact a failure of that infrastructure would create. Severity of impact is based on the number or criticality of customers losing service or experiencing a sewer backup into a structure, the volume of wastewater handled lost, and the environmental sensitivity of the location where the wastewater is lost.

There are several methods used to estimate the amount of wastewater that is lost in an overflow. Methods include estimating volumes pumped, size of opening and amount of flow that can escape, use of meters, estimating size of overflow area, and so on. In Figure 6.4, there are some basic methods that a utility uses, along with a newer idea that is becoming of interest, that is, the use of pictures that have been calibrated to show the amount of flow that can escape.

SSO Flow Estimation Methods

Volume of the SSO can be determined using a variety of approaches. The following sections will discuss two methods that are often employed. The person preparing the estimate shall use the method most appropriate to the SSO in question. Every effort shall be made to make the best possible estimate of the volume.

Method 1 Measured Volume

This method can be used on small spills if it is not raining.

Step 1: Sketch the shape of the spill that is contained.
Step 2: Measure the length and width.
Step 3: Measure the depth in several locations.
Step 4: Convert all dimensions to feet.
Feet = inches/12
Step 5: Calculate the area using the following formulas.

Rectangle Area = Length x Width

Circle Area = Diameter x Diameter x 0.785

Triangle = Base x Height x 0.5

FIGURE 6.4 Sanitary sewer overflow flow estimation (cu ft × 0.028 32 = m^3; ft × 0.3048 = m; gal × 3785 = m^3; 7.5 gal = 1 cu ft = 0.0284 m^3; in. × 25.4 = mm; and sq ft × 0.092 90 = m^2) (City of West Sacramento, 2008).

Step 6: Multiply the area times the depth to get the volume.
Volume ft^3= Area x Depth

Step 7: Multiply the volume by 7.5 gallons/ft^3 to convert it to gallons.
Gallons = Volume x 7.5 gallons/ft^3

Method 2 Duration and Flow Rate

Duration: The duration is the total elapsed time from when the SSO started until it stops.

Flow rate: The rate at which the SSO is flowing. Usually expressed as gallons per second (GPS) or gallons per minute (GPM) or gallons per hour (GPH).

Open channel flow: Often overflows run into nearby dry ditches or street gutters. Total volume, gallons, of flow can be quantified by measuring the cross-sectional area and speed of the flow. Measure a set distance paralleling the SSO flow route. Measure, in inches, the midway width and depth of the flow over this distance. Then measure the time, in seconds, it takes a float to travel the set distance. Record total time of the SSO flow.

Calculate the total SSO volume of the following example:

Example: After measuring off a set distance of 20 feet, it was determined that the float took 20 seconds to travel this 20 feet. The width and depth at the midway point of the flow was 28 inches and 3 inches, respectfully. The total time of the SSO flow was 20 minutes. What is the total volume, gallons, of this SSO event?

Total Volume (gal) = Velocity (ft/sec) x Area (ft^2) x total time (seconds) x 7.5 gal/ ft^3
= 20 ft/20 sec. x (28 x 3)/144) ft^2 x (20 min. x 60 sec./min.) x 7.5 gal/ ft^3
= 1 x 0.58 x 1200 x 7.5 gal
= 5,220 gallons

Pump stations: SCADA systems can provide flow or pump run time data for sewer and storm water pump stations. Pump curves may need to be obtained to determine flow rates. The flow rates can be used to determine flow volumes. Contact the city's Treatment Plant Mechanics to obtain SCADA data.

SSO flow estimation pictures (see next page): Provides pictures of sewage flowing from a manhole cover at a variety of flow rates. Observations by the responding utility maintenance crew are used to select the appropriate flow rate from the chart.

FIGURE 6.4 (Continued)

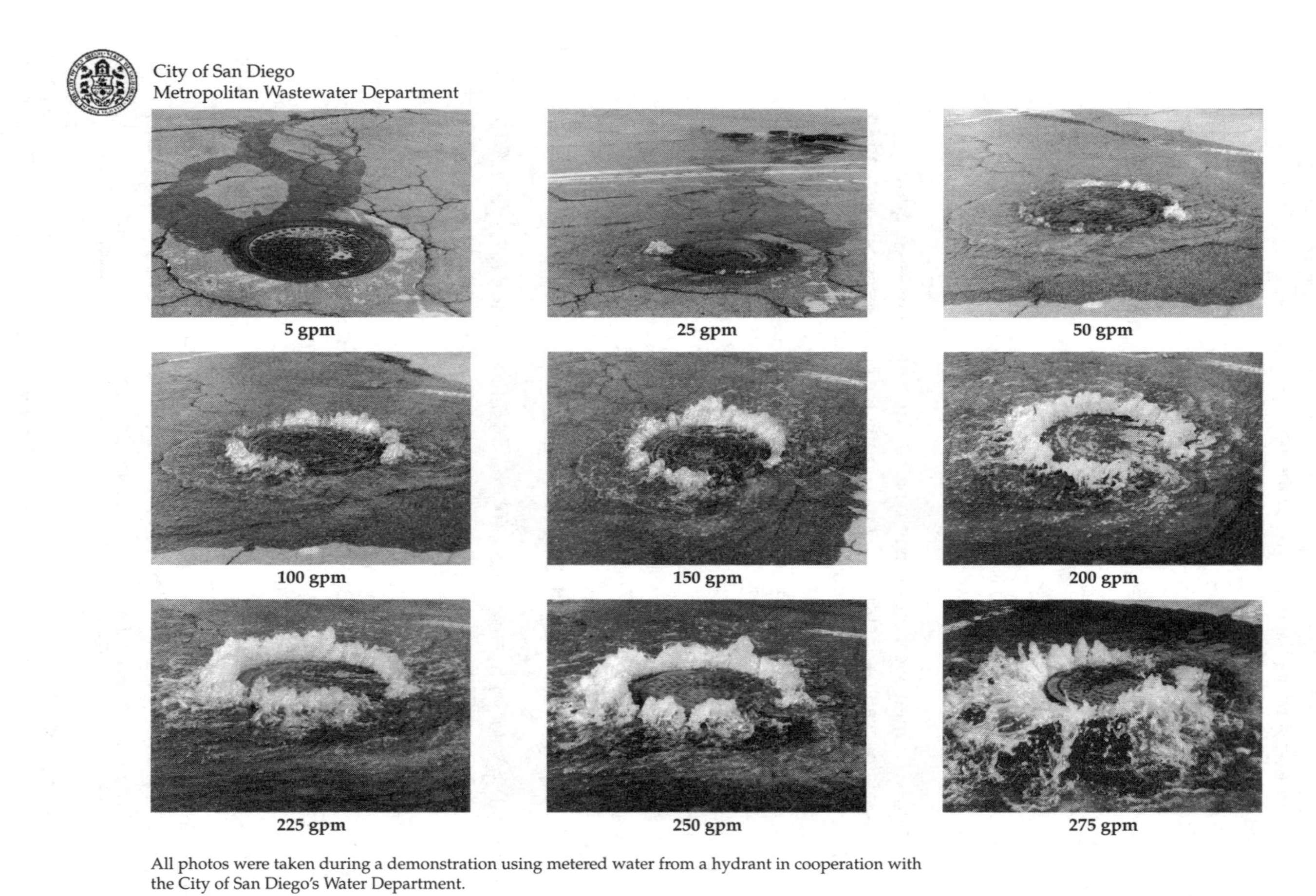

All photos were taken during a demonstration using metered water from a hydrant in cooperation with the City of San Diego's Water Department.

FIGURE 6.4 (Continued)

3.5 Staffing

Crew safety is the first thing to address when looking at staffing requirements. Some jobs will take more employees because of the safety factors involved. Staffing levels should be determined by comparing the cost of adding additional employees to the return of resources invested. For instance, running rods is typically a two-to-three-person job depending on the type and automation of equipment used, while jetting a line may only require two people. One should review the productivity of the crews with different staffing levels and determine if the added expense of additional employees is worth the return on investment. That additional employee may be more productive to the organization by working on another task.

3.6 Reporting Requirements

The National Pollutant Discharge Elimination System (NPDES) is a permit required to own and operate a wastewater system. Provisions within this permit require permit holders to report overflows, as discussed in Chapter 3.

Standard operating procedures should be created for overflow reporting. This is especially important for operating staff in the field. Procedures should include the following:

- Person(s) to notify and time frame required,
- Location of the overflow,
- Receiving water (if there is one),
- Duration of the overflow,
- Estimated volume of the overflow,
- Reason for the overflow,
- Corrective action, and
- Pipe age and material.

Figure 6.5 is an example of a report that can be used by field crews to document overflow information that can then be forwarded to the person that will eventually respond to federal and state regulators. Figure 6.6 shows a computerized reporting system that can be used in the field to get started and then added to in the office. The computer version also allows for pictures to be easily attached. Figure 6.7 shows one locality's decision matrix that can be used to decide the steps necessary to report and respond to an overflow.

Sanitary Sewer Overflow (SSO) Report

EPD Use Only			
SSO Category	________	ID #	________
Latitude	________	Longitude	________

Date	________
Equipment Hours	________
Worker Hours	________
Work Order No.	________

Called: ____________________ am pm

Arrived: ____________________ am pm

1. Est. SSO Start ____________________ am pm **2. Est. SSO Stop** ________________ am pm

3. Est. Duration of SSO ________________

4. Est. Volume __________ Gals Dimensions L ________ W __________ D ________

5. Location: Address ____

6. Description: ◎ Street ◎ Apartment ◎ Commercial ◎ Parking Lot
◎ Residential ◎ Intersection ◎ Other ________________

7. Source: ◎ Manhole ◎ Cleanout ◎ Ground ◎ Pump Station
◎ Other ____

8. Cause: ◎ Blockage ◎ Grease ◎ Roots ◎ Debris ◎ Other ________

9. Route of Flow: ____
____

10. Destination: ◎ Street ◎ DI ◎ Channel ◎ Surface Water
◎ Building/Private Property ◎ Other ________________

11. Contractor responsible? Y / N Name: ______________________________

Address/ Phone ____

12. Clean Up / Recovery (Actions): ____
____

13. Remediation: ____
____

14. Ways to Prevent Recurrence: ____
____
____

15. Was the blockage / problem in the City owned collection system? Yes ◎ No ◎

FIGURE 6.5 Example of a report that can be used to report overflows to Federal and State regulators (City of West Sacramento, 2008).

SSO Category 1 Determination	Yes	No
Was the volume 1000 gallons or more?	☐	☐
Was there a discharge to surface water?	☐	☐
Was there a discharge to a storm drain/RD-900 drainage channel that was not fully captured and returned to the sanitary sewer system?	☐	☐
If any "Yes" response, complete the rest of this form. If all are "No", skip to item #19.		

16. SSO Volume: to surface water ________ Gallons
to storm drain/drainage channel ________ Gallons
not recovered from storm drain/drainage channel ________ Gallons

17. Response:
☐ Cleaned Up ☐ Inspected to Determine Cause (CCTV)
☐ Contained ☐ Returned All/Part of SSO to Sanitary Sewer
☐ Restored Flow ☐ Other ________________

18. Samples taken? Y / N

19. Notification: (* If category 1 SSO, agency notification required)

Contacted				Left Message	Initial
☐	* OES (800) 852-7550	Date: ________ Control #: ________	Time: ________	☐	________
☐	* Regional Board (916) 464-4761 fax (916) 464-4645	Date: ________	Time: ________	☐	________
	Certification Notification Required? Y N	Date: ________	Time: ________		________
☐	* Yolo County Env. Health (530)666-8646	Date: ________	Time: ________	☐	________
☐	Utility Supervisor (916) 617-4849	Date: ________	Time: ________	☐	________
☐	Env. Program Division (916) 617-4825	Date: ________	Time: ________	☐	________
☐	Mechanics (916) 799-3890	Date: ________	Time: ________	☐	________
☐	RD - 900 (916) 371-1483	Date: ________	Time: ________	☐	________
☐	Business/Contractor	Date: ________	Time: ________	☐	________

Send this report to Environmental Program Division within 24 hours.

Prepared by: ________________________ **Date:** ________

FIGURE 6.5 (Continued)

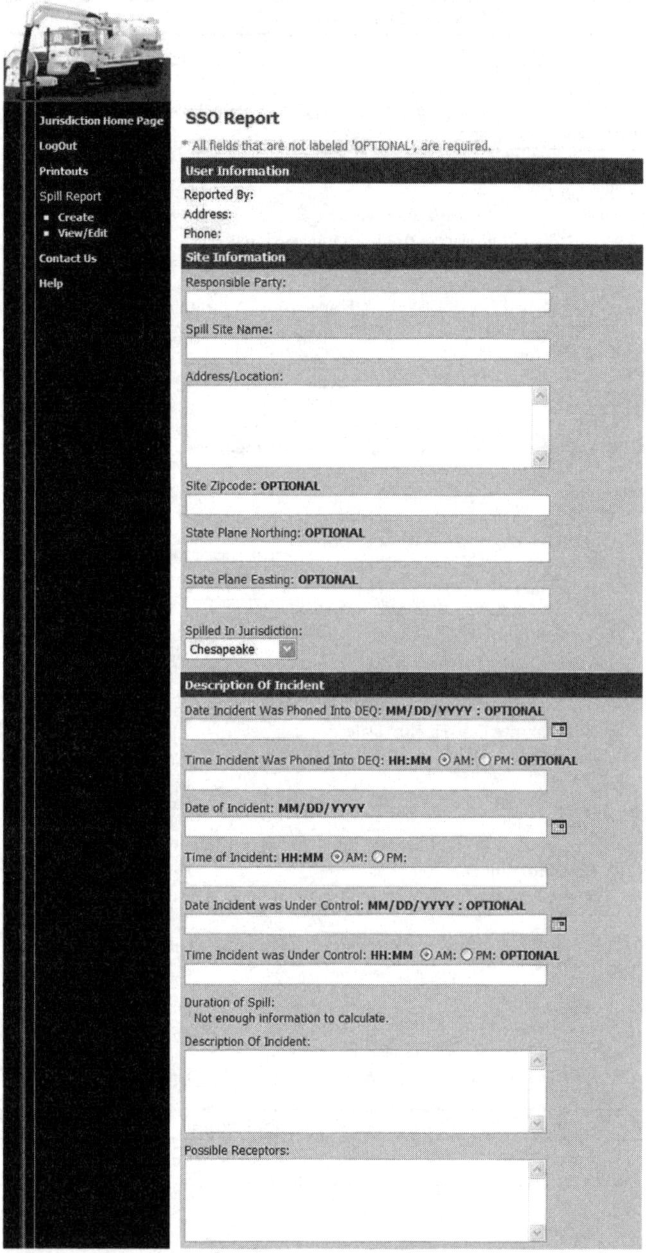

Jurisdiction Home Page
LogOut
Printouts
Spill Report
- Create
- View/Edit

Contact Us
Help

SSO Report

* All fields that are not labeled 'OPTIONAL', are required.

User Information

Reported By:
Address:
Phone:

Site Information

Responsible Party:

Spill Site Name:

Address/Location:

Site Zipcode: **OPTIONAL**

State Plane Northing: **OPTIONAL**

State Plane Easting: **OPTIONAL**

Spilled In Jurisdiction:
Chesapeake

Description Of Incident

Date Incident Was Phoned Into DEQ: **MM/DD/YYYY : OPTIONAL**

Time Incident Was Phoned Into DEQ: **HH:MM** AM: PM: **OPTIONAL**

Date of Incident: **MM/DD/YYYY**

Time of Incident: **HH:MM** AM: PM:

Date Incident was Under Control: **MM/DD/YYYY : OPTIONAL**

Time Incident was Under Control: **HH:MM** AM: PM: **OPTIONAL**

Duration of Spill:
Not enough information to calculate.

Description Of Incident:

Possible Receptors:

Figure 6.6 Sanitary sewer overflow reporting system for the department of environmental quality (courtesy of Hampton Roads Planning District Commission).

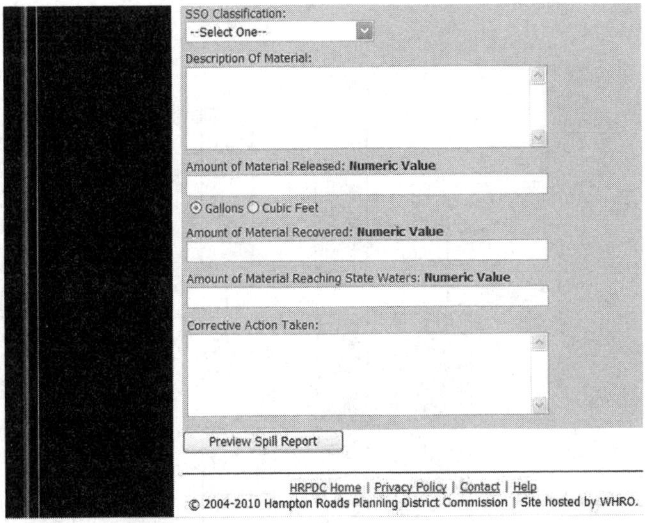

SSO Classification:

--Select One--

Description Of Material:

Amount of Material Released: **Numeric Value**

Gallons Cubic Feet

Amount of Material Recovered: **Numeric Value**

Amount of Material Reaching State Waters: **Numeric Value**

Corrective Action Taken:

Preview Spill Report

HRPDC Home | Privacy Policy | Contact | Help

© 2004-2010 Hampton Roads Planning District Commission | Site hosted by WHRO.

FIGURE 6.6 (Continued)

4.0 PREVENTIVE MAINTENANCE

Preventive maintenance includes proactive actions taken by those who operate sewer systems to inspect, clean, and do routine maintenance to stop or minimize those problems that will occur in the sewer system. Problems found during preventative maintenance that require a repair project are considered corrective maintenance. Many of the problems found in a collection system are caused by others. For instance, the issue of excessive FROG is typically introduced into the sewer system by businesses or residents. Many localities have adopted public information programs to educate the public to not pour grease down the drain. Other agencies within a locality may also inspect restaurants, for instance, to ensure they are using and maintaining properly sized grease traps.

Maintenance crews may know where grease is likely to build up and require regular cleaning to remove. Regular viewing of the inside of sewers to identify accumulation of FOG and FROG, sand, and other materials, followed by cleaning, can help to greatly reduce this cause of overflows. FOG and FROG-related blockages are a primary cause of overflows in gravity systems, as shown in Figure 6.2.

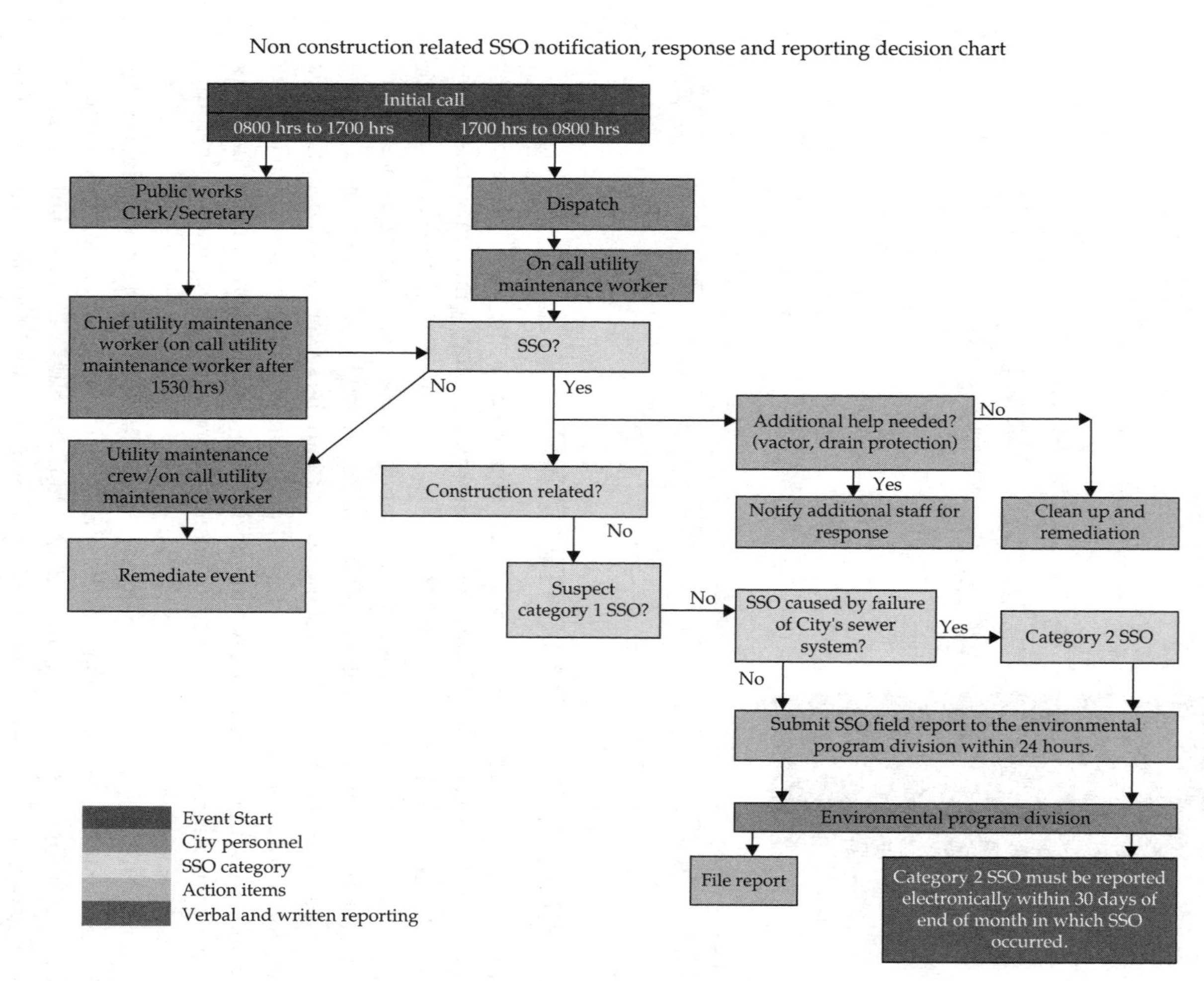

FIGURE 6.7 Example of response decision matrix for an SSO (City of West Sacramento, 2008).

Non construction related SSO notification, response and reporting decision chart

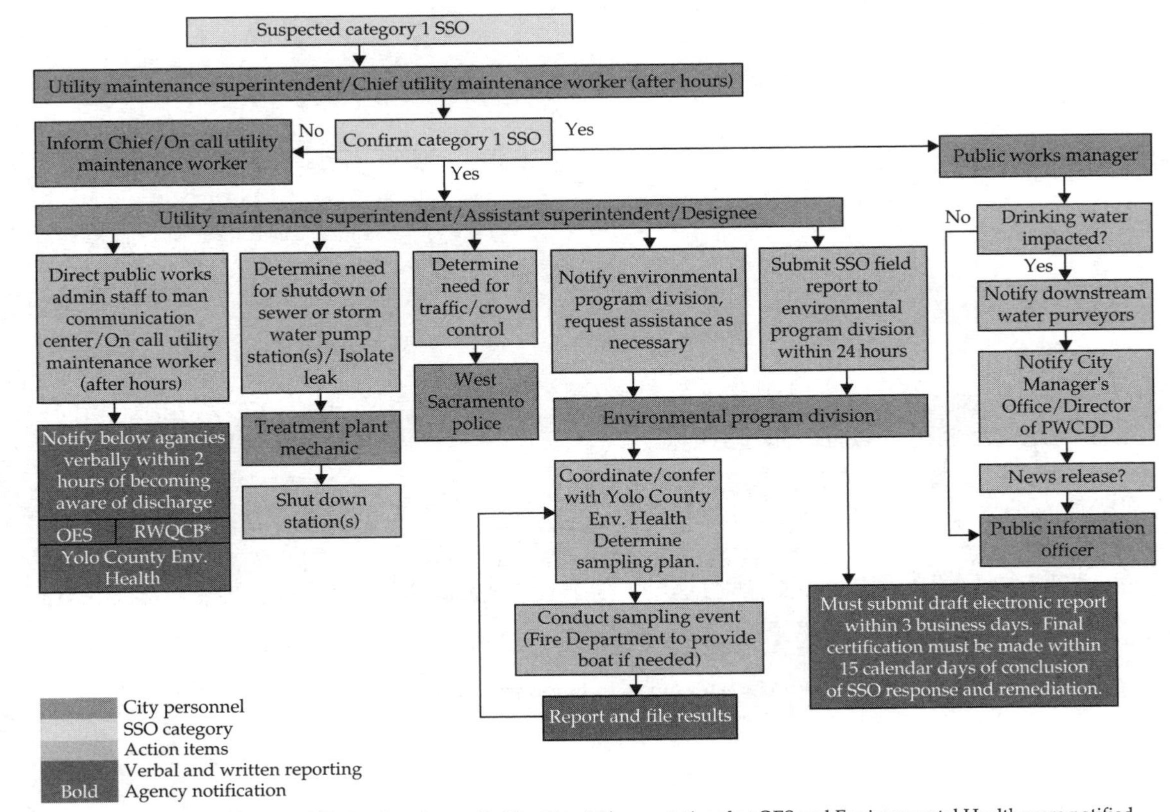

* Additionally, must submit certification (see Appendix F) within 24 hours stating that OES and Environmental Health were notified.

FIGURE 6.7 (Continued)

Construction-related SSO notification, response and reporting decision chart

Construction related SSO

Notify engineering construction management

Engineering construction management

Work with contractor on clean up and remediation

Is the SSO contained in trenches or temporary structures?

Yes

Notify environmental program division

Environmental program division

Determine if external notification is required

Upon completion of clean up and remediation

Utility maintenance superintendent

Seek reinbursement for costs incurred

City personnel
SSO category
Action items
Verbal and written reporting

Figure 6.7 (Continued)

There are several articles and publications that discuss sewer evaluation and rehabilitation programs. *Trenchless Technology* magazine's Web site http;/trenchlessonline.com contains articles regarding new sewer and manhole evaluation and rehabilitation programs and projects. Water Environment Federation has two publications related to sewer and manhole evaluation and rehabilitation programs, *Trenchless Technology Pipeline and Utility Design, Construction, and Renewal* (Najafi, 2004) and *Existing Sewer Evaluation and Rehabilitation* (WEF et al., 2009).

Most sewers are located in public rights-of-way. To improve the aesthetics of streets and boulevards, trees and shrubs are planted in major rights-of-way. Tree and shrub root structures seek sources of water to thrive, whether the source is natural groundwater or wastewater exfiltrating from sewers through small cracks or open joints. As roots extend to find water sources, they often find their way into sewers through these openings. Root intrusion can lead to blockages that either impede wastewater flow or widen the opening to allow soil inflow, causing a washout condition and, ultimately, sewer failure. Either situation can lead to stoppage of wastewater flow and may cause wastewater backups.

Closed-circuit television is one tool that is widely used to help maintain sewer systems. Through the use of these tiny cameras, the inside of piping systems can be viewed to determine their physical state of repair. Establishing a routine inspection period with CCTV can greatly increase knowledge of the condition of the inside of the sewer and identify problems such as corrosion, cave-in, leaking joints, FOG and FROG problems, debris, or intrusion of other pipes that can then be scheduled for cleaning or repair. The National Association of Sewer Service Companies (NASSCO) has developed methodologies for recording data associated with CCTV inspections called the *Pipeline Assessment Certification Program* and the *Manhole Assessment Certification Program*, which allow for more uniform descriptions of the status of the pipes and manholes, respectively (NASSCO, 2001, 2006). A modified version of CCTV inspection is to simply use a pole-mounted camera to view the inside of manholes. Often, pipes entering and leaving the manhole can be viewed for more than 100 meters (several hundred feet) to provide an indication of any problems. Observation of standing water inside the manhole can also be an indication of a blockage.

Another simple tool used to evaluate gravity sewers is smoke testing. This is an inexpensive, quick way to see where significant inflow problems exist. Smoke is pumped into the system through manholes and allows breaks, interconnects with storm systems, open cleanouts, and illegal connections such as roof drains to be easily viewed. Repairs can then be made that often reduce the amount of inflow tremendously.

4.1 Sewer Inspection

Documentation of the condition of sewers is critical to assess changes in their condition from one inspection to the next. This will assist the utility in prioritizing any corrective maintenance or major repair activities. A uniform method of rating the condition of sewers needs to be used so that the assessment will be consistent from one inspection to the next (e.g., NASSCO).

There are many different tools for evaluating sewers. One problem recognized in the industry is that, without a standardization of terms and methodologies, the data from the work could be subjective. There have been programs established to standardize the recording and evaluation of inspection data. Many firms have created their own methods and some software packages, such as computerized maintenance management systems (CMMSs), support several different methods. The main tools for the general public are discussed in the following subsections.

4.1.1 *Water Research Centre*

The Water Research Centre (WRc) developed their *Manual of Sewer Condition Classification* (WRc, 2004) based on more than 30 years of work. The WRc standardized inspection procedures, coding principles, header details, and condition codes. The codes were adapted by the United Kingdom to become the country's national equivalent codes. The WRc program includes codes for linings and repairs. The WRc has assigned ratings of 1 through 5 to the individual defects, with "1" being the best and "5" being the worst.

4.1.2 *National Association of Sewer Service Companies*

Building on the work of WRc, NASSCO developed the *Pipeline Assessment Certification Program* to standardize the performance of pipeline inspections in the United States. Additionally, NASSCO has established standardization requirements for software including the requirement that the software programs must be able to share data files. The *Pipeline Assessment Certification Program* includes a condition grading system to assign a condition rating to pipelines. The program also provides codes for linings and spot repairs, including defects. This grading system only considers the internal conditions obtained from the inspection; it does not consider pipe material, depth, soils, and surface conditions. These must be considered in the final evaluation of an individual pipeline. The NASSCO system uses a grading system from 1 through 5. A grade of "1" is considered excellent and a grade of "5" requires immediate attention. The National Association of Sewer Service Companies does provide some general guidelines relative to the rate of deterioration. NASSCO's *Pipeline Assessment Certification Program* cautions that actual rates are highly dependent on local conditions (NASSCO, 2001). They also indicate that these guidelines should be verified by actual research under prevailing local conditions. The National Association of Sewer Service Companies provides training and certifies both trainers and users for the *Pipeline Assessment Certification Program*. They have also developed similar programs for evaluating manholes (e.g., *Manhole Assessment Certification Program*) and sewer laterals (e.g., *Lateral Assessment and Certification Program*).

4.1.3 *Smoke Testing*

Smoke testing has been used as an inexpensive test to identify primarily inflow sources in the sewer, and can also be used to detect overflow points in a sewer system if the groundwater is below the sewer. After notification of adjacent property owners and police and fire crews, a test section of the sewer is isolated and chemical smoke is introduced into the sewer. A blower is used to pressurize the pipe and force the smoke out through potential leaks. Smoke testing has its limitations, and cannot be used on sewers containing suspected sags or water traps as faulty conclusions may result. Additionally, smoke testing cannot be used on rainy or snowy days; when the ground over the pipe is saturated, frozen, or snow-covered; or on pipes that are carrying maximum flow. A positive finding during smoke testing will pinpoint infiltration and inflow sources. However, a negative finding does not necessarily negate the existence of infiltration and inflow sources.

4.2 Manhole Inspection

Manholes can also be a source of infiltration and inflow. Flooding can allow flow to enter the top of the manhole. Special inserts can be installed that will allow the manhole to breathe, but will not allow water to enter. Infiltration and inflow can also enter the sides and bottom of manholes; as such, inspection and maintenance of manholes is very important. An easy way to inspect manholes is to use a pole camera. A similar coding system of data is available for manholes to standardize the inspection data.

4.2.1 *Water Research Centre*

The Water Research Centre introduced coding for manholes in their *Manual of Sewer Condition Classification* (WRc, 2004). Part B of the manual outlines their coding principles and details for manholes and inspection chambers. The rationale for their program is that a manhole or inspection chamber is a vertical pipe. The codes are used to describe the survey location in the chamber. As part of their program, the header sheet is revised. The WRc program does not include linings in manholes.

4.2.2 *National Association of Sewer Service Companies*

Like WRc's program, NASSCO's *Manhole Assessment Certification Program* was established based on the concept that a manhole is a vertical pipe (NASSCO, 2006). And, like WRc's program, the *Manhole Assessment Certification Program* revises the header sheet used for the *Pipeline Assessment Certification Program* to be specific to manholes. The *Manhole Assessment Certification Program* includes a condition grading system to assign a condition rating similar to that used for pipelines. This grading system does not consider manhole material, depth, surrounding soils, or surface conditions; these must be considered in the final evaluation of an individual manhole. The NASSCO system uses grades 1 through 5. A grade of "1" is considered excellent and a grade of "5" requires immediate attention. The National Association of Sewer Service Companies does provide some general guidelines relative to the rate of deterioration. Their program cautions that actual rates are highly dependent on local conditions. They also indicate that these guidelines should be verified by actual research under prevailing local conditions. The National Association of Sewer Service Companies provides training and certifies both trainers and users for the *Manhole Assessment Certification Program*. The NASSCO program was developed concurrently with the WRc program. *Manhole Assessment Certification Program* also addresses linings and lining failures.

4.2.3 *Existing Sewer Evaluation and Rehabilitation Manual*

This Manual of Practice (WEF et al., 2009) indicates that data should include the following:

- Exact location of the manhole;
- Diameter of the clear opening of the manhole;
- Condition of the cover and frame, including defects;
- Whether the cover is subjected to ponding or surface runoff;
- Potential drainage area tributary to the defects;
- Type of material and condition of the chimney, corbel, and walls;
- Condition of the steps and chimney and frame chimney joint;

- Configuration, diameter, and type of incoming and outgoing lines, including drops;
- Signs of frame chimney leakage;
- Type of material and condition of apron and trough;
- Observed infiltration sources and rates;
- Indications of surcharge; and
- Depth of flow and flow characteristics.

The manual also addresses safety and data recording.

4.3 Pumping Stations

Pumping stations are an integral part of any sewer system and must be maintained to avoid overflows. Pumps, motors, variable frequency drives (VFDs), valves, electrical systems, and generators all must be kept in excellent condition to avoid the possibility that their loss may cause significant problems within the sewer system. Corrosion, which is a significant problem in pumping stations, can quickly destroy concrete, steel, copper, and aluminum components of a pumping station and its associated equipment. Other routine maintenance is required on the moving parts of pumps and motors to keep them operational. Another, sometimes less visible, part of a pumping station is the wet well. Corrosion can be a significant cause of problems there as well. Regular cleaning to remove grit and grease is recommended to ensure the suction of the pumps are free and clear and that the concrete or steel present is still intact. The loss of power in a pumping station can also quickly result in an overflow. Maintaining emergency generators or emergency pumps is also important.

Entrained air from pumping stations that is pumped into discharge piping is another significant cause of corrosion and pipe failure. Care should be taken to ensure flow entering a wet well does not splash and cause air entrainment. In a wet well, this agitation can also release hydrogen sulfide that can attack the concrete and steel in the wet well. The Hampton Roads Sanitation District (Virginia Beach, Virginia) has produced a video entitled, "Protecting Our Pipelines—Eliminating Air Entrainment", which is a useful guide to reducing pipeline corrosion and failure caused by air entrainment (available at http://www.hrsd.com/protectingpipelines.htm).

4.4 Force Mains

A force main is difficult to maintain once it is in service because it is typically full of wastewater and under pressure. A force main is designed to operate under specific flow conditions to maintain enough velocity to keep solids suspended; in reality, however, the conditions are never perfect. Force mains are designed to have changes in grade, which can result in problem areas at the transition points. Solids can build up in the low areas and air and gas can build up in the high points. Air is the main enemy of a force main, and keeping it out requires some of the best maintenance possible. Air causes two problems. In sewers, it mixes with naturally occurring hydrogen sulfide and forms sulfuric acid that can destroy ferrous and concrete pipe in a short amount of time. Air and gas can also cause a blockage in a force main.

It is important to ensure air does not get into the force main either through air-release valves or as a result of entrained air in pumping stations. Any air that gets into a force main needs to be allowed to discharge out of air-release valves that should be critically positioned at high points in the system. It is important to keep air-release valves unplugged and operable. This requires a routine venting program to ensure gases are being released on a regular basis. Grease and debris can often plug the air-release valve, so a rod may be necessary to remove the blockage.

Allowing a force main to become a vacuum is not recommended. This will pull massive amounts of air into the sewer, causing corrosion and air blockages and, possibly, the pipe to collapse. Operating the system to maintain a positive pressure is, therefore, extremely important. Some utilities actually use automatic adjustable valves to keep a back pressure in force mains to maintain a positive head.

Water hammer is another problem that can occur in force mains, especially when pumping stations are turning on and off in hilly terrain. This constant pounding can eventually lead to breaks. Dampening systems and using variable-speed controllers on pumps can help minimize this problem.

Having shut-off valves spaced throughout the force main to isolate portions during a problem is important, but the valves must be operated regularly to ascertain they are going to work when needed. Knowing the number of turns to completely shut the valve is important to being confident that the valve is completely closed when required.

A significant issue with a force main is inspecting it inside when it is full of wastewater. Because of corrosion problems, checking the crown of the pipe is desirable. There are several methods used to try to determine wall thickness and whether corrosion is occurring because CCTV cannot be used while the pipe is in service. Of course, if the pipe is down for any reason, CCTV is a good idea. Non-destructive testing such as ultrasonic testing can be performed from the outside, but it requires digging down to the pipe. This can be dangerous because the cover on the pipe may be the only thing holding it together. It is also possible to do core samples, but they must be repaired and can weaken the pipe or create a problem where there was none before.

Force mains that lack enough flow to maintain the required velocity to keep solids suspended can begin to have deposition. Many operators believe that deposits will occur in force mains until the opening begins closing up and the velocity increases. As the velocity increases and lifts up the sediment, the opening stabilizes. Other operators prefer to clean the deposits before they become hard as concrete. One method used to do so is "pigging". This requires a port to enter the abrasive "pig" and allow it to travel down the pipe and clean it. It is best to perform pigging during peak flows so the sediment will be carried down the system to a pumping station or plant to be removed. It is also important to have a recovery station to get the pig out of the system. If the pig gets stuck, it is a significant problem.

Flow metering or sensing is also a way to determine what is or is not going on in a force main system in the event of a break or plugged pipe. Another method to help control corrosion in metallic pipe is cathodic protection, although this only protects the outside of the pipe from external corrosion. This method requires constant maintenance to replace cathodes or to check the electrolysis of the system to ensure the pipe is not corroding.

Where a force main enters a gravity system at a manhole, it is important to ensure the manhole is protected from corrosion with a coating or lining. All of the sulfides that have been brewing in the force main will be released in the manhole.

4.5 Gravity and Combined Sewers

The purpose of routine preventive maintenance of gravity and combined sewers is to maintain the integrity of the sewer and to maximize its capability to transport flows to a treatment facility. Preventive maintenance consists of regularly scheduled inspections of sewers and sewer cleaning, as necessary, to remove solids sedimentation or accumulation of other debris. Combined systems have both wastewater and stormwater entering the same system, which creates for a difficult system to control and maintain.

For small-diameter sewers, CCTV and archival equipment are used to inspect and document conditions. For larger-diameter sewers, visual inspection is typically used. If entry into a sewer is deemed hazardous (due to its depth below grade or if hazardous gas or an oxygen-depleted atmosphere exists), alternate methods may be used. Technology for inspecting the inside of sewers is quickly advancing. Radar, sonar, and other sound devices are often being used for internal sewer inspection. While some of these devices require an empty pipe, others can be used with the pipe in service.

Hot spots may be areas in gravity sewers with known recurring maintenance issues. Hot spots may be caused by a partial or total collapse of the sewer; reverse grades; buildup of solids through deposition (due to low, dry weather flow velocities); localized chemical and mineral buildup due to illegal industrial and commercial discharges to the sewer; development of root problems; or the accumulation of grease. If not rectified in a timely manner, hot spots may lead to premature discharges of untreated dry or wet weather flows into receiving streams or will cause wastewater backups into residences or other structures, creating health hazards and possibly resulting in litigation. A *hot spot* may also be someplace in the system where something must be done before a storm, for instance, so that basements do not get flooded or a discharge does not occur. These are known because of problems in the past. Finally, a *hot spot* may be a section of sewer that may back up weekly requiring routine flushing until permanent repairs can be made. A list of hot spots should be maintained to make sure those locations get the attention required in some timely manner, whether it be weekly, monthly, or before a storm so that a backup, discharge, flood, or other consequence can be avoided.

Routine inspection of the sewer infrastructure can identify potential localized problems that need immediate attention. However, the following are other indicators that can identify problem areas:

- A depression in the ground surface in the vicinity of a sewer may indicate a solids washout, which may result in a partial or total collapse;
- A high number of flooded basement complaints in an area may be an indication of a sewer blockage or sewer laterals that may have become disconnected from a trunk sewer (it is important to note that this condition may also be due to a heavy, intense wet weather event); and
- Continuous high solids (with a high level of inorganic material) loading at the treatment facility may be an indication of solids entering through cracks or open joints in a sewer section.

Documentation of the condition of sewers is critical to assess changes in their condition from one inspection to the next. This will assist the utility in prioritizing any corrective maintenance or significant repair activities. A uniform method of rating the condition of sewers needs to be used so that the assessment will be consistent from one inspection to the next (e.g., NASSCO).

Gravity and combined sewers often have joints that represent a weak point. Open joints can occur from lateral or vertical soil movement, the freeze–thaw cycle in northern climates, or from poor workmanship during construction. Open joints can lead to infiltration of groundwater, which can also bring soil into the sewer. Open joints can also allow wastewater to flow out of the sewer, contributing to the loss of supporting soil. Over time, loss of supporting soil will allow increased movement of the sewer and, if left unaddressed, will cause a partial or total collapse. Loss of soil below the invert will result in settlement of the sewer. Loss of supporting soils along the sides of the sewer may result in failure of the sewer line by buckling. Regularly scheduled inspection of sewers should detect this situation before it becomes a problem.

Cracks may occur in sewers due to questionable workmanship during pipe manufacturing or sewer construction or because of the material and methods used during construction. For instance, the tunneling and "cast-in-place" method of combined sewer construction in the 1970s resulted in cracks in secondary liners. While the cracks may have initially been small, they have a tendency to expand over time due to constant exfiltration and infiltration. If the surrounding soil contains fine particles, they can infiltrate through small cracks and may cause the cracks to expand, allowing larger soil particles to enter the sewer.

Loss of sewer grade may be an indication of a loss of soil surrounding the sewer. Weight-bearing soil may be washed into the sewer through open joints or cracks, creating voids underneath the sewer. If the void(s) become so significant that the remaining soil cannot support the sewer, the pipe will shift.

Regularly scheduled street cleaning activities is one of several best management practices in controlling solids and debris entering a combined sewer system (CSS) during wet weather periods. The schedule depends on the amount of debris being deposited. Dirt, automobile coolant, and other leaked automobile fluids left on roadways and sundry debris will be washed into combined sewer systems (CSSs) during wet weather events. During the fall, smaller tree leaves can fall through catch basin grates and enter the sewer; larger leaves will clog catch basin grates and create ponding on streets.

If a street cleaning program is not implemented, a catch basin cleaning program can reduce solids buildup in the sumps, which otherwise can lead to solids carryover into the sewer during heavy wet weather events. Removal of leaf buildup on catch basin grates will reduce ponding of water on streets. Regular inspection of catch basins will reveal damage such as tipped frames, which will impact street drainage or cause significant cracking of the concrete substructure, and which can, in turn, increase infiltration or inflow of surrounding soil and water.

After several wet weather events, overflow control structures accumulate a wide assortment of dirt and debris. Accumulation of debris can lead to

- Failure of the proper operation of mechanical equipment, such as debris preventing the operation of flow regulators;
- Deposition of solids in outfalls which, over time, can reduce the hydraulic capacity of the structure and may cause backup of wastewater into the system; and
- Inflow of receiving stream water after a wet weather event if larger debris becomes lodged in inflow-preventing structures (e.g., backwater gates), preventing them from properly sealing shut.

Regular inspection of overflow control structures and end-of-pipe structures and timely maintenance activities, as necessary, will minimize problems related to the proper operation of these structures.

5.0 CORRECTIVE MAINTENANCE

Corrective maintenance is action taken to repair or replace portions of the system on a planned basis before it becomes an emergency. Often, these problems are found during preventative maintenance actions such as inspection or cleaning. These can be scheduled in a capital improvement plan (CIP) for larger projects or be part of an asset management program. This type of maintenance activity is proactive and is done to avoid emergencies and, thus, overflows. Many localities have chosen to reduce corrective maintenance budgets, but often find that having to make repairs under emergency conditions is far more expensive and disruptive to the community.

Most localities require some sort of utility locating be done when working on anything in the ground to ensure utilities are not disturbed. Many firms now provide a service called *subsurface utility engineering* using different methods to identify what is actually underground near planned work areas. Other localities may still rely on as-built drawings to show what is in an area. It is important that a utility provide this service to ensure contractors do not cause service disruptions and/or overflows by inadvertently hitting sewer or other utility lines.

5.1 Pumping Stations

Pumps stations can become "bottlenecks" in the sewer system. If the number of pumping units available to control flows is insufficient, levels will rise upstream of the station and may contribute to avoidable wet weather discharges and basement flooding. While preventive maintenance activities and equipment replacement schedules will minimize pump outages, there will be circumstances where equipment failures are unavoidable. Corrective maintenance activities are required when preventive maintenance inspections or normal observations indicate the need for prompt action and before correctable maintenance becomes impossible and emergency maintenance is necessary.

Three significant design requirements for pumps are flow, total dynamic head, and wet well size. As pump run times increase, rotating parts begin to wear and impact the pumping characteristics. Raw wastewater is abrasive, due to grit and other material in wastewater, and wears on pump impellers and bowls, reducing the hydraulic capacity of the units. Pump impellers and bowls have a close tolerance, and exceeding the tolerances will cause the pump's operating characteristics to drift from the design point on the pump curve. A manufacturer's recommendations should be followed for pump disassembly and inspection after the specified number of operating hours. Impellers and bowls can be rebuilt (if minor wear is found) or replaced (if significant wear is found).

Pump shafts and bearings should also be inspected. Worn bearings can score the shaft and lead to premature failure. Bearing cooling and lubrication systems should be checked to ensure proper flow of cooling water or oil for lubrication. Components should be replaced if indicated. Testing for radial movement of the shaft will identify problems with the pump that will affect reliability. Irregularities in the trueness of the shaft will lead to eccentricities (e.g., wobble) in the pump rotation, will cause excessive pump wear, and may result in premature failure of the pump. If indicated, the pump shaft should either be straightened or replaced.

Motors are inspected, cleaned, and tested as part of the preventive maintenance program. Motor testing reveals problems that affect the overall efficiency of the pump and motor. A pump must operate at, or near, its pump curve rating to maximize transport of wastewater. A loss of motor efficiency can cause a pump to operate below design parameters.

Motor testing will indicate if a relatively simple fix is required (e.g., replacing brushes) or if a more complicated action is needed (e.g., motor rewind). It is sometimes more cost-efficient to replace a small horsepower motor than to repair it. As such, it is important to check pricing before taking action.

Control panels must be inspected and attended to. Panels must be cleaned to remove any accumulation of dust. Unless there is positive pressure ventilation inside of a control panel, dust can build up on the electrical contractors and cause arcing and, in some instances, severe damage in the panel. Hydrogen sulfide, which is common around pumping stations, can lead to quick failure of copper and brass components in control panels. Coating connections can be helpful, but regular inspection is also required. Panels located outdoors and subject to the elements should be constructed of proper corrosion-resistant materials and have an appropriate rating from the National Electrical Manufacturers Association.

Heat buildup inside of panels, especially those that generate a substantial amount of heat (e.g., VFD panels), can lead to premature failure of electronic components. If such panels have air-moving equipment with air filters, the filters must be checked and replaced as necessary to keep airflow at a constant rate.

Supervisory control and data acquisitions (SCADA) systems monitor and control wastewater treatment and sewer systems. For operations staff to make informed decisions based on SCADA data, the information must be accurate and timely. Supervisory control and acquisition data can also be used to re-create events for operators to better understand how their system reacts to different wet weather events and how to better control flows in future events. Frequent calibration of instruments is necessary to ensure the accuracy of information. If replacement of SCADA equipment is necessary, prompt action is required. End devices such as level sensors must be checked for zero and span. For instance, if a pumping station's wet level information is used, either by the operator or as part of an algorithm to start and stop pumps, an error in the level data can cause pumps to be started or stopped too late. Pumps that start late may cause avoidable overflows upstream of the station; late pump stops may cause cavitations and damage to the units.

If operators are dependent on SCADA information for monitoring and control, the data communication system between the operator control room and the facilities' end devices must be constantly checked. Loss of data means that operators would not know the ever-changing conditions in the sewer system during a wet weather event. Without real-time data acquisition, SCADA programs would not function to start and stop pumps or open and close regulators or gates.

Pumping station buildings also require maintenance due to mechanical vibrations and humid environments. Periodic inspection of the structure and prompt corrective action of deficiencies will prolong the life of the facility. Structural components such as leaking roof systems, cracked concrete walls, floors and foundations, or structural steel problems can impact the service life of the building and equipment if not promptly addressed. Wet wells can often deteriorate quickly from hydrogen sulfide. Regular inspection is necessary to make certain they are physically sound.

5.2 Force Mains

Force mains in sewer systems can comprise a portion of sewer from the discharge of a pump lifting the wastewater to a higher elevation or they can be a larger network of sewers common to low-lying areas with little elevation change. Force mains typically remain full of wastewater under pressure and are difficult to take out of service and examine. Numerous attempts have been made to establish ways to determine their integrity while in service, which is obviously difficult. It would be easier if wastewater was homogeneous like water or oil, but, of course, it is not. A break in a significant force main can be a catastrophe.

If it was noted during a previous inspection that a portion of a force main was in need of repairs or replacement, it is always prudent to schedule those repairs so they can be done during normal hours and not during torrential rains. Doing the repairs under controlled conditions when proper notice can be given to all concerned is the more efficient and cost-effective way to get the job done.

Planning is critical in force main repair or maintenance projects. When working on force mains, it is often necessary to isolate a portion of the flow. Typically, valves are not located where they are needed, but several contractors can provide line stop valves where necessary within the force main system. Force main projects should always be planned in advance. Diverting flow around the problem areas is often easy to do and, as always, planning not to have an overflow is essential.

Asset management tools represent one way to determine when a force main needs repairs or replacement. Comparing the age, type of material, and operating conditions to other force mains in the system that have needed repairs or replacement may be the best way to attempt to be proactive. If experience has determined that air-release valves will only last a given amount of time, then setting up a program to replace them at that recommended timeframe would be a good example of corrective maintenance. Being proactive in repairs and replacements is necessary to obtain the goal of no SSOs.

5.3 Gravity and Combined Sewers

In gravity and combined sewers, the first indicator of a problem may have been a cave-in or a blockage that had to be fixed but that led to further investigation and a planned approach for further repairs. A CCTV inspection may have spotlighted internal corrosion before it became an emergency, thus allowing time for a more economical, planned fix. When a break occurs, it is always prudent to do some CCTV work in the area to assess the extent of damage. Even a line damaged by a contractor provides a valuable opportunity to look inside. Planned repairs of gravity systems can often mean a complete replacement of pipe or in-place rehabilitation using the many methods available. Timely corrective maintenance of gravity and combined sewers is critical before a situation becomes catastrophic (i.e., the sewer collapses). Corrective maintenance is less costly than emergency maintenance because the latter will require an exorbitant expenditure of resources to correct. Another adverse effect to such a situation would be a potential health hazard from wastewater backups into residences and buildings because of a collapsed sewer.

As discussed earlier in this chapter, open joints may cause voids not only in the soil near the open joint, but also at locations significant distances from the suspect joint. The entire extent of soil loss will not always be apparent at the repair, and extreme care must be used to determine the extent of soil loss before correcting the problem. Corrective maintenance may require grouting of the voids prior to sealing open joints.

The discovery of several parallel, diagonal cracks is indicative of significant voids that have formed beneath the sewer and result from the sewer pipe being supported by soil only on either end of a significant void. Immediate corrective action is required to fill the void(s) and repair the cracks before failure occurs.

Soils above the sewer may also shift to fill the new void created by the sewer shifting. The failure of the sewer will cause wastewater backups or avoidable, untreated discharges. Corrective action is critical. If the sewer has already shifted, the sewer section may have to be excavated and replaced after the void(s) have been filled to correct the misalignment. The key here is that repairs be planned so that disruptions are minimal.

Chemical corrosion in sewers is caused from either the buildup of hydrogen sulfide or industrial discharges of corrosive material. In slow-flowing sewers, the deposition and decomposition of organic solids creates hydrogen sulfide, which is released and mixes with oxygen, creating sulfuric acid that is extremely corrosive and attacks sewers. For instance, if concrete sewers or manholes in areas that are susceptible to hydrogen sulfide attack do not have corrosion-resistant amendments in the concrete or a coating applied to the surface, the chemical attacks the concrete, softening the cement paste that binds the aggregate. As the softened paste is washed away during periods of high flow, aggregate is lost and results in a decrease in the sewer wall thickness. If this situation is detected, corrective maintenance may include grouting sections of the sewer to restore structural integrity and coating or relining the walls with a corrosion-resistant material.

Other sources that impact the structural or hydraulic integrity of gravity and combined sewers may be from illegal industrial discharges. Corrosive discharges will attack sewer walls much the same way as hydrogen sulfide, and can even destroy the invert of the sewer if the corrosive chemicals are heavier than water and travel along the bottom of the sewer. Other industrial discharges can cause mineral deposits to build up in sewers, causing an obstruction that can result in a reduction of hydraulic capacity. If these situations are observed through inspection, corrective activities should include the investigation of the industrial dischargers' tributary to that section of sewer to halt this type of discharge.

Routine inspection of sewers will reveal the presence of root intrusion. Such conditions should be addressed in a timely manner (i.e., root removal via chemical or manual means) and corrective action taken as necessary (i.e., repair and replacement of damaged sections of sewer) before a catastrophic event such as a sewer collapse occurs.

6.0 EMERGENCY MAINTENANCE

Preparing for a storm is often like preparing for a battle, in that the enemies are excessive infiltration and inflow and possible overflows. It is essential that maintenance and operating groups work together closely during emergencies to ensure the problems are fixed expeditiously and with as little effect to the environment as possible. A goal to keep wastewater in the pipe, if at all possible, is the best way to start. Who, what, when, where, and how are the questions to be prepared for. Additionally, the operator should be prepared for the unknown and should not be afraid to call in help. Have people on standby with leaders selected, and have material and equipment ready to go.

Perhaps when the problem occurs it is small and can be handled with a sump dug into the ground using a sump pump to pump the flow back into the pipe downstream until the final fix can be made. Having contractors on call that can be brought in to fix the problem can be helpful because many localities are short-staffed and can use their own employees to fight other fires rather than get bogged down in a significant project. Sometimes in a force main break an item as simple as a wooden plug can get the geyser under control until the break can be properly repaired.

At times, a pump-and-haul operation is essential until the final fix can be completed. A sump, manhole, or wet well may serve as a place to pump out the flow into tank trucks and haul it to other areas or a plant until the repair is made. Possibly, the flow can be pumped via a temporary pipe to another area and maybe bypass the area to be repaired. This can be done for gravity systems and force mains. Having portable pumping is always a good idea. Having access to vacuum trucks can help dewater the area being repaired and can be helpful in a cleanup after the break.

One thing to keep in mind in an emergency is to get other utilities in the area located and marked as soon as possible. Breaking underground optical phone lines or gas lines can cause additional problems and should be avoided. A hydraulic model that is kept up-to-date with current flows and conditions can be a vital tool in an emergency. The amount of flow being dealt with and other conditions such as pressures when a portion of a force main is shut down can be important information. Closing a valve could cause potentially disastrous problems elsewhere in the system.

It is always good to have a system set up to keep regulators informed of emergencies. Regulators typically prefer to hear the news from the locality instead of seeing it on television. Informing the police, the department of transportation, fire, public transit, and radio stations can also be essential, especially if the break is in a roadway.

What can be done to fix a force main that goes under a river, for instance? This requires a contingency plan. There is sometimes no alternative to working on a force main "hot" (i.e., full of wastewater and under pressure), even though the flow may be escaping, resulting in an SSO. It is necessary to get the section of pipe that needs repair open and to get the flow away from the area being fixed by any means possible. If it is found that the whole top of the force main is gone for a distance, then an emergency contract may be necessary. It is good to have these contracts already in place so crews can get on the job quickly to repair the problem.

Blockage is a common problem encountered with a gravity line. Blockage can disrupt service for a large area and can cause backups of wastewater into homes and businesses. The utility will have to deal with the wastewater and take a proactive and responsible approach to protecting public health in the wake of the overflow or blockage. In short, planning ahead will pay off in the end.

7.0 EFFECTS OF DESIGN AND CONSTRUCTION ON MAINTENANCE

The design of wastewater collection systems can have a crucial effect on maintaining the system properly. Besides ensuring the correct size, other design issues such as pipe material and coatings can affect the long-term life of the facilities. Controlling the grade of a gravity system is critical and can also be important for force mains. In the past, force mains were often constructed at a constant distance below grade, and not much attention was paid to high points that could occur. At every high point, there is increased risk of air pockets accumulating and then the accompanying corrosion ensuing. Planned slopes of force mains with designed high points with air valves and internal coatings can extend force main life tremendously and be a long-term benefit to maintenance crews.

With the use of more plastic pipe, it is important that bedding be properly installed. Many breaks are often caused by a misplaced rock under a plastic pipe. Monitoring the bending of pipe during construction can also be important to avoid stress breaks later.

A problem often seen in gravity systems is a sag in the pipe. This can be the result of not preparing the bedding properly, especially where the pipe crosses another utility causing the pipe to settle. When making repairs to a pipe, preparation of the bedding needs to be a significant concern to ensure sags do not occur. Sags result in accumulations of grease and debris that may cause other blockages in the future. Excessive height of flow falling into manholes can cause hydrogen sulfide to be released that can result in the manhole being severely corroded. Drops need to be kept to a minimum and piped to ensure a smooth transition to avoid problems.

The design used for pumping stations can also be detrimental to the collection system. Breaks where the pipe leaves the station are common and air entrainment caused by falling wastewater can mean the quick deterioration of discharge piping due to corrosion. Accumulations of grit and grease in wet wells can lead to overflows; therefore, it is important to make wet wells accessible and easy to clean. New self-cleaning wet wells are being constructed to aid in this process. Variable-speed pumps and design to soften water hammer can also contribute to the operator's attempt to keep the pipe together. Of course, the capability to lift equipment in the station and connections for emergency generators and portable pumps are also nice to have when needed. Additionally, the ability to pump the wet well into a tank truck for pump-and-haul operations and the ability to valve off the flow coming into the station can be helpful in emergencies.

Designing redundancy into a force main system is always beneficial whether it be parallel pipelines to allow flow to be taken in different directions around a problem or adding mainline valves to the system that will allow portions of the system to be taken out of service. Adding valved outlets will provide the capability to easily add bypass piping in the case of a problem. Making it easier to take a section of force main out of service and allowing a path for the flow to continue either for repairs or inspection can make the life of the service crews much easier. Some technology even utilizes air vents for internal inspection, so having them well placed is definitely advantageous.

8.0 EFFECTS OF OPERATION ON MAINTENANCE

The way in which a utility's operators run a system can significantly affect the maintenance required. Some systems have the capability to divert flows in different directions, especially from force mains. Balancing the flows to ensure proper velocities are maintained helps with deposition, which can cause backups. Ensuring the pumps are pumping correctly is essential. Pumps slamming on and off can lead to quick failure, not only of the pumps, but of the piping system as well.

Making sure emergency power and pumping are working can prevent an overflow. When venting air valves, it is wise to keep an eye on gas accumulations in the force main system, which may indicate problems. When systems do not have enough flow, especially when they are newer, wastewater often sits and digests, causing extreme amounts of gas. This gas can cause the system to back up. Chemicals can be used to keep the wastewater fresh and slow down the gas problem.

Operators can inadvertently make changes in the system that can affect the operating level either in pumping stations or in the gravity collection system. Keeping the level of wet wells low and allowing the wastewater to fall an excessive distance will cause corrosion problems in the wet well; gases can also affect the inside of the pumping station and corrode electrical and mechanical equipment. Allowing the wet well to get too high may back up the wastewater into the collection system and surcharge the pipes unknowingly. If a manhole is seen to be at a higher level than others during routine maintenance, it should be investigated.

9.0 MANAGING THE SYSTEM

Proper management of the wastewater system is crucial to making it all work, especially during a disaster. U.S. EPA's *Guide for Evaluating Capacity, Management, Operation, and Maintenance (CMOM) Programs at Sanitary Sewer Collection Systems* (U.S. EPA, 2005) provides an excellent framework for ensuring everything is in place to properly manage a system. Even if U.S. EPA is not threatening to take enforcement action, it is wise to have such a system in place, in addition to operations manuals and standard operating procedures. Benchmarking other utilities can provide ideas that may prove worthwhile to implement.

The advent of the GIS represents a great management tool for the utility. Storing maps and as-builts for easy viewing is helpful, along with the massive capability to store system-related information about maintenance, breaks, and more. Many hydraulic models tie directly into GIS systems and, although hydraulic modeling has traditionally been a great planning tool, it can be an invaluable aid in operations also. Keeping the GIS updated to reflect the actual operating system is helpful when problems arise. Estimating how much flow must be handled during a break and seeing how to divert flow or establish bypass pumping can easily be done with an operational model.

Contingency planning is another management tool, well worth the time investment, in which scenarios are set up to determine how they would be managed under emergency conditions. Additionally, having the needed equipment, material, and maybe even contractors lined up to assist will help limit the amount of overflows and show that the utility is indeed running the show.

Other management tools, such as CMMSs, can help the utility make certain the system is being maintained properly and begin storing those records that can then assist with an asset management program. Used in conjunction with GISs, these programs can be valuable tools for maintaining the sewer system and minimizing overflows. The goal is for the utility to get the point where they can actually begin proactively replacing old facilities instead of always being in the reactive emergency mode. Being proactive and getting ahead of the problem is the main way utilities can hope to reduce SSOs.

9.1 Recordkeeping

Comprehensive and up-to-date operations, maintenance, and regulatory records must be kept. It is recommended that a complete set of all records for each facility be kept on-site and that another set be kept at a central repository site for all facilities. Federal and state regulatory agencies have the "right-of-entry" to inspect facilities and have been known to conduct such inspections, which includes verification of on-site recordkeeping practices.

Consideration should be given to converting records to a common reference system and requiring future inspections be performed following this same system. In this manner, the development of defects over time can be tracked. The data can also be used to project a system-specific service life.

9.1.1 *Operations*

Operators must keep an up-to-date and accurate log for each event at each site that includes information such as time of arrival to the facility, inspection of equipment, chemical inventories, problems encountered during the event, postevent cleanup, and all other activities prior to, during, and after an SSO event. The date and time of each activity and descriptive activity documentation are necessary so that a review of activities during a postevent debriefing session will be beneficial in improving the operation of the facility during future events.

9.1.2 *Maintenance*

A complete and comprehensive record of all maintenance activities must be kept at the facility, with a copy kept at a central repository. Information on the preventive, corrective maintenance, or replacement for each piece of equipment (i.e., equipment name and number, date of service, description of maintenance activity, etc.) must be duly recorded. Whether the utility uses a CMMS or a manual maintenance management system, all maintenance activities related to each facility must be made available to the facility via remote computer terminal or hard copy.

As part of the operator's duties during non-wet weather periods, inspection of the facility and its equipment is necessary to ensure, to the extent possible, that the equipment and systems will be operable during the next event. It is a good idea to note when a piece of equipment is found to be inoperable so proper maintenance can be scheduled. It should be noted that SSOs can occur anytime, not just during wet weather.

Housekeeping is a significant undertaking after an overflow event. Figure 6.8 shows one utility's method for cleaning up an area after an overflow. At the end of one wet weather event, clean-up activities of basins and equipment must be done so that the facility is ready for the next event.

Sanitary Sewer Overflow Clean up & Remediation

SOP No: EPD-2

Page

Issued: 12-3-08

Revised:

Responding Staff

Dispatch and arrival on scene.

5. Note the time of dispatch and arrival on scene.
6. Upon arrival promptly investigate the cause and stop the sanitary sewer overflow. **Appropriate safety procedures shall be observed.**
    a. Note the start and stop time that the overflow occurred.
7. Contact additional help as necessary to aid the clean up/remediation efforts.
    a. The Vactor truck(s) should be dispatched immediately.
    b. If necessary, contact City Mechanics to shut down sanitary sewer pump stations.
    c. If necessary, contact staff to bring appropriate protection mechanism(s) from Corp Yard.
8. Determine the direction of flow.
    a. Determine whether overflow has entered storm drain collection system, ditch or canal.
        i. Determine the final destination.
    b. Install appropriate mechanism to divert or contain the flow to protect storm drain inlets.
        i. Make every effort to contain overflow above ground

Spill contained above ground

(Streets, Sidewalks, Driveways)

1. Protect public from the area.
2. Immediately begin collecting overflowing/overflowed material with Vactor Truck.
3. Wash down the impacted area.
    a. Collect all wash water.
    b. Dispose of wash water into sanitary sewer.

(Soil, Equipment)

1. Collect any debris.
2. Wash equipment with bleach solution.
    a. Collect wash down water and dispose of into sanitary sewer.
3. Remove or decontaminate contaminated soil/plants.
    a. Collect wash down water and dispose of into sanitary sewer.

FIGURE 6.8 Sanitary sewer overflow cleanup (City of West Sacramento, 2008).

Sanitary Sewer Overflow Clean up & Remediation

SOP No: EPD-2

Page

Issued: 12-3-08

Revised:

Spill entering/entered storm drain collection system

1. Protect public from the area.
2. Immediately begin collecting overflowing/overflowed material with Vactor Truck.
3. Install mechanism to prevent SSO from further entering storm drains or surface waters.
4. Determine how far downstream the overflowing/overflowed material has reached.
    a. Once determined, go to next manhole downstream.
5. Turn off storm water pump station(s).
    a. This item might not be possible during rain event.
6. Install pipe plugs at the determined downstream location.
7. Collect all material contained within the impacted storm drain collection system.
8. Wash down the impacted area.
    a. Collect all wash water with Vactor truck.
9. Once overflow has ceased and clean up is complete remove all plugs and dams used to contain flow.

Spill has entered a waterway

1. Protect public from the area.
2. Make every effort to stop the flow from entering the waterway.
3. Begin collecting the wastewater.
4. Contact the Environmental Program Division.
5. Remove debris.
6. Remove or decontaminate contaminated soil/plants.
    a. Collect wash down water and dispose of into sanitary sewer.
7. Place warning signs around the impacted area.
8. Collect water samples for analysis. See sampling guidelines contained within the SSO Response Plan.
    a. This item will usually be conducted by Environmental Program Staff.

FIGURE 6.8 (Continued)

Regulatory reporting for wet weather facilities is similar to regulatory reporting for publicly owned treatment works. Although each state or region may have differences in what to report, an event report, which includes volume, duration, effluent quality, and analytical data, is typically required. Some states also require a monthly summary of events.

9.2 Performance Measurements

It is prudent to have some sort of measure to make sure the efforts being put forth to reduce overflows are paying off. U.S. EPA listed several goals and measures in their *Performance-Based Strategy for the Sanitary Sewer Overflow (SSO) National Compliance and Enforcement Priority* (U.S. EPA, 2004a). These fall in the categories of protecting public health and water quality and protecting public investment. Basically, U.S. EPA's goals are enforcement-related. However, in their discussion of measures, they do touch on some areas the utility can draw on to help in determining performance measures.

One possible measure is the number of SSOs per 160 km (100 miles) of pipeline. U.S. EPA also discusses measures such as beach closures and shellfish bed closures.

If the utility is under enforcement and has developed a long-term control plan (LTCP), then measures associated with meeting the plan may be appropriate. If a CMMS is in use, then measures associated with maintenance of the system would be good. Are pipe inspections being completed every 5 years as planned? Is the number of emergency hours used to deal with SSOs going up or down? Measures of the age of pipe could be an indicator that things are improving. Benchmarking what another utility is doing may lead to some good ideas.

Once a good measure is decided on, it is important to begin using it as soon as possible and then examine how it changes over time. Changes in direction can then be an indicator that something may need more attention. The old adage of "how do you know anything if you do not measure it" is something to keep in mind.

Readers may also refer to *Core Attributes of Effectively Managed Wastewater Collection Systems* (2010), which was developed by the American Society of Civil Engineers, American Public Works Association, National Association of Clean Water Agencies, and Water Environment Federation.

9.3 Planning

Planning helps put the utility in the proactive mode where its efforts can be most cost-effective. Short- and long-range sewer master plans, development plans, and comprehensive plans can help the utility identify current problems and see what effect longer-range development and growth can have on the system. With a long-range plan, anything that happens in the interim can be guided to fit into the future puzzle. If a future condition is predicted, then regular updates to a plan (including hydraulic modeling) may be helpful to see when something absolutely must be done. This can be a part of the priority system used to define capital improvement projects. If the utility has already been placed under a consent order or other regulatory action for CSO and SSO, then the LTCP is a tool typically required to define what the locality must do to fix the overflow problem and to lay out a schedule tied to funding to get it done. Typically, this plan looks out 15 to 20 years or more.

As discussed earlier in this chapter, contingency planning makes operators think about "what if" questions and forces them to make plans to handle the most difficult situations. Sooner or later there is going to be a problem with a pipeline that cannot be taken out of service or is located under a river. In these instances, long-range planning can even incorporate facilities to help, such as a parallel system.

Asset management is the use of existing records and history to look forward and plan for the replacement, rehabilitation, or upgrade of facilities. With aging infrastructure, this allows the utility to start planning ahead to fund these projects. At some point, the utility has to define when it is more advantageous to replace a facility rather than to keep fixing it. For instance, how many patches can a deteriorated pipe have before it makes sense to finally replace it and stop the extremely expensive process of dealing with it on an emergency basis?

Then there is the CIP, an annual or biannual plan that puts funding in place to actually get started with a project. With information from all the other aforementioned plans, CIP provides a format to list and prioritize projects, identifies funding needed and its source, and provides a timetable and justification for the project. The CIP is the final management tool that completes the planning process and moves the utility forward to implementation.

9.4 Conclusion

Stopping SSOs and CSOs completely is the goal all utilities should be working toward. Proactive maintenance and planning with supportive management will help achieve the goal. Preventive maintenance will keep a utility on the cutting edge of knowing where potential problems exist so they can be taken care of before a problem occurs and will help keep the system operating at its peak. Unfortunately, emergencies are going to occur. However, corrective maintenance, scheduled when needed, should result in decreased numbers of emergencies and allow for more cost-effective repairs. Being in front of pending problems instead of always reacting to emergencies helps the budget, minimizes employee turnover, and, most importantly of all, keeps SSOs and CSOs to a minimum.

10.0 REFERENCES

American Society of Civil Engineers; American Public Works Association; National Association of Clean Water Agencies; Water Environment Federation (2010) *Core Attributes of Effectively Managed Wastewater Collection Systems;* NACWA: Washington, D.C.

City of West Sacramento (2008) *Sanitary Sewer Overflow (SSO) Response Plan;* Environmental Program Division, City of West Sacramento: West Sacramento, California.

Najafi, M.(Ed.) (2004) *Trenchless Technology Pipeline and Utility Design, Construction, and Renewal;* McGraw-Hill: New York.

National Association of Sewer Service Companies (2001) *Pipeline Assessment and Certification Program;* Reference Manual Version 4.3.1; National Association of Sewer Service Companies: Owings Mills, Maryland.

National Association of Sewer Service Companies (2006) *Manhole Assessment and Certification Program;* Reference Manual Version 4.3; National Association of Sewer Service Companies: Owings Mills, Maryland.

U.S. Environmental Protection Agency (2004a) *Performance-Based Strategy for the Sanitary Sewer Overflow (SSO) National Compliance and Enforcement Priority;* U.S. Environmental Protection Agency: Washington, D.C.

U.S. Environmental Protection Agency (2004b) *Report to Congress: Impacts and Controls of CSO's and SSO's;* EPA-833/R-04-001; http://www.epa.gov/npdes/pubs/csossoRTC2004_chapter04.pdf (accessed June 2010).

U.S. Environmental Protection Agency (2005) *Guide for Evaluating Capacity, Management, Operation, and Maintenance (CMOM) Programs at Sanitary Sewer Collection Systems;* EPA-305/B-05-002; U.S. Environmental Protection Agency, Office of Enforcement and Compliance Assurance: Washington, D.C. http://www.epa.gov/npdes/pubs/cmom_guide_for_collection_systems.pdf (accessed June 2010).

Van Buren, M. E.; Safferstone, T. (2009) *The Quick Wins Paradox;* Harvard Business Publishing: Cambridge, Massachusetts.

Water Environment Federation; American Society of Civil Engineers; Environmental and Water Resources Institute (2009) *Existing Sewer Evaluation and Rehabilitation,* 3rd ed.; WEF Manual of Practice No. FD-6; ASCE Manual and Report on Engineering Practice No. 62; McGraw-Hill: New York.

Water Environment Research Foundation (2003) *Effective Practices for Sanitary Sewer and Collection System Operations and Maintenance;* Water Environment Research Foundation: Alexandria, Virginia.

Water Research Centre (2004) *Manual of Sewer Condition Classification,* 4th ed.; Water Research Centre: Swindon, Wiltshire, U.K.

11.0 SUGGESTED READINGS

Andrews, M. E. (1998) Large Diameter Sewer Condition Assessment Using Combined Sonar and CCTV Equipment. APWA International Public Works Congress: NRCC/CPWA Seminar Series Innovations in Urban Infrastructure; Las Vegas, Nevada.

City of Mill Valley, California (2008) *Collection System Maintenance and Management, Sewage Spill Reduction Action Plan;* Paragraph 3.5.1.2; Alto Sanitary District: Mill Valley, California.

Jeong, H. S.; Abraham, D. M.; Lew, J. J. (2004) Evaluation of an Emerging Market in Subsurface Utility Engineering. *J. Construction Eng. Manage.,* 130 (2), 225–234.

U.S. Environmental Protection Agency (1991) *Handbook, Sewer System Infrastructure Analysis and Rehabilitation;* EPA-625/6-91-030; U. S. Environmental Protection Agency, Office of Water: Washington, D.C.

U.S. Environmental Protection Agency (2002) *Fact Sheet, Asset Management for Sewer Collection Systems;* EPA-833/F-02-001; U.S. Environmental Protection Agency, Office of Wastewater Management: Washington, D.C.

Chapter 7

Overflow Mitigation Technologies

(continued)

(continued)

1.0 BACKGROUND AND PURPOSE

This chapter discusses the utility and design of system components and facilities to control discharges from combined sewer overflows (CSOs) and sanitary sewer overflows (SSOs). Dry weather overflow control techniques, such as measures that help limit maintenance problems, are also considered. Operation and maintenance practices, such as fats, oils, and grease programs, are discussed in Chapter 6 of this manual and in other Water Environment Federation publications. This manual focuses on ultimate facilities or measures that will provide a high level of control, consistent with current requirements of regulatory enforcement actions.

The range of technologies used for controlling overflows act in different ways. These practices may reduce the amount of flow generated, store the flow during wet weather events, or provide adequate treatment for excess flows, either in the collection system or after conveyance to the wastewater treatment plant (WWTP). Sanitary sewer system (SSS) overflows, under the terms of the Clean Water Act (CWA), are deemed illegal. As such, wet weather SSO control technologies are more limited, and consist primarily of flow reduction or wet weather storage and conveyance to treatment. Under certain peak flow conditions, the application of advanced high-rate treatment technologies other than secondary treatment may be supported through a no feasible alternatives analysis (WEF, 2006). As such, these technologies may be relevant to the control of SSOs as well.

1.1 Short-Term Controls (Nine Minimum Controls)

The U.S. EPA's 1994 combined sewer overflow (CSO) control policy (CSO policy) required that combined collection systems implement the nine minimum control technologies by January 1, 1997. The nine minimum controls listed here are specifically related to CSOs, although SSO control planning must often apply the same concepts (particularly numbers 1, 2, 4, and 8) until such time as remedial measures are implemented to control the discharges. Aspects of the nine minimum controls are contained in many long-term control plans (LTCPs), as they may provide a cost-effective component to the control of CSO discharges.

The nine minimum controls are

1. Proper operation and regular maintenance programs for the sewer system and the CSOs;
2. Maximum use of the collection system for storage;
3. Review and modification of pretreatment requirements to ensure CSO effects are minimized;
4. Maximization of flow to the publicly owned treatment works (POTW) for treatment;
5. Prohibition of CSOs during dry weather;
6. Control of solid and floatable materials in CSOs;
7. Pollution prevention;
8. Public notification to ensure that the public receives adequate notification of CSO occurrences and CSO effects; and
9. Monitoring to effectively characterize CSO effects and the efficacy of CSO controls.

Aspects of the nine minimum controls that are also elements of longer-term controls are addressed in various sections of this chapter.

1.2 Long-Term Controls

The CSO policy requires that LTCPs result in the accomplishment of water quality standards and the technological requirements of CWA. These technological requirements are defined consistent with nine minimum controls. The control of CSO discharges to meet water quality standards requires a significant reduction in the pollutant load from discharges, typically accomplished through volumetric or conveyance controls that allow more flow to receive treatment. Volume can be managed through reduction in flow volume generated, storage of excess flow during peak conditions, or by increasing the rate at which the collection system can transport the flow. Treatment of CSO discharges is described in the CSO policy as a minimum of primary clarification, solids and floatables disposal, and disinfection of effluent, if necessary.

Sanitary sewer overflow discharges must be controlled volumetrically as current law does not allow for partially treated discharges from separate sanitary systems. Thus, options for SSO control (such as are caused by wet weather conditions, rather than maintenance conditions) are in the range of global technologies that include flow reduction, storage, or transport to treatment. Capacity, management, operation, and maintenance (CMOM) programs are widely implemented in separated sewer systems to ensure reliability of capacity and operation. Operations and maintenance practices are discussed in Chapter 6 of this manual.

Technology controls typically applied in LTCPs and those discussed in this chapter are listed in Table 7.1.

2.0 SOURCE CONTROLS

The control of flow at the source helps to reduce flow that must be managed in the sewer system. Either alone or in conjunction with other technologies, source control is one component of flow management. Source controls address a variety of potential extraneous flow sources, including direct runoff connections, indirect transfer of flow between storm and sanitary sewer systems (SSSs), and groundwater and inflow from surface waters. Sources can also include water main leaks or abandoned services. Source control is applied broadly in separate sewer systems, where prevention and removal of such sources is necessary for the system to have reasonable response to wet weather events. *Existing Sewer Evaluation and Rehabilitation* (WEF et al., 2009) provides more focused information on source control in separate sewer systems. Source control is also beneficial in combined sewer systems (CSSs).

2.1 Removal of Direct Connections

Direct connections of runoff, either from public rights-of-way or private properties, will result in SSOs or basement backup if connected to sanitary systems. Such direct connections in combined systems accentuate peak flowrates in the system.

Typical investigatory techniques help to identify the location of various connections. Generally, a combination of investigatory efforts including physical inspection,

TABLE 7.1 Wet weather control technologies and application.

Technology category	Specific technologies (and chapter reference)	CSO applicability?	SSO applicability?
Source control flow, volume and pollutant load (various sections as noted)	Infiltration and inflow removal (2.0)	X	X
	Low-impact development (3.0)	X	Partial
	Pretreatment measures to reduce flows and/ or pollutants discharged into the collection system from industrial users (5.2)	X	NA
	Flow reduction (2.0)	X	X
	Source volumetric control (4.0)	X	NA
	Stormwater controls (5.1)	X	NA
Sewer system optimization (section 6.0 and 7.0)	Increasing capacity to WWTP	X	X
	Relief sewers	X	X
	Hydraulic modifications (regulators, other control structures)	X	X
	Flow rerouting	X	X
Sewer separation (section 8.0)	Complete sewer separation	X	NA
	Separation of specific portions of the combined collection system	X	NA
Storage (section 10.0)	In-system storage	X	X
	Pipeline storage	X	X
	Storage basins or tunnels	X	X
Treatment (various sections as noted)	Facilities for removing floatables from CSOs (9.0)	X	NA
	Facilities for providing primary treatment or advanced primary treatment (such as high rate treatment or ballasted flocculation facilities) (11.0, 12.0, 14.0)	X	Limited
	Facilities for providing disinfection and de-chlorination of CSOs (13.0)	X	NA

televising, smoke and dye testing, and plumbing investigations are necessary to develop a comprehensive understanding of sources. These investigatory techniques may also include flood testing of adjacent sewers to identify flow transfer.

Types of sources that constitute direct connections include catch basins (public or private); private drainage systems; roof drainage (directly connected downspouts, roof drains, or internal roof drains); driveway drains; stairwell drains; patio drains; drains from depressed loading docks; foundation drains; sump pump discharges; and other sources. Critical elements for effective removal of these sources involve locating the sources through comprehensive investigation, institutional programs to accomplish removal, and legal enforcement mechanisms to ensure that removal occurs. Technical aspects associated with removal generally involve redirecting the flow to either a pervious area or reconnecting to a storm sewer. This can be challenging if the source is complex (such as internal roof drains in multistoried buildings), lower in elevation than adjacent storm sewers (such as driveway or stairwell drains), is in highly dense locations (e.g., downtown districts or row housing), or collects flows that should not be directed to surface waters (such as loading dock drainage). In some instances, significant extensions of storm sewer systems or pumping may be required. Often, these types of sources are minimized as much as is practical, but the connection is left in place. Examples of such a situation would be installing an awning or roof over a stairwell drain or curbing around a depressed loading dock to limit the flow contribution to a small area.

While the initial reaction to implementation of private inflow removal is often one of concern related to feasibility, the ability to physically separate plumbing and drainage systems on private property is reasonably achievable for most properties. An investigation conducted in portions of Boston, Massachusetts, with construction dating from the late 1800s, indicated that 69% of the properties either currently had separated plumbing or could be readily separated (Recos et al., 2008).

2.2 Manhole Structures

Manhole structures may allow infiltration and inflow into a SSS. Excess flow can enter through the cover, frame, leveling rings, walls, or at pipe connections. Manholes in low areas where ponding occurs are particularly susceptible to inflow. Gravel-road base on top of poorly drained soils may effectively be drained by the manholes that penetrate the roadway structure. Deteriorated manholes, or ones with shifted components (such as an offset frame), are also particularly susceptible. A variety of rehabilitation methods are available for limiting the excess flow into manholes, including improved covers, liner pans, interior and exterior modifications to leveling ring portion and liners, and sealing techniques. These techniques are covered in depth in *Existing Sewer Evaluation and Rehabilitation* (WEF et al., 2009).

2.3 Below-Ground Sources (Groundwater and Rapid Infiltration)

Infiltration and inflow to sewer systems from groundwater sources may produce very high rates of flow. Many sources that actually enter the system in the ground result from rapidly infiltrating water directly associated with wet weather. An example of this is foundation (footing) drain response, which may result, in part, from rapid infiltration of rainwater into basement backfill areas rather than rising groundwater tables. Other below-grade sources include leakage into pipes through joints and transfer of flow between storm and sanitary sewers. Rates of pipe flow transfer can be very high, and the sanitary sewer, which is typically lower than the storm sewer, generally acts as the receiving pipe. In Grosse Pointe, Michigan, a 305-m (1000-ft) terminal section of sanitary sewer flowed full until the adjacent storm sewer was lined. The storm sewer was located below lake level and always contained water. Following completion of the storm sewer lining effort, the flow in the sanitary sewer dropped dramatically. Similar observations of flow reduction have occurred following rehabilitation of interceptors adjacent to rivers and streams, where infiltration is occurring at joints.

Removal of foundation (footing) drains is often accomplished by installation of pumping systems to manage the flow. Discharge of these sump pump systems may be to lawn or landscaped areas or directly to storm sewers. Discharge locations need to be selected to avoid nuisance conditions. Additionally, reliability issues must be addressed, including backup for pump or power failure. A single incidence of sump pump failure may result in the homeowner reconnecting the flow to the gravity sanitary system. A foundation drain removal program in Ann Arbor, Michigan, has accomplished the removal of several hundred foundation drains from the sanitary system. The vast majority of foundation drains in the program produced flows in the range of 0 to 65 m^3/d (0 to 12 gpm), with a majority between 15 and 25 m^3/d (3 and 5 gpm) (Nordstrom et al., 2009). A footing drain disconnection program (see Figure 7.1 and Table 7.2) in Auburn Hills, Michigan, reduced wet weather peak flows by 74% through a neighborhood footing drain disconnection program involving 526 homes (Westmoreland et al., 2006).

BEFORE

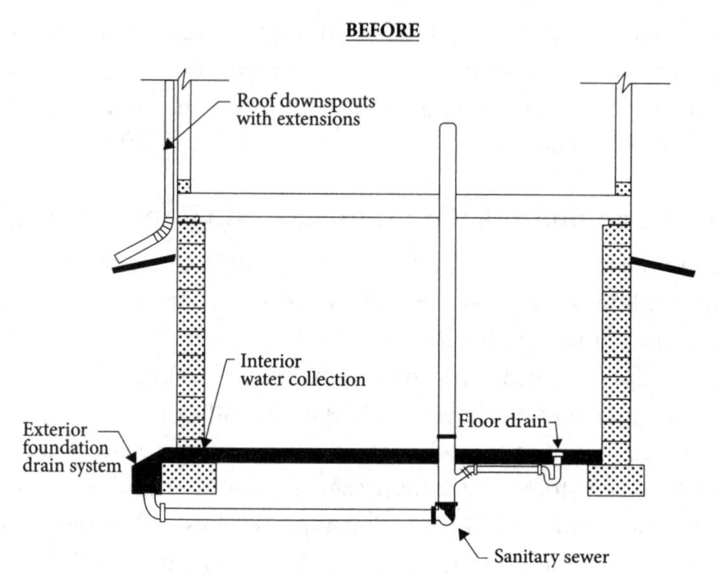

Existing foundation drain detail

AFTER

Roof downspouts
with extensions
1 ½"
Pipe plug
Flexible pipe
with adaptor
1½" Rigid pipe
To street curb
and outer
Interior
water collection
Rubber coupling
for pump removal
Check valve
Floor drain
Exterior
foundation
drain system
Sump pump
Sump
Check valve
Sanitary sewer

Typical sump pump installation detail

FIGURE 7.1 Footing drain disconnection (Westmoreland et al., 2006).

Table 7.2 Flow-reduction results in a footing drain removal program, Auburn Hills, Michigan (from Westmoreland et al., 2006) (cfs × 0.028 32 = m^3/s; gpcd × 3.8 = L/person·d).

FDD stage and year	Average dry weather				Peak wet weather			
	Average dry weather flow (cfs)	Total dry weather flow removed (cfs)	Percent of dry weather removed	Dry weather flow per capita (gpcd)	10-Year, 1-hour projected peak flow (cfs)	Total peak flow removed (cfs)	Percent of peak flow removed	Wet weather flow per capita (gpcd)
0% (1999)	0.55	0.00	0.00	135	3.43	0.00	0%	844
25% (2001)	0.48	0.07	12%	119	3.13	0.30	9%	770
50% (2002)	0.33	0.22	40%	81	1.37	2.06	60%	337
100% (2005)	0.23	0.32	59%	55	0.89	2.54	74%	220

Below-ground sources of excess flow include leakage into public and private sewers through trench infiltration. Recent programs have worked to address the deterioration of private laterals, which are more difficult to maintain and inspect than public sewers. These private lateral sewers often make up more than 50% of the length of pipe in the system. As a result, they may produce a comparable amount of excess wet weather flow. A variety of rehabilitation techniques have been used including pipe bursting, lining, and replacement. Typical defects in private laterals are depicted in Figure 7.2.

Additional below-ground sources include

- Power failures that allow sumps associated with foundation drains to overflow to the floor drain;
- Abandoned laterals that have ineffective plugs (particularly prevalent in areas where smaller lots have been consolidated over the years to allow for the construction of larger buildings); and

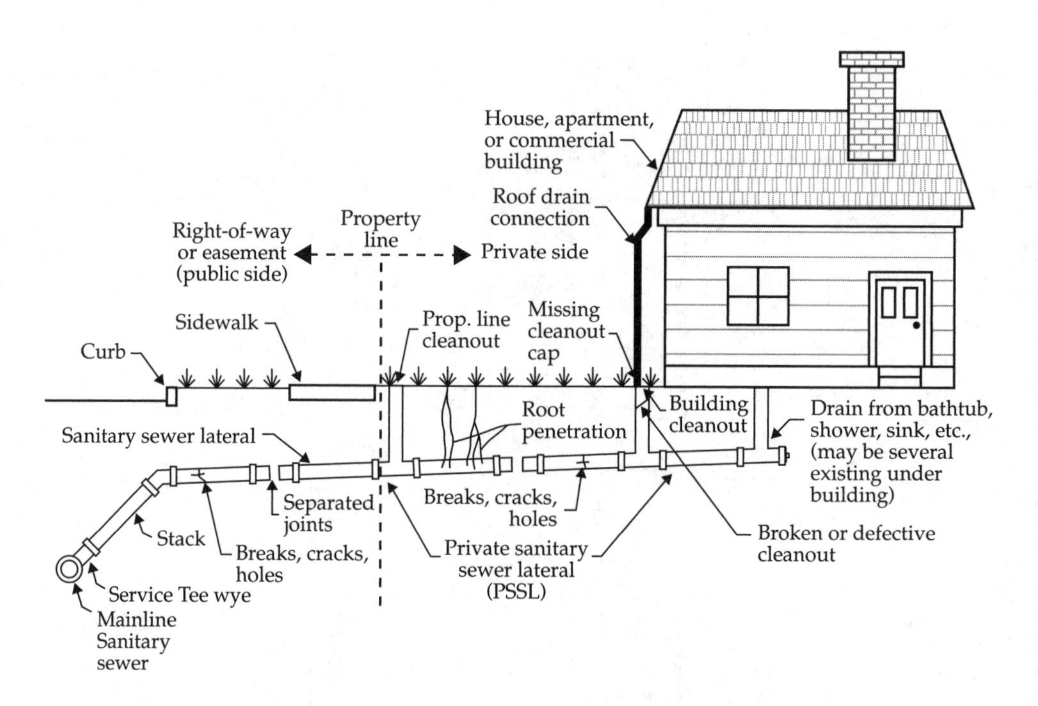

FIGURE 7.2 Private lateral defects (Bhattarai and Guthikonda, 2008).

- Poor connections between private laterals and the sewer main, such as "plumbers taps" that involved busting a vitrified clay pipe sewer main with a hammer and sticking the lateral in the hole.

2.4 Protection from Surface Water Infiltration and Inflow

Sewers that connect to surface waters (such as combined sewer outfalls), are adjacent to surface waters (such as interceptors), or were constructed through wet areas (such as wetland or hydric soil) may be susceptible to intrusion from the surface water source. This may include flow entry through outfalls that do not have adequate backflow protection, manhole or pumping station submergence during high-river stage, or joint and manhole infiltration into sewers located in such areas. Study of stream crossings in Huntsville, Alabama (see Figure 7.3) showed excellent results with focusing attention on defects in proximity to such surface waters. A single defect repair resulted in a 5% reduction in dry weather flows and a 44% reduction in peak flows.

Rehabilitation of such sewers should ensure that the systems are protected for conditions during flood stage. Actions could include

- Installation of reliable backflow protection in chambers that will protect backflow valves from boat traffic or ice damage,

FIGURE 7.3 Stream inflow source to sewer system (Enfinger and Cook, 2008).

- Raising or relocating manhole structures so that rims are above a design flood stage,
- Lining or replacing manholes to prevent wall infiltration,
- Lining of interceptor sewers with jointless pipe, and
- Providing watertight frames and covers if the manhole top is below flood elevation.

Structures that are subject to frequent flooding may act as sources of major inflow at times when they are inundated. Flooding of basements that result from stormwater drainage limitations, river or stream flooding, or sump pump failures related to electricity outages will result in direct connection of surface water through such system elements as floor drains. Flood protection for wastewater facilities may require that frequently flooded properties be isolated from the remainder of the collection system.

Some CSSs receive inflow directly from streams that are routed into the terminal (upstream) end of the pipe. Such situations occur where combined sewers were built to enclose the downstream end of surface streams while leaving the upper portions of the watersheds unsewered. Similar situations exist where separated storm and sanitary sewers in a newer portion of a city discharge into combined sewers in an older portion of a city. In either case, the opportunity exists to redirect larger stormwater or stream flow around the CSS.

2.5 Legal and Political Strategies for Addressing Work on Private Property

Work on private property requires effective legal and financing concepts. At a minimum, ordinances must be in place that require modifications to private property when inflow sources are connected to sanitary systems. More stringent ordinances may require correction of defective private pipes (house laterals and side sewers) that allow excessive infiltration to enter the system. Effective implementation typically requires a "champion" who will work with local politicians to explain the needs of the program. Programs have ranged from completely voluntary (limited effectiveness) to those enforced through legal mechanisms. Funding has ranged from completely private to completely public funding.

Water Environment Federation maintains the Private Property Virtual Library (WEF, 2010), which addresses all aspects of infiltration and inflow control from private property sources. The library can be accessed at http://www.wef.org/PrivateProperty/.

2.6 Flow-Use Adjustment Strategies

Wastewater authorities can work with residents and large discharges to reduce the amount of sanitary wastewater during wet weather. For example, the Milwaukee Metropolitan Sewerage District (Milwaukee, Wisconsin) encourages residents to use less water during heavy rains through its "Every Drop Counts" program. Large industrial dischargers may have operational flexibility to reduce water use during high flow periods or temporarily delay discharging by storing water in flow equalization tanks.

3.0 LOW-IMPACT DEVELOPMENT TECHNIQUES

The volume and rate of development of combined wastewater and, consequently, overflow from CSO discharges is directly affected by the hydrologic characteristics of the area served by the CSS. As such, changes to the tributary areas that reduce the amount of runoff, or defer and delay the rate of runoff, can be an important component of the control of CSO discharges. These approaches have quietly been implemented in CSO control programs for years. For instance, Manistee, Michigan, constructed bottomless catch basins to promote infiltration in sandy soil to reduce the amount of combined sewer area as early as the 1980s. Hydrologic modification is a significant component of the CSO control programs of Portland, Oregon, Seattle, Washington, Philadelphia, Pennsylvania, Louisville, Kentucky, Cincinnati, Ohio, and Kansas City, Missouri, among others.

While this manual addresses both CSOs and SSOs, the application of these practices to SSO control is limited. Implementation of techniques that include ponding or infiltration of stormwater should be spaced some reasonable distance from sanitary sewers and laterals so that increased infiltration and inflow into separated sanitary systems does not occur. Research performed for the Milwaukee Metropolitan Sewerage District indicated that a 3.05-m (10-ft) separation distance between residential rain gardens and private sewer laterals is adequate to avoid causing increased infiltration and inflow in the silty soils typically found in Milwaukee (Gonwa and Ellis, 2006). There are three primary groupings of hydrologic modification that can support control of CSOs. These are low-impact development (LID), stormwater best management practices, and stormwater flow management. All of these approaches support the control of CSO discharges. As communities have sought to increase the "green infrastructure" component of their CSO control programs, the label has been applied to each of the approaches. However, U.S. Environmental Protection Agency (U.S. EPA) intends the term *green infrastructure* to generally refer to systems and practices that use or mimic natural processes to infiltrate, evapotranspire (i.e., the return of water to the atmosphere either through evaporation or by plants), or reuse stormwater or runoff on the site where it is generated (Grumbles, 2007). The primary groupings of hydrologic modification are described as follows:

- *Low-impact development*—these approaches most directly align with the U.S. EPA's definition of *green infrastructure*. Low-impact development techniques attempt to maintain a site's predevelopment hydrology by minimizing runoff and treating and retaining stormwater at the site scale to the maximum extent practical (Landers, 2006). Low-impact development techniques generally include the use of plants and soils to provide treatment to the stormwater and to reduce the overall volume leaving the site. Low-impact development technologies include green roofs, rain gardens, expansion of tree canopy, porous pavement, and bio-retention systems such as rain gardens, curb bump-outs, and planter boxes, for example.
- *Stormwater best management practices (BMPs)*—these practices generally relate to the management of stormwater after flow has been collected and conveyed to a particular type of management system, such as a detention pond. As such, they can be implemented on larger sites or systems than LID techniques. Stormwater BMPs typically use sedimentation and detention to improve quality and reduce flowrates. In a CSS application, these practices primarily help to reduce the rate of flow, but may not have a significant total effect on the volume of flow generated. However, some stormwater BMPs that include infiltration components provide similar quality and flow benefits to LID techniques (University of New Hampshire, 2006).
- *Stormwater flow management*—management of stormwater flows solely from the perspective of attenuating peak flowrates discharged to the CSS is another approach that has been used with success in controlling CSOs. These techniques include reducing stormwater inlets on streets and street or subsurface storage of stormwater.

Depending on the amount of land that can be modified, such as whether public rights-of-way, private property, or both are included, and the aggressiveness of the program, significant reductions in overflow volume can be accomplished. Section 3.0 of this chapter addresses LID and stormwater BMP techniques. Section 4.0 addresses other measures that provide for temporary storage or reduction in peak rates without green infrastructure techniques.

3.1 Considerations

Use of hydrologic modifications to control CSOs requires that the stormwater component of the flow is separated from the sanitary wastewater at the scale of the control being implemented. In larger control facilities, such as constructed wetlands that are meant to serve larger areas, sewer separation may be required upstream of the facility. This could be accomplished through intercepting stormwater runoff before its entry into constructed drainage facilities (as is the case with curb extensions and catch basin retrofits), site-scale modifications, or collector storm drainage systems. In other systems, new storm sewer construction may be necessary.

An issue of debate in the industry is the level of control that can be accomplished with site-level green infrastructure controls alone. Capture (or delay) of the runoff generated from the first 1.3 to 2.5 cm (0.5 to 1 in.) of rainfall is the typical sizing criteria used for many LID practices. While larger systems can be constructed, the ability of stormwater reduction captured by these distributed systems may not be adequate to achieve regulatory targets at every CSO outfall. As such, most programs that are focusing on green infrastructure will consider which areas are most amenable to green infrastructure and also incorporate additional constructed measures to achieve the final level of control required in necessary areas of the CSS. The level of control that can be achieved through practices is being studied in various cities with combined systems, including Kansas City, Missouri (Struck et al., 2009), Portland, Oregon (Kurtz, 2007).

Typically, the rainfall captured by these systems can be 60% to 80% of the total annual rainfall runoff, producing little or no runoff from smaller storm events and capturing a portion of the larger storms. To eliminate overflow, hydrologic controls need to capture the majority of the runoff generated in the selected recurrence frequency storm.

Particularly true of the various stormwater management practices is the need for site-specific considerations and an evaluation of their ability to perform in local seasonal and soil conditions. These practices are particularly sensitive to the ability to store flow and also for rainfall to be intercepted and evapotranspired from vegetation, which may not occur during winter seasons.

Because of the distributed nature of most stormwater and LID controls, construction of these types of measures includes significant local construction near homes and businesses. Stormwater and LID controls also provide potential landscape beautification and other aesthetic value to the neighborhood through the implementation of naturally landscaped areas. The role of adjacent property owners in supporting and maintaining low-impact development facilities can be encouraged through educated outreach and involvement during the predesign and design phases. Some citizens may choose to dedicate some of their private property for stormwater controls. Several of the green infrastructure controls may also change parking or traffic flow. To achieve citizen support for programs, an intensive public outreach effort should be conducted along with the implementation.

Modeling approaches for LID and distributed stormwater controls may need to address much smaller tributary areas than traditional combined sewer hydrologic modeling. Portland, Oregon, has successfully predicted hydrology through modeling that uses highly detailed descriptions of parcels (Liebe and Collins, 2007). U.S. EPA is currently releasing the System for Urban Stormwater Treatment and Analysis Integration (SUSTAIN). This model provides a tool for optimum BMP placement and selection strategies based on pre-selected potential sites and applicable BMP types, including green infrastructure. The optimization criteria can include minimized costs, maximized pollutant flow and/or load reduction, or a combination of these (Shoemaker et al., 2009).

3.2 Land Use and Development Standards

Development ordinances and roadway design standards that promote the implementation of either LID or stormwater BMP controls, or otherwise reduce the percentage of directly connected impervious area, are essential for effective implementation of CSO controls that are supported by modification in hydrologic characteristics. Use of these techniques requires all jurisdictions that determine the methodology of land or transportation corridor development to contribute to the implementation of the program. An agency that has control only over collection facilities will need to gain the cooperation of local and regional governmental agencies such as planning agencies, public works, and, possibly, other departments to fully implement these practices.

Core elements of appropriate ordinances that address reduction in wet weather flow generation could include limits on peak flow generation, total flow generation, and total effective impervious area, culminating in a requirement that the site function comparably to undeveloped conditions. Sample ordinances and libraries of ordinances, particularly those focused on LID, can be accessed through the Center for Watershed Protection Website at www.cwp.org. Ordinances and supplemental literature typically would identify requirements for redevelopment and new development. Design standards, ownership, and maintenance of the facilities in perpetuity are other concepts that are addressed. Generally, ordinances evolve through a progression of concepts from stormwater peak flow control to stormwater quality control to more comprehensive LID implementation, all of which are beneficial to CSO flow management.

3.3 Specific Low-Impact Development and Stormwater Best Management Practice Measures

Good reference material for LID and BMP technologies is available through U.S. EPA and low-impact development center Web sites. A summary of available technologies along with considerations for application is presented in the following sections.

3.3.1 Permeable Pavement

Permeable pavements are available in a variety of surface types (U.S. EPA, 2009b). Permeable pavements typically include a porous layer (pavers, permeable concrete, or porous asphalt) overlaying a stone reservoir for temporary storage of stormwater before infiltration into the soil. The permeable pavements' structure should be placed above the natural groundwater table. Where native soils are not supportive of infiltration, underdrains may be used for ultimate collection of runoff. Permeable pavements are best applied in parking areas with minimal truck traffic. In areas subject to freezing conditions, the stone reservoir should extend below the frost level.

Permeable pavement capacity is determined by the storage capacity of the stone reservoir, the permeability of the surface pavement, and the porosity of the soil. The ability to size the stone layer and its means of dewatering should make it possible for most locations to effectively use permeable pavement for storage (at a minimum). Permeability rates for asphalt and concrete pavement are very high, and the capacity of the underlying soil to infiltrate is more the limiting factor in its use.

The long-term effectiveness of permeable pavement has been uneven. Regular vacuum sweeping (three to four times a year) of the pavement to avoid clogging of pore spaces is the minimal recommendation to prolong effective life. A typical permeable pavement cross section is shown in Figure 7.4.

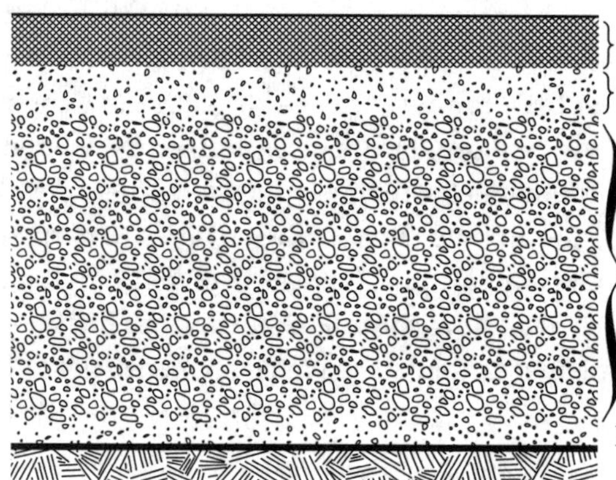

FIGURE 7.4 Permeable pavement cross section.

3.3.2 *Green Roofs*

Green roofs are roof areas that are covered with plants and help control the runoff volume from rooftop areas. A green roof may include soil substrate and a variety of plants. Green roofs are typically encouraged on private or publically owned buildings through development ordinances or financial credits. Interest in, and implementation of, green roofs is increasing because they are beneficial for smaller event stormwater control, they can counteract urban heat island, and they can reduce heating and cooling costs for the building.

As a design consideration for CSS, the amount of rainfall volume that a green roof can capture or detain during a relatively large storm event is significant. Work by the Pennsylvania State Green Roof Center (U.S. EPA, 2009a) indicated that green roofs reduced by approximately 25% the total runoff associated with an approximately 2.5-cm (1-in.) rainfall. The Seattle Green Roof Evaluation Project (Construction Innovation Forum, 2008) showed green roofs as capturing between 65 and 94% of total rainfall as measured over an 18-month period in Seattle, with excellent performance during smaller-intensity events. Study of a green roof in Portland (Kurtz, 2007) showed an annual reduction of 56% of the rainfall volume. Summer retention was 87% and winter retention was 47%. During winter conditions, the roof continues to provide a significant hydrologic control because of the soil media.

3.3.3 *Roof Runoff Capture*

On the single residential property scale, the implementation of rain barrels or cisterns to capture flow from downspouts has been promoted as both a volumetric control and a public engagement activity. Rain barrels may also be beneficial to a homeowner who has experienced seepage into their basement because of poor drainage characteristics around the house. Use of rain barrels also promotes rainwater harvesting for such purposes as gardening, helping the homeowner to engage in the water management process, and reducing the need to use potable water for such activities. Rain barrels are typically fashioned from a 210-L (55-gal) polyvinyl drum or manufactured in a comparable size. As such, one rain barrel can store the equivalent of 0.9 cm (0.35 in.) of runoff from 23 m^2 (250 ft^2) of roof area. Capture of larger rain events or larger roof areas would require multiple barrels or the use of larger-capacity cisterns.

Roof runoff for smaller properties may also be directed to an area where infiltration can occur. This may be as simple as directing runoff to a lawn area or installing a subsurface infiltration well or trench. Avoiding direct connection of roof areas to the CSS can have a significant volumetric and peak rate benefit.

3.3.4 *Bio-Retention (Rain Gardens)*

Bio-retention areas and rain gardens are landscaping features adapted to provide on-site treatment of stormwater runoff. They consist of shallow, landscaped depressions. Generally, they are intended to retain and infiltrate runoff from smaller storms with provisions to discharge excess runoff from larger storms to the storm sewer system. The beds of rain gardens are typically native soil, while bio-retention areas are constructed with an engineered soil consisting of compost and sand underlain by a stone or sand storage layer. Bio-retention areas may contain an underdrain system if the natural soils are not suitable for infiltration.

The landscaped depressions include a significant mulch blanket. Runoff will pond above the mulch and soil during the rain event and seep through the mulch and soil media during and after a rainfall.

The tributary area to a bio-retention system should be limited to no more than 2 ha (5 ac), avoided in steeply sloped areas, and placed above the groundwater table. A typical bio-retention area is about 5% to 10% of the tributary impervious area. A photo of a rain garden is shown in Figure 7.5.

FIGURE 7.5 A rain garden constructed at an office building (courtesy of Dan Christian).

3.3.5 *Infiltration Practices*

Infiltration practices that can be applied for larger tributary areas include infiltration swales, basins, and trenches. Other than swales, which are vegetated with grasses, these facilities are not landscaped and do not function as a visual enhancement. The facilities are depressed and often include structured layers of soil, sand, pea gravel, or rock to promote rapid infiltration below the surface. Infiltration practices require good permeable native soils for eventual infiltration into the ground. The facilities are sensitive to the amount of solids and other fine materials; as such, stormwater pretreatment is important. Generally, the facilities require relatively flat slopes, the ability to drain to native soil in approximately 24 hours (which defines the size or area required), and a distance of approximately 1 m (several feet) between the bottom of the infiltration practices to the groundwater table or bedrock. They may raise concerns relative to contamination of the groundwater.

Infiltration practices can be supplied with underdrain systems so that they function as more of a retention/detention/filtration facility. In these cases, the native soil permeability is not as significant; however, pretreatment to prevent clogging of the filter media is important.

Level of runoff control in support of CSO discharge reduction is variable and dependent on size and soil permeability. Typical design practice is often based on capturing the runoff associated with 2.5-cm (1 in.) of rainfall.

3.3.6 Detention and Retention Facilities

Detention and retention facilities for stormwater control peak flowrates and improve water quality, but are not intended to have a significant effect on the total volume. While not typically considered a part of distributed green infrastructure controls, they may be part of regional flood controls. Types of facilities include dry and wet pond constructed wetlands, underground storage chambers, and "blue roofs". Blue roofs use controlled flow roof drains to temporarily pond water on flat rooftops. The amount of volume that can be stored is significant. These facilities may require significant area to construct, that is, up to 2% to 3% of tributary area for wet ponds and up to 3% to 5% of tributary area for wetlands. Construction of these facilities requires available land and relatively moderate slopes. Groundwater table or native soils generally do not limit the implementation of the ponds, but a clay lining may be required for sites with high groundwater (to prevent groundwater contamination) or permeable soils (to maintain a permanent pool). These larger-scale facilities are particularly likely to require modified conveyance of stormwater from upstream areas (partial sewer separation).

3.3.7 Street Rights-of-Way Modifications

Modifications of the hydrology of street runoff can be accomplished through a series of measures that intercept flow as it leaves the street surface and stores it in bio-retention areas (either behind the curb or in curb bumpouts) or infiltration trenches. Other street modifications include narrowing of streets in residential areas to reduce the amount of impervious area. A similar effort would involve replacing parking area pavement with pavers or permeable pavement. Other concepts include implementing an increase in street trees, both for the benefits of the canopy interception of rain and the volume of water that can be absorbed by the tree.

Street rights-of-way modifications can reasonably be expected to control as much as 2.5 to 5 cm (1 to 2 in.) of local runoff. In some locations of lesser traffic, narrower streets with increased greenways have been implemented. Narrower streets have the side benefit of resulting in both traffic calming and an improvement in aesthetic appeal of the neighborhood. Examples of street rights-of-way modifications are included in Figure 7.6.

FIGURE 7.6 Street rights-of-way control—curb cuts and bio-infiltration area (Owen, 2007).

4.0 SOURCE VOLUMETRIC CONTROL

Source volumetric control is a practice that helps to reduce the volume and rate of flow directed into the CSS.

Most existing CSSs have some degree of this practice in place as a measure to limit surcharging and prevent basement backup. This may be either as an intentional effort to control basement backup or reduce overflows or an unintentional result of design standards from the era where the system was constructed, or the result of street deterioration, resulting in poor drainage effectiveness. It is important to understand the overall efficiency of the drainage system in a combined sewer area. Changes to the drainage system, that is, either future increases in capacity to resolve local flooding issues or intentional restrictions in capacity to reduce peak rates, will have consequences in the local area that need to be understood. In communities where storm drainage design as part of street improvements is routinely enhanced, the additional stormwater volume directed to the combined sewer may result in a deterioration in the overall level of CSO control unless the natural and planned limitations in the existing drainage system are understood in the planning effort.

4.1 Street Storage

Street storage provides for use of street rights-of-way for storage of stormwater runoff during wet weather events. This requires the use of inlet controls (catch basin restrictions) to force the storage. The design of inlet controls may require that multiple catch basin structures be connected so that the size of the control opening is small enough to control the amount of flow generated, yet not so small that it becomes a maintenance issue. Street storage is only feasible in streets that have occasional depressions and relatively low traffic volumes. Street storage should be carefully designed to prevent problematic flooding of parked or moving vehicles and to ensure that there is a positive overland flow outlet before risk of building flooding and that public safety is protected. Street storage should be designed as a temporary occurrence of ponding, which is relatively limited in duration, reasonably infrequent, and provided with positive drainage so that street deterioration does not occur.

4.2 Flow Slippage

"Flow slippage" is the practice of preventing stormwater entry to existing catch basins where the option exists to allow the flow to pass along the gutter to a downstream structure. Upslope catch basins are restricted or eliminated and the flow is conveyed through the gutter rather than through a pipe. This may enable the conversion of upstream sections of combined sewer for use as a sanitary sewer or may otherwise help to reduce the rate of flow into the combined system. As with street storage, locations should be limited to those where traffic is light and safety issues will not occur.

4.3 Localized Subsurface Storage

In locations where more storage is desired than can be accomplished on the street surface either due to inappropriate surface slopes or safety concerns, localized subsurface storage can be used to provide volume for stormwater before connection to the combined sewer. The city of Dearborn, Michigan, implemented controls of this type to reduce the amount of basement backup during large events. A series of catch basins were tied together into a large sewer pipe that was sized for storage. An outlet control provided for release of the stored stormwater back to the combined system at a controlled rate. This concept should be applied to the separate stormwater component of the flow only, and access to the storage pipe for cleaning should be provided.

4.4 Catch Basin Restrictions

Catch basin restrictions are the physical devices or methods that result in a hydraulic restriction of the flow at the catch basin. Methods include

- Limiting the open space in a catch basin grate. This can be accomplished by attaching a solid plate to the grate surface, filling in grate openings with concrete (which has a limited lifespan), or by replacing the grate with a constructed restricted cover;
- Installation of flow limitations below the grate, as an orifice or other restrictive device to reduce flow capacity;
- Reducing the size of the outlet pipe from the catch basin to the storm sewer; and
- Installation of outlet control on the catch basin lead.

Use of catch basin restrictions of any type can result in plugging and surface ponding that may become problematic flooding. Cleaning of debris becomes difficult when there is significant depth of water to be traversed before removing material. Any opening sized smaller than 15.2-cm (6-in.) clear is difficult to maintain without plugging with a variety of debris, especially in areas where leaves accumulate. Difficulty in maintenance or nuisance flooding may result in the removal of the device by operations crews. These factors need to be considered when selecting design type and placement. Operations and maintenance crews should be included in the discussions.

5.0 POLLUTANT-LOADING REDUCTION PRACTICES

Control of pollutant loads at the source is an element of the nine minimum controls program. These activities generally strive to reduce the potential pollutant load in wet weather discharges from CSOs before the implementation of long-term controls.

Control of pollutant loads applies to pollutants generated in stormwater, sanitary, and industrial components of the flow. The effectiveness of these controls is directly related to the persistent efforts to implement their use.

5.1 Stormwater Controls

Stormwater controls address those pollutants that are inherently present in stormwater entering the CSS. These types of practices may include structural practices, maintenance practices, and efforts of the public that are encouraged by public education practices. A number of practices that relate to construction or development standards are required through local ordinances.

Structural practices could include a variety of approaches to reduce solids and floatables loadings in the waste stream. These include

- Prevention of solids entry into the collection system through on-site practices (such as erosion control on construction sites or on-site stormwater treatment systems),
- Planned capture of solids in the collection system (catch basin sumps),
- Planned capture of floatables in the collection system (catch basin hoods), and
- Installation of flushing capabilities in the collection system (to prevent solids deposition that would be re-suspended in wet weather).

5.2 Industrial Waste Minimization

Businesses and industries can help reduce the pollutant load of stormwater to CSSs through a variety of structural and non-structural practices. Significant industrial users may be able to alter their practices during periods of wet weather to reduce the potential for wet weather discharge of industrial waste. Activities that could be implemented include control of batch discharges, separation of on-site wastewater from stormwater systems, allowing detention of the wastewater component. Wastewater collection authorities should consider whether industrial discharges can be given priority access to express sewers or interceptors, potential through direct connection, or rerouting of the flow to a sanitary area.

6.0 REGULATOR AND FLOW CONTROL STRUCTURES

6.1 Overview

Regulator structures are intended to control flow in collection systems. Purposes include preventing overloading of the interceptor system, shifting flow from one portion of the system to another, maximizing flow to treatment, making use of in-system storage volume, and providing system relief (overflow) during wet weather events. Each regulator location has at least two flow outlets, one to convey normal dry weather flow to the interceptor and one to convey wet weather flow to the diversion location, typically to a receiving water. Some regulator structures are also used to shift flow to an alternative flow direction.

6.2 Considerations

Regulator structures are hydraulic structures and are influenced by system capacities and behaviors outside of the structure itself. Operation of the regulator structure is influenced by the capacities of the influent and effluent sewers and the conditions present in the surrounding system. The overall hydraulic capacity of the structure to divert flow to treatment may be limited by the return (dry weather flow) sewer rather than the in-structure control. The overall hydraulic capacity of the outfall may be limited by the outfall pipe and the receiving water elevation, as opposed to the theoretical weir capacity. Design of regulators must, therefore, consider the entirety of the adjacent system.

As collection systems have sought to increase flow to treatment in conjunction with the nine minimum controls, flow-restrictive devices may have been removed or modified relative to the original design of the structure. As such, the structure itself may no longer be functioning as the hydraulic control. The control location may have shifted to adjacent sewers. In cases of interceptor overloading, surcharging may cause the regulator to function as an interceptor relief point, with flow occurring in the reverse to the original design, backflowing through the return line, and discharging through the outfall.

The majority of regulator structures that are used in current designs are static structures without moving parts. Experience has been that older installations that relied on movable components were not maintained sufficiently to function reliably as designed, which could increase the frequency and volume of wet weather overflows or even cause dry weather overflows. Many of the designs previously used that incorporated moving parts essentially functioned in an "open-and-shut" mode, where higher flows triggered a closed condition and 100% of the flow would be diverted to receiving waters. The primary objective in combined sewer flow control that is in current practice is to maximize flow to treatment while avoiding a condition where the interceptor is overwhelmed. To accomplish this, current regulator design typically provides some level of flow restriction without using a "closed" mode.

Regulators should be designed to avoid plugging and operate with as little maintenance as possible. Pipe openings and flow channels should be smooth and small-diameter pipes and sumps should be avoided. Self-cleaning designs are encouraged.

6.3 Types of Regulator Structures

Regulators of various types are summarized in Table 7.3. Regulator structures fall into a variety of categories based on location, means of hydraulic control, and intent of flow control. Structures may be located either on a tributary area, on the interceptor sewer, or at a junction of sewers. Method of control can include static or operable mechanisms. Intent of hydraulic control can be based on passing a design flow, controlling an hydraulic grade line, or functioning in either a dry or wet weather mode (basically an open-and-shut operation).

TABLE 7.3 Types of flow regulator structures.

Regulator type	Features	Cautions
Weirs/dams	Weir directs flow to collection system Excess flow exceeds weir crest and enters receiving stream	May not function as weir for monitoring purposes May be impacted by backwater from interceptor or receiving waters
Orifice	Commonly implemented with a weir or dam Orifice may be a plate or a small (relative to influent capacity) pipe	Variable discharge capacity with depth May be prone to blockage
Vortex valve	Flow-limiting device to control rate of discharge to the interceptor. More constant flow capacity in comparison to orifice	Air vent needed to prevent air blockage
Leaping weir/orifice	Intended to limit flow to the interceptor. As flow increases, more flow will enter receiving waters	Because capacity reduces as flows increase, not directly amenable to NMC for "maximizing flow to treatment"
Mechanical devices (historic)	Historic devices tended to operate in "open" and "closed" modes, where the flow to treatment would be essentially closed off during wet weather	Historic mechanical devices (particularly float/gate) tended to be difficult to maintain and have generally been removed
Mechanical devices (current technology)	Allows variable flowrates downstream and modulated off of downstream hydraulic grade line	Proprietary devices requiring evaluation of reliability prior to implementation

6.3.1 *Offline Control*

Many regulator structures are located at the connection of a combined sewer to an interceptor. The regulator determines the amount of flow that can pass to the interceptor. This configuration typically has a combined sewer flowing into the structure and a return sewer and an overflow sewer discharging from the structure. Many of the regulators used to control flow from offline areas include a means of flow limitation and an overflow weir or dam. These are discussed relative to the weir and dam design and the flow control design.

6.3.2 *In-line Regulators and Diversions*

In-line regulators or diversions allow for the discharge of overflow from the intercepting combined sewer. In this configuration, the amount of flow through the system is determined by the intercepting sewer capacity, the effective hydraulic gradient, and an in-line regulating devices.

Overflow chambers along an interceptor are most typically composed of overflow weirs that are located in a side configuration to the flow. In this case, the weir design can have a significant effect on the capacity of the weir and the amount of flow that is directed toward treatment.

7.0 REAL-TIME CONTROLS

7.1 Primary Concepts and Considerations

Real-time control is based on the concept of making use of the existing collection system to maximize conveyance to treatment and take advantage of large sewers to provide temporary storage during wet weather events. Real-time controls can be used to reduce the frequency and volume of overflow. Real-time controls can be more effective at maximizing storage in the collection system or distributing flow between several WWTPs than static controls.

Real-time control systems include monitoring of the sewer system to guide the decision-making process to manage the wastewater collection system. Data measured in the system are used to operate pumping stations and regulators within the system to achieve better performance during the ongoing process. Data may be transferred to either a central or distributed location where decisions are made for operation of regulators, pumps, and valves with respect to current (and anticipated) future states of the collection system. Different degrees of real-time control may be implemented, ranging from operator-determined decisions regarding operation of major collection system components, to dynamic control of local components, to system-wide control based on complex computerized algorithms and decision-making. The most complex installations incorporate precipitation monitoring and forecasting in conjunction with an algorithm to convert rainfall into sewer flow. In all circumstances, operators perform supervisory functions and can, at any time, take over control manually. In the event of a loss of communications, a default (fail-safe) mode would be implemented.

A variety in the level of automation is discussed by Stinson and Vitasovic (2006). Elements required to address various levels of complexity are included in Table 7.4.

Implementation of real-time control or in-system storage requires a detailed understanding of the system capacities under a variety of conditions, travel times in the system, and available storage volumes and treatment capacities.

Locations where real-time controls can provide the best opportunities for success contain the following characteristics:

- Excess interceptor capacity during most wet weather events;
- Large, relatively flat combined sewers that generally do not flow full; and
- Diversions structures that regulate flow between two treatment plants.

TABLE 7.4 Real-time control components required for varying system complexity (Stinson and Vitasovic, 2006).

Control Mode	Instruments	PLC's	SCADA/ Communications	Central SCADA Server	Active Operator Input, Monitoring	Central RTC Server	Rainfall Forecasting	On-Line Model
Local Manual Control	✓				✓			
Local Automatic control	✓	✓						
Regional Automatic Control	✓	✓	✓	✓				
Supervisory Remote Control	✓	✓	✓	✓	✓			
Global Automatic–Rule Based	✓	✓	✓	✓		✓	✓	
Global Automatic–Optimization	✓	✓	✓	✓		✓	✓	✓

Real-time controls generally will work to provide control of smaller storm events and a portion of larger events. As such, they will typically be combined with other long-term controls to meet regulatory objectives.

7.2 Maximizing Interceptor Capacity

Real-time control systems that maximize use of interceptor capacity rely on the ability to modify the settings of regulators, increasing or decreasing the conveyance capacity to the interceptor. The goal is to fully use the interceptor capacity during most storm conditions. Regulators are replaced with operable gates or valves. South Bend, Indiana, estimated a decrease in overflow volume between 5 and 10% for rainfall events of approximately 1.27 cm (0.5 in.), with the existing treatment capacity of 3 m^3/s (70 mgd). With an increase in WWTP capacity to 4.4 m^3/s (100 mgd), a reduction of approximately 16 to 24% versus fixed regulators was projected (Ruggaber, 2009).

Use of volume provided in storage facilities can be maximized in conjunction with maintaining full flow in the interceptor system. Many low-intensity overflow events can be fully captured by use of the basin and interceptor as a system. The Acacia CSO basin in Oakland County, California, successfully contained CSO events up to 23 000 m^3 (6 mil. gal) in a 15 000-m^3 (4-mil. gal) basin by continually operating dewatering pumps when interceptor capacity was available (Oakland County Drain Commissioner, 2000).

7.3 Maximizing Hydraulic Gradient in Interceptor and Tributary Sewers

The ability to maximize the hydraulic capacity of the interceptor and improve effectiveness of storage in tributary combined sewers can be accomplished on a passive or partially passive basis by raising overflow weirs to their maximum effective height. Columbus, Ohio, used this as an approach to reducing overflows as a nine minimum controls measure. Evaluation of weir modifications required careful assessment of hydraulic conditions upstream in an effort to avoid potential basement backup. Additionally, the available capacity of the reduced regulator opening required evaluation. Anticipated benefits following detailed design were less than projected in the initial plan. Nevertheless, both frequency of activation and overflow volume were expected to decrease by approximately 30% (Domenick, 2007).

The Milwaukee Metropolitan Sewerage District uses real-time controlled sluice gates to control the maximum grade line in its interceptors. In this manner, they maximize the hydraulic gradient in the interceptor while protecting against flooding conditions.

7.4 Maximizing Storage in Large Sewers

Large-diameter combined sewers may have sections with adequate volume to provide significant storage for flow generated during runoff events. In-system storage is accomplished through installation of a hydraulic control that will result in a backwater condition in the sewer, resulting in storage. This can be accomplished with inflatable dams, operating gates, or pumping station controls. Considerations for implementation of hydraulic controls also involve an assessment of the existing sewer condition (i.e., the ability to install new control without new structure or pipe replacement), flow maintenance requirements during construction of control, surface conditions (i.e., placement relative to surface use, such as roadway), and depth and effectiveness of storage. The control system must ensure that water levels do not exceed critical elevations. Slow operation of hydraulic controls could result in unacceptable surcharging of the upstream sewer during rapidly rising flow conditions. Overly rapid operation could result in transient conditions causing unacceptable surge waves.

Detroit Water and Sewerage evaluated multiple types of controls and summarized advantages and disadvantages (DWSD, 2000b) of both inflatable dams and various types of operable gates. These are presented in Table 7.5.

TABLE 7.5 Hydraulic control devices for use in in-system controls (DWSD, 2000b).

Hydraulic control element	Advantages	Disadvantage
Inflatable dam	No cover requirement Technology proven Overflow device Manufacture of devices limited	Full cross-sectional area slightly reduced with dam deflated Sludge buildup upstream Single source supplier (or not available) Sewer must be in good condition
Double leaf gate	Gates fully retractable Can operate as overflow or underflow device Proven technology	Cover required for retracted gates/operators Gate may not seat with sludge buildup Rollers/slides may freeze up Poor flow modulation with underflow operation Jet under throttle gate may cause downstream erosion

(continued)

TABLE 7.5 (Continued)

Hydraulic control element	Advantages	Disadvantage
Single leaf gate	Gates fully retractable Proven technology	Cover required for retracted gate/operator Less operational flexibility than double-leaf gate concept Rollers/slides may freeze up Poor flow modulation with underflow operation Jet under throttle gate may cause downstream erosion Rollers/slides may freeze up Gate may not seat with sludge buildup
Radial control gate	None	Operator may impede flow for high flows Large structure required for gate/operator Debris may collect on gate Cover required for retracted gate Poor flow modulation with underflow operation Jet under throttle gate may cause downstream erosion Unproven technology in sewers
Bascule gate	Overflow device	Gate in dry weather flow in lowered position Sludge buildup in gate pocket Cover required for operator Large structure required for gate/operator Height limit of 3.05 m (10 ft) Unproven technology in sewers
Weir wall with orifice	Passive operation–fail-safe Overflow device Proven technology	Orifice may plug with debris Large structure required for sufficient weir length Sludge buildup upstream
Bending weir/orifice	Passive operation–fail-safe Overflow device	Orifice may plug with debris Large structure required for sufficient weir length Sludge buildup upstream

Inflatable dams can be installed directly in sewer pipe in good condition. These dams have been installed in communities, including Detroit and Cleveland, to take advantage of large flat sewers, economically maximizing in-pipe storage.

8.0 SEWER SEPARATION

Combined sewer overflows occur during storm events. Therefore, separating combined sewers into sanitary and storm sewers is the most direct means of reducing sanitary pollutant loads in wet weather discharges. To effectively complete sewer separation, the entirety of the collection system, both public and private, must be considered. This translates to ensuring that private stormwater sources and private sanitary sources are properly connected following completion of the project.

Sewer separation, when properly implemented, eliminates human and industrial waste pollution from being combined with stormwater flow. It does not remove pollution from stormwater sources. Where stormwater pollution could result in violations of water quality standards or degradation of the receiving water from existing conditions, the program should contain stormwater quality measures to be comprehensive.

Sewer separation has been implemented in small and large communities, including congested areas. Most LTCPs include some aspect of sewer separation, particularly for small tributary areas. Costs for sewer separation, based solely on comparison to other CSO control costs, may or may not be competitive. However, sewer separation often includes significant infrastructure renewal (a benefit not provided by "end-of-pipe" controls), is more effective at controlling basement backup, and has lower long-term operation and maintenance costs than storage and treatment facilities.

A good understanding of how the existing system is configured, which is essential to effective implementation of sewer separation, should be obtained. Existing systems generally fall into three categories:

- Combined trunk sewers with separated tributary sewers—in these areas, the main trunk sewer is combined, but most individual sewers connected to the trunk sewer are either sanitary or storm sewers. In these locations, separation is primarily achieved by constructing a new storm or sanitary trunk sewer.
- Two-pipe systems with cross-connections, where both pipes may function as a combined sewer—research of these areas to determine if the second pipe was constructed as a relief sewer or if it was intended to perform a separation function will help to identify the feasibility of separation and the extent of new construction.

- Truly combined systems—one sewer is predominant in these areas, and most sewers clearly function as a combined sewer. To separate these areas, extensive new construction is required.

8.1 Considerations

Sewer separation is most applicable when the following conditions apply:

- Small to moderately sized area tributary to outfall (≤ 121 ha [300 ac]);
- Existing infrastructure in poor condition, needing major rehabilitation;
- Interceptor and treatment capacity will be available for all of the anticipated post-separation sanitary flow (may be considered "wet sanitary"). Suggested hydraulic capacity minimum of four to six times dry weather flow, dependent on tributary area size;
- Existing combined sewers are accessible for construction or could be used for sanitary flow collection satisfactorily; and
- Alternative controls (storage and treatment) would be extremely costly to implement.

Sewer separation is least applicable when the following circumstances apply:

- Area tributary to outfall is extremely large;
- Interceptor and treatment capacity is limited and expansions are not required under other alternatives being considered; and
- Existing combined sewers are very difficult to access (such as in 4.5-m [15-ft] alleys) and could not function satisfactorily as sanitary sewers (as a result of flat slopes, oversized, etc.).

A comprehensive list of considerations for evaluating sewer separation is included in Table 7.6.

8.2 Partial Separation (Inflow Reduction)

Comprehensive sewer separation including all private property sources is difficult to accomplish. In some instances, partial separation is used to redirect stormwater and thereby reduce the frequency and volume of overflow to established targets without comprehensively completing separation. Generally, new sewers collect street drainage

TABLE 7.6 Sewer separation comparative evaluation issues (adapted from Vander Tuig et al., 2009).

Evaluation topic	Evaluation question
Infrastructure investment	What is the resulting system useful life, and what OM & R costs will be required in the long term (50+ years). How might changes in technology and discharge limits affect the selected alternative?
Level of service	What is the status of the existing system in terms of basement backups, structural integrity and functional reliability? What improvements are required to provide a uniform level of service for different CSO control alternatives?
Water quality effects	What are the anticipated pollutant characteristics of the discharge on both an annual and event basis? What are the in-stream impacts that result from those pollutants in terms of attainment of water quality standards? Are there legacy pollutant issues in the sewershed (such as contaminated industrial sites) or particular land use practices (such as highly urbanized area with excess refuse/litter/etc. potential for impact with stormwater)? Is there a particular concern (e.g., swimming beaches) related to occasional discharge of flows containing raw or partially treated sewage of human origin?
Location and correction of direct dischargers	Is there a commitment to confirming that the dry weather flow from each property is properly connected to the sewer system and that there are no inadvertent direct discharges to the receiving waters (either in retained combined system or following sewer separation)?
Available transport and treatment capacity	Under a separation alternative, what flows are anticipated following completion of construction? Are all private inflow sources going to be removed as part of the program, including foundation drains, or will these be incorporated into the design of the system that will be in place following the project? Are the existing interceptors and WWTP facilities adequate for flows that would result following separation?
Public disruption	Is the potential disruption to the public consistent with what would be required for other infrastructure efforts in the public rights-of-way (such as water main replacement/ street repair/ reconstruction)? Is the construction activity an opportunity to implement additional stormwater quantity and quality controls that will result in less pollutant discharge and reduced stormwater discharge volume than what might otherwise occur?
Schedule	What schedule results are required/can be achieved with the various options under consideration? Would sewer separation be granted additional time to implement? Would this favorably affect affordability?

or convenient private property drainage into a new stormwater system, while retaining the existing combined sewers for flows that originate on private property, both sanitary and storm.

New storm sewer construction is often used for picking up local catch basins, even in projects where new sanitary sewer construction is predominant. New storm sewers are typically shallow and do not require relocation and reconnection of building services. For large service areas, large storm sewers are required, which may be difficult to locate in existing rights-of-way with other utility conflicts. New storm sewers must extend to pick up not only street drainage, but also private roof and parking lot drainage, particularly from non-residential properties.

Fundamental elements of this approach require investigation of the project area to establish all public drainage sources and determine if any existing sewers can be used in the new drainage system.

8.3 Separation with Construction of New Sanitary Sewers

Most existing combined sewers are old and have served most of their design life. They may be in generally poor condition, have poor grade (slope) characteristics, and be too large to effectively serve as sanitary sewers. Additionally, when large sewers (e.g., previously combined sewers) are used for sanitary flow transport, additional new large storm sewers are required to replace their capacity, often in congested rights-of-ways. Construction of new sanitary sewers provides the benefit of a well-functioning sanitary system from a reliability perspective.

New sanitary sewer construction can be more difficult and may be determined to be infeasible under certain conditions. Existing combined sewers may be located in narrow alleys, under buildings, or swimming pools. Replacement of such sewers is extremely difficult, yet new sanitary sewers must be placed in the same general location to reconnect the building services. The following are items to consider:

- Deep construction may be required to establish a grade that meets current pipe slope requirements and allows for connection of local sewers under the existing combined trunk sewer;
- Identification of all sanitary building connections is necessary for proper reconnection to the new sanitary sewer. This is particularly difficult with homes with multiple connections, each of which serves a different purpose (i.e., one carries foundation drainage while another carries sanitary wastewater), and on corners and whose laterals are not intercepted during construction; and

- Proper abandonment of former sanitary building connections, to prevent future failure of a portion of abandoned leads, which could result in voids under pavement and pavement failure.

9.0 FLOATABLES AND SOLIDS CONTROL SCREENING

Floatables and solids control technologies are intended to remove materials of a size that can be seen visually from the waste stream. The sixth minimum control is intended to "reduce, if not eliminate visible floatables and solids using relatively simple measures" (U.S. EPA, 1995). Higher control requirements for floatables and solids consider the removal of solids for both public health and pollutant-load reduction. These ranges of controls include Michigan Department of Environmental Quality's objective to control the discharge of "sanitary trash and identifiable sanitary solids (floating, sinking or vertically stationary)" (Rouge Watershed CSO Workgroup, 2001), or reducing solids discharge by 50% (Washington State Legislature, 2000). The complexity of facilities necessary to achieve high-end solids controls objectives typically requires major screening facilities or other process configurations.

9.1 Short -Term Measures

Nine minimum controls-type measures can be implemented in relatively short time frames and are grouped into the following three categories: measures applied at outfalls, measures applied in the receiving waters, and measures to control floatables control at or near the source. These floatables controls are typically designed to capture larger materials that are visually detrimental to the aesthetics of the receiving water, such as litter and other unnatural solids materials.

9.1.1 Baffles

Underflow baffles consist of a transverse baffle mounted in front of, and perpendicular to, an overflow conduit. During a storm event, flows to the outfall pass under the baffle, while the highly buoyant floatables are retained behind the baffle. Baffles were installed as retrofit features in Louisville, Kentucky, and at Sanitation District No. 1 in Northern Kentucky based on requirements for floatables control to be placed at all outfalls. Retrofitting locations must consider both minimizing hydraulic loss through the altered structure and providing velocities that encourage trapping floatables (Courter et al., 2008; Guthrie et al., 2008). Figure 7.7 shows a baffle constructed under the Louisville, Kentucky, floatables program.

FIGURE 7.7 Baffle constructed under the Louisville, Kentucky, floatables control program (Guthrie, Melisizwe, Sydnor, and Kraus, 2008).

9.1.2 *Trash Racks*

Trash racks installed in CSO outfalls as retrofits typically consist of relatively widely spaced bars (2.5 to 7.5 cm [1 to 3 in.]) that are manually cleaned. These screens will trap large debris, including rags and branches, and large litter. A provision should be included to allow for bypass of the screen during wet weather events when blinding occurs.

9.1.3 *Static Screens*

Static screens are placed at regulators or diversion chambers where the overflow will pass through the screen, typically in an upflow orientation. Static screens are either manually cleaned or provided with a cleaning mechanism that washes captured solids back to the sewer system. Static screens are available commercially in a variety of opening sizes. Hydraulic design needs to consider the effect of a blinded screen. As screen-size opening decreases, chamber size may need to increase to provide the necessary screening area to limit head loss. As such, smaller sizes may require construction of a new structure and eliminate the ability to be implemented as a retrofit. Figure 7.8 presents a static screen as constructed in the Louisville floatables control program.

FIGURE 7.8 Static screen as constructed in the Louisville, Kentucky, floatables control program (Guthrie, Melisizwe, Sydnor, and Kraus, 2008).

9.1.4 Nets, Cages, and Booms

Nets, cages, and booms are intended to capture debris either before discharge or following discharge. The devices are often used in conjunction with each other and may consist of a manufactured system or a simple system constructed by utility staff. The basic concept of the system is to place a net or cage system on the overflow pipe downstream of a regulator to capture floating materials in fabric bags. These netting systems have been widely implemented in certain regions of the country, particularly New York and New Jersey. Plan and section views of in-line versions of these systems are shown in Figure 7.9. A photo showing a floatables control netting structure in Ridgefield Park, New Jersey, is included as Figure 7.10.

Design information required for netting systems includes peak flowrate, total design event volume, and the amount of materials anticipated. Sites must be accessible and reasonably level, and in-water sites require adequate minimum water depth (U.S.EPA, 1999). These criteria establish the necessary netting volume. Installations are more effective when the velocity is minimized approaching the nets. Operational staff for the Northeast Ohio Regional Sewer District indicated that the floating systems are generally more effective for this reason (Courter et al., 2008). Reports of

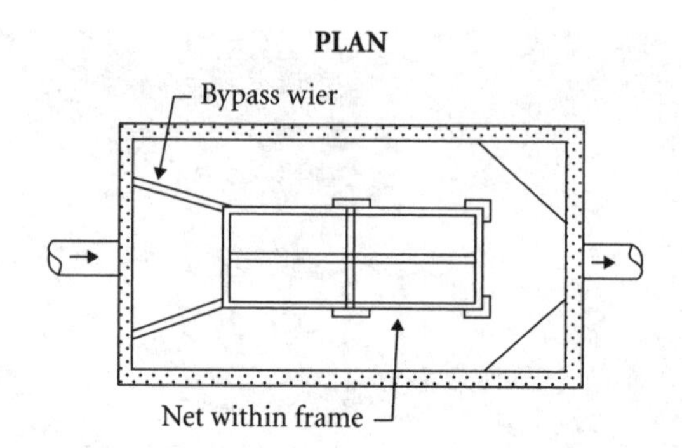

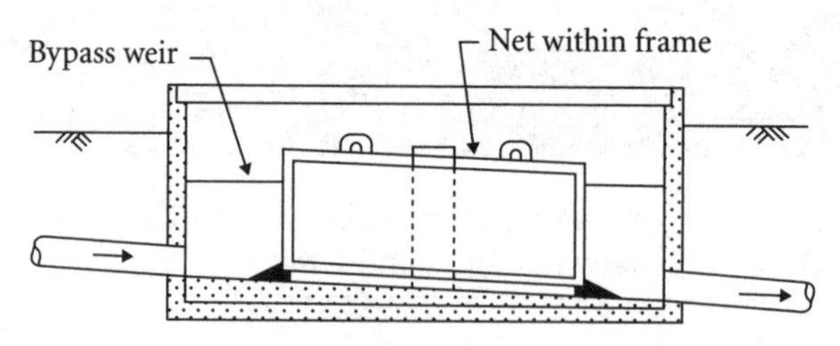

FIGURE 7.9 Plan and section views of typical in-line netting installations.

FIGURE 7.10 Ridgefield Park, New Jersey, floatables control netting structure (courtesy of Michael Lorenzo, HDR).

effectiveness range from 65% to 95%. Netting systems are labor-intensive to maintain and, in a number of locations where they have been implemented (e.g., New Jersey and Cleveland, Ohio), the operations effort is performed through contractual services.

9.1.5 *Skimmer Boats*

Skimmer boats and vessels collect floating debris after it has entered the receiving water. They collect litter and debris that originates at a number of sources. Skimmer boats are most typically deployed in relatively open water.

9.2 Long-Term Measures

Control objectives for long-term measures are typically related to the objective of removing objectionable floatable material. While CSOs contain wastewater-related waste items, they also contain litter from street runoff and natural materials, such as leaves or other plant matter. These items include those that float, settle, or have neutral buoyancy.

9.2.1 *Screen Types*

Screening can provide high-rate separation of coarse solids from wastewater. Screens used for CSO treatment include mechanically cleaned bar screens, raked horizontal and vertical bar screens, traveling screens, and drum screens. Screens are designed to remove a given particle size and are effective in removing solids of the design size or larger. Additionally, some of the solids with sizes smaller than the design size can be removed as the screen becomes solids-laden. A disadvantage of screening is the relatively high cost of maintenance that is required. Screens are also susceptible to clogging and mechanical difficulty, particularly as the design solid removal size is decreased. When small particles are removed, the quantity of solids and difficulty of disposal increases. Table 7.7 provides a description of screening devices used in CSO treatment. These are as follows:

- *Bar screens (various types, mechanically cleaned)*—bar screens vary in opening size from approximately 4 to 50 mm (0.16 to 1.97 in.). One hundred percent of the flow passes through bar screens. Solids handling must be provided. These screens provide relatively low head loss and high capacity for the footprint.
- *Combed screens (horizontal and vertical orientation)*—Two combed screens that have increased in use over recent years include a horizontal orientation screen manufactured by COPA (Marden, Kent, U.K.) and a vertical orientation screen

TABLE 7.7 Screen types and considerations.

Screen type	Description
Bar/catenary screen	Fine-to-medium screen. Vertically oriented screen extends above flow channel and generally requires screening building. Screenings disposal to dumpsters or via conveyor or re-introduced into flow. Head loss of 8 cm (3 in.) or more, depending on blinding.
Vertical-raked bar screen (e.g., Romag)	Fine screen. Small spaced bar screen with height of approximately 1.2 m. Bars are aligned in horizontal direction. Bar spacing at 5 mm. If screen is blinded, flow overtops screen as emergency relief. Head loss of 61 cm through screens typical. Installation in Detroit, Michigan. United Kingdom Industry Research Limited (1999) reported 31% solids retained value (% of objects > 6 mm retained).
Horizontal-raked bar screen (e.g., COPA)	Fine screen. Similar to Romag in size and head loss. Flow passes up through screen. Rake pushes flow to size of screen, then to chamber below. Emergency overflow is provided by adjacent weir. United Kingdom Industry Research Limited reported 22% solids retained value.
Perforated/ traveling screen	Vertically aligned perforated screen. Head loss of approximately 61 cm (24 in.) at design flowrate. Screen extends above flow channel, requiring screening building. Perforated opening allows for better capture of solids.
Drum screen	Fine screen. Drum screen partially submerged in flow. Requires screening building to house equipment. Head loss of approximately 30 cm (12 in.). May require coarse screen upstream from it.
Horizontal static screen	Medium-to-coarse screen. Fixed screen of varying size openings. No mechanical cleaning mechanism, although a wash-down mechanism could be provided. Flow could bypass screen via adjacent weir if blinded. Can be manufactured to project specifications; proprietary screens in perforated configuration also available.

manufactured by ROMAG AG (Ddingen, Switzerland). The screen is "system-constructed" and entails modular standardized elements of poststressed thin stainless steel plates with 4- to 5-mm (0.16 to 0.20 in.) spacing for flow passage. Combed screens require that a portion of the flow entering a screening structure be directed to treatment to transport solids away from the screen face. ROMAG screens are shown schematically in Figure 7.11.

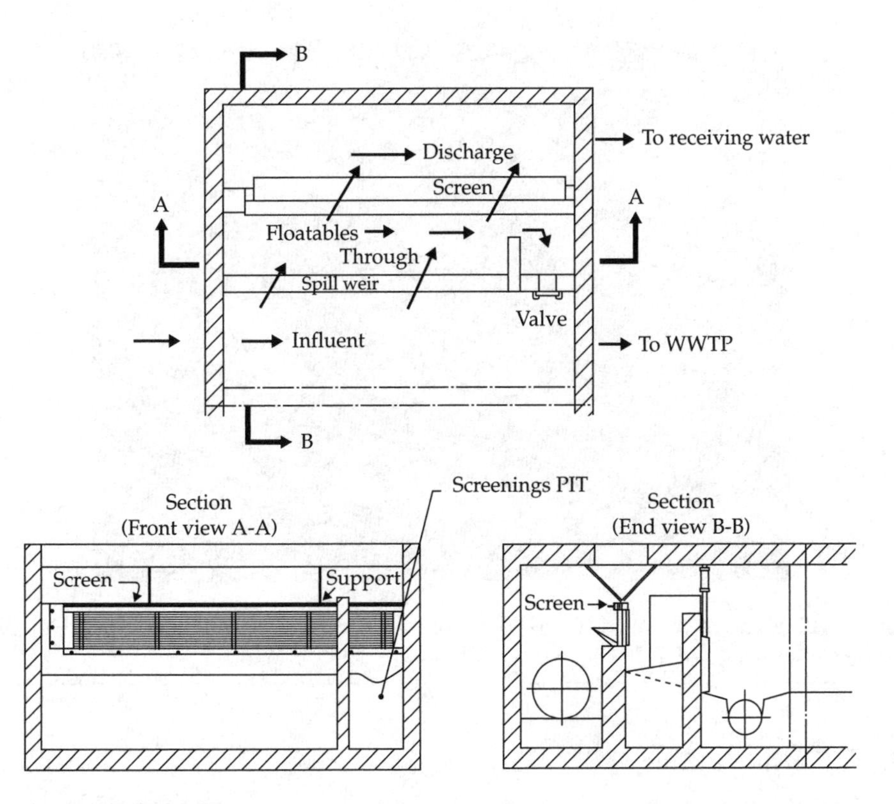

FIGURE 7.11 ROMAG screen.

- *Traveling screens*—traveling screens provide the option for a perforated plate screen in a vertical orientation. On these screens, the entire screen surface moves to the location where it is cleaned. The screen room at the Baby Creek CSO facility in Detroit, Michigan, is shown in Figure 7.12.
- *Sansep units*—these are proprietary units made of a perforated screen in a circular configuration. Flow is introduced tangentially to the screen, resulting in tangential cleaning action along the screen face. Solids are returned to treatment via an underflow or collected in a sump. The units are frequently confused with swirl concentrators, but actually function as a screen.

FIGURE 7.12 Screen room of the 5200-cfs Baby Creek CSO Control Facility, Detroit, Michigan (courtesy of Tetra Tech).

9.2.2 *Considerations*

A variety of issues should be considered relative to screen selection. These include the following:

- Amount of screen height and materials handling space required above flow surface and crane equipment requirements to lift and maintain screens should be considered.
- Head loss through screens can be significant and is not reported consistently by manufacturers. The total difference in necessary water surface elevation from upstream to downstream should be identified to support design.
- Flow stream toward treatment; some screen types require a constant flow stream toward treatment to transport debris away from the face of the screen. These screens will not work properly without this capability. The percent of return flow required may overload conveyance sewers.
- Solids management requirements increase as screen opening decreases. Bar screens may be able to discharge to dumpsters, while screens that generate more

material may require a screen collection area (as is done in Atlanta, Georgia), or may use sluicing troughs for transport of collected solids to treatment.

- Solids generation rates for small screen openings are not well defined in literature for wet weather flows. Use of dumpster containers for large facilities may not be practical.
- For fail-safe operation, a means of bypassing the screens should be evaluated.

10.0 STORAGE FACILITIES

10.1 Introduction

Storage temporarily holds excess flow. Once the collection system and WWTP have available capacity, volume is released to be conveyed to and treated at the WWTP. Most storage facilities are "offline" facilities, that is, basins, tunnels, or pipelines. Such basins or pipelines only receive flow during wet weather events and are dewatered during dry weather periods.

10.2 Pipeline Storage

Storage pipelines are typically newly installed sewers or box sewers that provide storage volume in a linear alignment. They may be constructed as either online or offline storage facilities. Pipeline storage may provide small to large volume, comparable to tunnels (Strause et al., 2008).

The principal operation of pipeline storage is similar to tunnels in that flow is diverted to a large pipeline. Key differences with pipeline storage are that the depth is shallow, eliminating any drop related hydraulics. Methods will be required to divert flow into the storage pipeline. Ability for overflow from the facility (or adjacent control structure) and for the facility to dewater must be provided. Hydraulic design of the facility needs to account for the associated flow management. Pipeline storage will be sloped over its length to allow for dewatering and cleaning.

A principal consideration in design of pipeline (linear) storage includes the routing of the pipeline. Due to the large size of the potential pipeline, adjacent utility conflict potential should be minimized. Pipelines may be able to be installed conveniently in linear parks located adjacent to waterways. Support of large pipeline in the event of poor soils may require piles. Floatation risk should also be considered and prevented.

Most pipeline storage facilities have some provision for flushing. Augusta, Maine, constructed an 1128-m (3700-ft) -long, 3-m × 1.8-m (10-ft × 6-ft) precast box sewer for transport and in-line storage. This in-line storage was provided with seven flushing chambers (Strause et al., 2008), using flushing gates to generate a scouring wave. Massachusetts Water Resources Authority constructed the BOS019 storage conduit as twin 3-m-wide × 5-m-high (10-ft-wide × 17-ft-high) box sewers 85-m (280-ft) long. This storage facility was also provided with flushing gates for cleaning (MWRA, 2006).

10.3 Outfall Storage

Locations where the outfall pipe downstream of the regulator is long, mildly sloped, and large provide an opportunity for pipeline storage during smaller events. This configuration sometimes occurs when combined sewers discharge into what effectively is an enclosed creek. Installation of an outfall control structure at the downstream end and an ability to dewater the enclosure provides an opportunity to use the existing pipeline for storage. Louisville, Kentucky, has incorporated an outfall (Sneads Branch in-line storage site) of this type into their real-time control system (Akridge et al., 2006).

10.4 Basin Storage

Storage in basins is a reasonable option when significant volumes of flows need to be controlled at a single location. Basins provide a relatively cost-effective means of providing storage with limited construction risk. While public reaction to such facilities may initially be negative, most constructed facilities are not objectionable and can be designed to complement surrounding land uses. In some instances, the concrete slab on top of a basin may provide beneficial uses such as tennis or basketball courts. Basins require land for construction, although they can be located either above or below ground. Basin construction costs, on a unit basis, decrease as the volume increases. As such, maximizing the amount of area controlled at a single location helps to make the cost of storage more efficient and reduces operation and maintenance expenses.

A series of considerations relative to basin storage design is discussed by Walker and Heath (2008) and in the city of Toledo, Ohio, LTCP (City of Toledo, 2005). The flowrate on which the facilities' hydraulic design is based has an effect on the overall cost of facility construction. Combined sewer storage facilities typically need to have hydraulic design for larger peak flowrates, which result in larger channels, piping, and potentially pumping facilities. These facilities tend to have a higher cost than SSO control facilities of comparable size because of hydraulic considerations.

In concept or preliminary design for a storage facility size, hydraulics, site selection, and geotechnical and environmental issues are the most significant. Lead time to address these issues should be included in the schedule.

A summary of basin storage components is presented in Table 7.8.

10.4.1 *Facility Sizing and Hydraulic Profile*

Volume and peak conveyance capacity must be determined for storage facilities. Collection of additional flow monitoring data to support this objective may be appropriate if data were limited in the study phase. The hydraulics of the discharge from the tributary area to the interceptor need to be evaluated.

The hydraulic profile through the facility will be controlled by design river-stage elevations on the downstream end and critical in-system hydraulic grade elevations on the collection system side of the facility. If pumping of the process flow is required, influent pumping can help reduce excavation requirements. The preference is for the facility's process flow to operate by gravity. Pumping is best suited for dewatering because dewatering has smaller and less variable flowrates than filling. The hydraulic profile needs to account for losses through the facility, including channels, weirs, and other hydraulic structures. If the facility is compartmentalized (as is typical), this is typically accomplished with weirs to separate the cells. The weir may be placed so that its elevation becomes submerged when the facility is full, or may be placed so that it provides a true separation between compartments.

10.4.2 *Site Selection*

Site selection must be performed with care to avoid projects that are delayed. A public outreach effort is essential to ensure that stakeholders and the general public have an opportunity to weigh in on site use and appearance. Sites must also be investigated to ensure that property ownership is well understood. Many locations used for storage facilities are in proximity to rivers, and may have unclear property ownership or may be difficult to acquire if a purchase is involved.

10.4.3 *Geotechnical and Environmental Considerations*

Early evaluation of geotechnical conditions will help to evaluate overall options in facility construction. An environmental screening of the site will define the potential environmental risks associated with a site.

Table 7.8 Storage basins design considerations.

Facility Component	Description/Requirements
Facility size criterion	Facility sized to limit overflows to design frequency
Basin construction	Concrete or earthen basin, generally below-grade and covered (depending on site), although open basins may be acceptable in some locations. Compartmentalized to reduce cleaning requirements.
Influent flow	Flow can enter a facility via gravity or pumping. Gravity influent flow conditions are preferred over pumped influent and depend on basin elevation relative to the sewer system, river flood stage and local topography.
Flow mode	Basins may be operated either to provide first-flush capture or to overflow from the storage facility after full
Bleedback capability	Gravity or pumping facilities to enable sending flow back to the interceptor when interceptor capacity is available maximize flow to treatment
Hydraulic design	Basin should be able to operate during design flood-stage conditions
Dewatering	Dewatering facilities should generally be capable of emptying the facility within 24 to 48 hours. Actual dewatering timing would be based on flow conditions in the interceptor and at the treatment plant.
Floatables control	Floatables control suitable for design objectives should be included at the facility or adjacent to the facility
Flushing system	Flushing should be provided to clean the facility. Supplementary flushing hoses for channels, wet wells, and so on should be provided.
Odor control	Odor control sized for air volume displaced as storage capacity is filling has been successful
Ventilation	Ventilation per 10 States Standards (GLUMRB, 2004) or other design guidance
Monitoring	Flowrate or level monitoring in influent, effluent, and dewatering sewers. Sampling capabilities for influent and effluent. Level indicators in basin compartments. Flow depth monitoring in adjacent interceptor.
Buildings	Building requirements depend on type of equipment provided with the basin and amount of staff facilities needed

10.4.4 *Basin Construction*

A majority of basins are constructed as concrete tanks, using either cast-in-place or precast technology or a combination of both. For the type of structural and geometric configurations under consideration, pertinent aspects of the dimensional requirements need to be evaluated to ascertain whether the application is appropriate for the site. For example, height of roof domes for circular precast tanks should be determined and evaluated for suitability in the storage location.

In some locations, particularly at WWTP sites, storage has been provided in lined earthen cells. Potential lining materials include low-permeability clay excavated on-site, imported bentonite clays, and geotextiles. The Metropolitan Water Reclamation District of Greater Chicago is in the process of converting the Thornton Quarry into a 30 million cubic meter (8 billion gallon) CSO storage reservoir and will ultimately have more than 57 million cubic meters (15 billion gallons) of CSO storage in three reservoirs.

10.4.5 *Tank Placement (Elevation)*

The depth of a storage facility below ground can have a significant effect on its cost and construction complexity. The actual tank depth may be similar to the working depth or the tank may have significant head space for operator access. In all instances, sufficient head space for ventilation and structural clearances would be provided. The amount of excavation required will be equal to the structural components (i.e., slabs), plus slope allowances for the floor, plus the working depth of storage, plus the distance between the maximum water surface elevation and existing ground. Geotechnical conditions at the elevation of tank placement would need to be considered, particularly such items as rock excavation, dewatering, and need for pile support due to either support on poor soils or for floatation protection.

10.4.6 *Flow Routing*

Routing of flow through the facility during routine and peak wet weather conditions needs to be evaluated. Common options for flow routing include capture of first flush independent of later flows and sequential compartments or multiple compartments filling concurrently. Capture of the first portion of flow in an initial compartment facilitates later cleaning. Other flow routing concepts that need to be considered are flow splits or controls. Most flow splits use passive controls such as weirs.

10.4.7 Odor Control

Air displacement during filling and natural venting may result in objectionable odors. If the storage basin is located in an area frequented by the general public, measures to control odors may be required.

10.5 Tunnel Storage

Tunnels provide an option for storage in a linear alignment. Tunnels can be a good option for wet weather flow control when sites for storage are not available or when multiple sources of flow need to be consolidated. Many wastewater collection authorities that have selected tunnels for wet weather control have based the decision on the lack of sites for surface facilities or concern about the public acceptance of visible wastewater facilities. Tunnels may result in lower operation and maintenance requirements than multiple individual facilities. As with basins, tunnels become more cost-effective as the total volume and the tunnel diameter increases because of the large mobilization and equipment costs associated with tunnel construction. Tunnels have the benefit of providing additional conveyance capacity along with storage volume. They can serve as relief sewers. Milwaukee, Wisconsin, and Chicago, Illinois, rely on their tunnel systems for both storage and conveyance.

Key considerations relative to tunnel design include the following:

- *Geotechnical evaluation and placement of tunnels*—tunnels can be placed in soil or rock; determination of the best depth for a tunnel is dependent on the suitability of the various soil strata. Rock tunnels are typically more cost-effective than soft ground tunnels. As tunnel depth increases, the tunnel drop structures become more complex.
- *Flow routing through tunnels*—tunnels can function in first-flush capture, flow-through, or partial-capture modes. The mode of functioning affects the design of the system in the following ways:
    - *First-flush capture*—drop shafts are sized to accept peak hydraulic flow rates; once the tunnel is full, additional flow bypasses the tunnel.
    - *Flow-through*—drop shafts are sized to accept peak hydraulic flowrate. After the tunnel is full, additional flow enters the tunnel, displacing previously stored volume that is discharged at a designed overflow location. Flow-through capacity must have sufficient hydraulic capacity for the design flow condition.

    - *Partial capture*—drop shafts are designed for flowrates up to a selected event. Tributary flows that exceed the drop shaft capacity result in overflow at the previous discharge location. Once the tunnel is full, discharge may occur at either prior overflow locations or from a tunnel overflow location. Drop shafts under this condition may be smaller than those that are intended to convey peak flow.

- *Location of regulators relative to the tunnel*—the location of drop shafts relative to former regulator chambers is a key element of tunnel layout. The tunnel can intercept either the outfall pipe from the regulator or can be located upstream of the regulator. In the latter situation, the need to avoid diverting flows to an offline tunnel during dry weather poses challenges and may result in the need to modify or relocate the regulator. A diversion located downstream of the regulator provides some protection against dry weather overflows.
- *Hydraulic capacity of tunnel elements*—drop shafts, consolidation conduits, and tunnel overflows need to be able to transport the design flow. Some tunnels have been installed with limited drop shaft capacity, if the drop shaft capacity is exceeded; higher flowrates are discharged to the overflow. In this design, overflow can occur from the system even if the tunnel is not full. However, costs associated with the larger hydraulic capacity add to the overall facility cost without adding usable volume.
- *Surge control*—tunnels can experience undesirable performance through surges that result from the inability to vent air or from hydraulic surging. Modern tunnel design includes a number of elements that allow for de-aeration of flow, venting of air, and surge control volume.
- *Easements*—subterranean easements are required along tunnel and connection-tunnel alignments. Care in alignment is required, and locations where support for structures may exist (e.g., bridges, buildings, etc.) need to be avoided.
- *Dewatering*—tunnels typically include pumping facilities for dewatering. By their nature, tunnel dewatering pumping stations tend to be high-lift facilities. The required dewatering capacity for the tunnel pumping station will affect the overall cost of the facility and the complexity of design.

- *Infiltration and exfiltration control*—tunnel infiltration during construction and afterwards can lower the groundwater table and cause building settlement or rotting of old timber pilings. Tunnel exfiltration during operations could potentially contaminate groundwater.

Tunnel construction techniques influence the overall cost of the facility. Factors considered in construction include safety concerns; main tunnel and connection-tunnel construction techniques; shaft construction techniques; power availability; handling and disposal of tunnel and shaft spoils; handling, treatment, and discharge of water present during tunnel and shaft construction; and protection of existing structures (Trypus et al., 2008).

Conveyance of flow to the tunnel is accomplished through drop structures. These structures typically include the components of inlet structure, drop shaft, vertical bend, horizontal conduit with or without a de-aeration chamber, and an outlet structure. Primary components are shown in Figure 7.13. The primary types of drop structures are vortex-flow drop structures and plunge-flow drop structures. A summary of tunnel considerations is presented in Table 7.9.

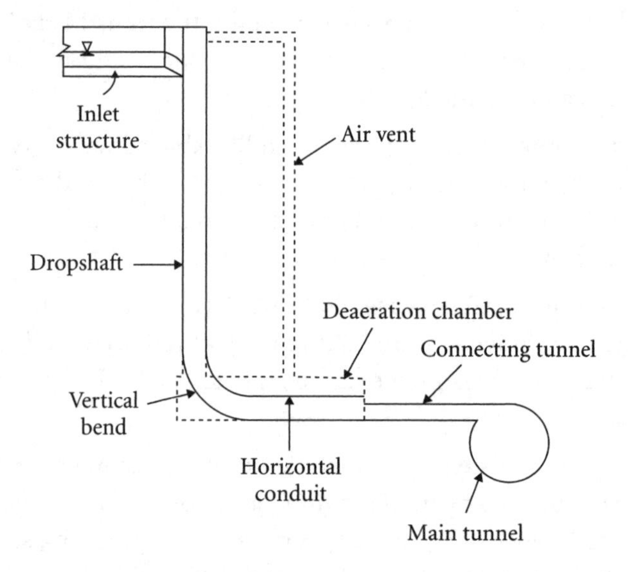

FIGURE 7.13 Tunnel components.

TABLE 7.9 Tunnel considerations.

Facility or component	Description/requirements
Facility size	Facility is sized based on selected design condition.
Tunnel description	Underground large sewers generally 3.66 m (12 ft) or greater in diameter are mined through selected layer (either soft ground or rock).
Influent flow	Flow enters the facility through gravity drop shafts. Drop-shaft capacity either sized for 100% of tributary flow or a lesser component of the flow.
Flow mode	Tunnels may act as first-flush capture or flow through to discharge. As noted under "Influent flow", this may be for 100% of the influent flow or a lesser amount. Overflows may occur from pre-existing outfalls or a tunnel overflow, or both.
Placement relative to regulator	Tunnels may intercept flow upstream or downstream of the existing regulator. In the event that the tunnel alignment connects upstream of the existing regulator, the regulator could be relocated upstream, hydraulic modification at dropshaft connection could be placed to force flow to the tunnel, or another approach to balance flow between the interceptor and the tunnel could be provided.
Bleedback capability	Pumping facilities should be able to send flow back to interceptor when capacity is available at flow rates determined by location.
Hydraulic design	Generally, flow enters a tunnel by gravity. Tunnels should be protected from river backflow, but are generally not designed with pumping of peak wet weather flow. Flow control measures would be necessary to prevent surges and other adverse hydraulic phenomena associated with tunnel hydraulics. Hydraulic relief points (surge tanks) may be required.
Dewatering	Pumped dewatering rates typically capable of emptying the facility within 24 to 48 hours. Flexibility provided for dewatering should allow for variable interceptor capacity.
Floatables control	Floatables control should meet technology standards of nine minimum controls.
Flushing system	Isolation gates or flushing gates could be installed to allow for sequential downstream release of retained wastewater to accomplish some scouring of the tunnel deposits. Receiving pit at downstream end may be required.
Odor control	Odor control has generally not been provided in tunnel systems

(*continued*)

TABLE 7.9 (Continued)

Facility or component	Description/requirements
Ventilation	Ventilation generally not provided for tunnels. Ventilation likely required for routinely occupied areas of dewatering pumping stations.
Monitoring	Monitoring may include flowrate or level monitoring in influent, effluent, and dewatering sewers. Sampling capabilities at influent and effluent. Level indicators in the tunnels. Flow depth in adjacent interceptor also included.
Buildings	Building requirements would be driven by the pumping station design; floatables control facilities and ancillary equipment (generators, odor control, etc.).

10.6 Cavern and Other Subsurface Storage

Under certain conditions, construction of an enlarged cavern or a deep, large-diameter shaft may provide a better solution than an above- or below-ground basin storage. Factors that may make such a solution preferable include favorable geology and little land availability. Atlanta implemented such a facility as part of their Underground Linear Storage Facility (Gray et al., 2006). This approach was more cost-effective than constructing either a tunnel or an at-grade basin.

Gray and Hedin (2008) evaluated the concept of storage in abandoned coal mines in the Pittsburgh, Pennsylvania, area. The concept was limited due to the potential commingling of mine drainage with combined wastewater.

10.7 In-Receiving Water Storage

The "flow-balance technique" uses baffles and curtain walls to temporarily store flow in receiving waters until capacity is available in the collection system. This practice has received minimal traction in the United States because regulatory agencies typically do not allow converting waters of the state into wastewater treatment units.

10.8 Flow Routing Options

Storage facilities may be provided as in-line or offline facilities and may be installed remotely from the POTW or at the POTW itself. In-line facilities use the same facility for both dry and wet weather flows. Offline facilities remain dry unless in use for a wet weather event.

11.0 GENERAL COMBINED SEWER OVERFLOW TREATMENT SYSTEMS

Facilities that are commonly used for wet weather treatment include retention treatment basins and screening and disinfection facilities. Retention treatment basins (RTBs) combine the treatment components of screening, settling, and disinfection. Screening and disinfection facilities are intended to remove gross solids (through screening) and provide disinfection in a tunnel or existing outfall. Treatment shafts function similar to a RTB (i.e., they provide screening and disinfection in an equivalent volume), but the volume is provided in a vertical (shaft) alignment and has different settling characteristics than a basin.

11.1 Retention Treatment Basins

Retention treatment basins capture small storm events and provide primary clarification, control solids and floatables, and disinfect discharge during larger-volume events. Treatment includes screening at the influent or effluent, skimming, settling through compartmentalized basin storage, and disinfection. Other process elements may include de-chlorination and aeration. Retention treatment basins may be above- or below-ground facilities, with open or closed tops. Typically, at least the first compartment to receive wastewaters is covered to control odors.

Retention treatment basins may be sized based on a desired capture volume, a required surface area and depth to achieve primary treatment, and the desired detention time for disinfection during the design storm. The basin component may provide for settling and disinfection contact in the same compartment. The 10 States Standards (GLUMRB, 2004) establish surface loading rates for achieving settling. Disinfection contact time is typically based on a period of 15 to 30 minutes during the design event, but can be shorter if satisfactorily demonstrated to regulatory agencies. A first-flush compartment that functions as capture only is sometimes provided to comply with state requirements or to facilitate cleaning. Some states (such as Illinois) require capture of the first-flush volume for full treatment. Compartmentalization helps with cleaning following events. A discussion of the various components of RTBs is included under each process component in this chapter. Key design concepts are included in Table 7.10. Flow diagrams for RTBs are shown in Figure 7.14. The Grand Rapids Market Avenue Retention Basin (Grand Rapids, Michigan) is shown in Figure 7.15.

TABLE 7.10 Retention treatment basin design concepts.

Facility or component	Requirements
Facility size	Facility sized based on necessary volume to accomplish settling of flow and disinfection contact time. Settling based on surface loading rates and weir loading rates per design guidance. Credit for surface loading rates generally applied to facility as a whole rather than in individual compartments.
Disinfection	Where provided, disinfection uses sodium hypochlorite or alternate chemical with contact time provided of 15 to 30 minutes of contact time. Chemical dosage rates (design) range from 15 to 20 mg/L chlorine, typically using 5% solution. Extensive flexibility in pumping systems to support variable chemical demand rates and various flow rates. De-chlorination, using sodium bisulfate, would be required for any disinfected overflows.
Chemical storage	Chemical storage is necessary.
Flushing system	Flushing as with storage basins.
Odor control	Odor control is supported by the storage of chlorinated wastewater, which is less odor-producing relative to storage. May reduce the need for odor control.
Ventilation	Ventilation in the screening facility is provided per standards in the industry. Special ventilation requirements apply to chemical feed and storage areas.
Monitoring	Flowrate or level monitoring in the incoming sewer(s), with alarms for high flowrate or level, and flowrate and water quality of discharges to the receiving stream. Sampling capabilities for influent and effluent. Influent flow monitoring needs to support chemical dosage. Oxidation–reduction potential monitoring should be provided.
Buildings	Chemical feed and storage facilities are housed in a building. Basic flow control and diversion structures direct the overflow to the facility. Provisions to bypass the screens if they should become blinded or otherwise cause upstream flooding.

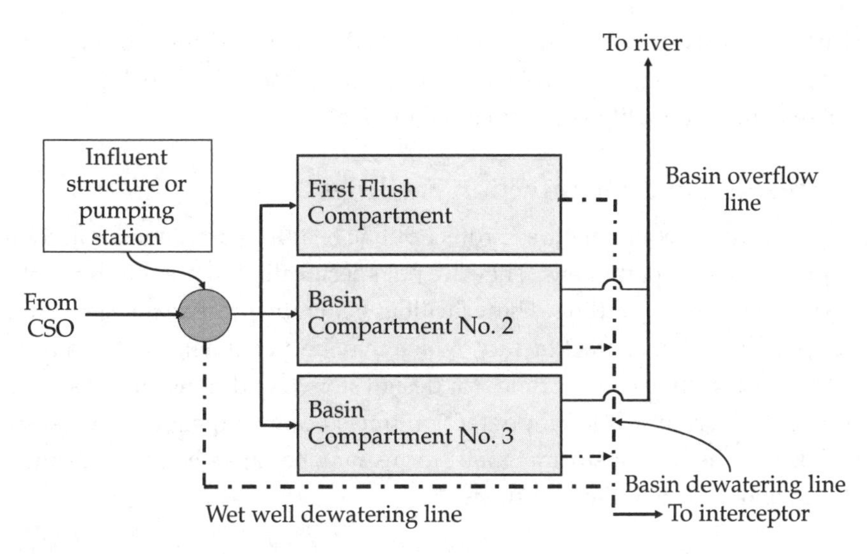

FIGURE 7.14 Sample basin configuration.

FIGURE 7.15 Grand Rapids Market Avenue Retention Basin, Grand Rapids, Michigan (courtesy of Tetra Tech).

Effluent quality from RTBs was evaluated as part of the Rouge National Wet Weather Demonstration Project (Rouge Watershed CSO Workgroup, 2001). Summarized effluent quality is shown in Figure 7.16.

11.2 Screening and Disinfection Facilities

Screening and disinfection facilities protect public health by removing objectionable solids and eliminating pathogens. They are not specifically intended to promote pollutant reduction through settling. These facilities generally provide fine-screening followed by a disinfection contact facility, which may be a channelized basin or a long outfall pipe. Disinfection contact times for design storm conditions are between 5 and 15 minutes, although much longer detention times occur in practice as most storms are smaller than the design storm. Small storms may be captured in the disinfection contact area and dewatered to treatment.

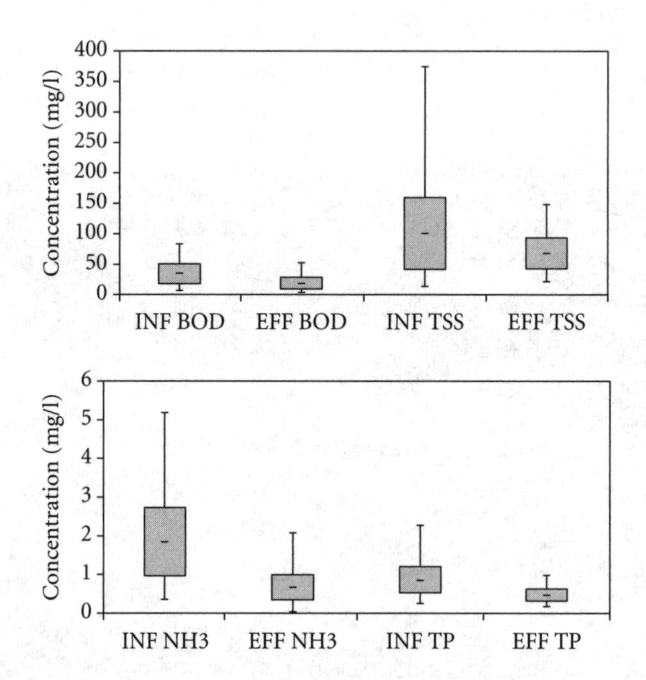

FIGURE 7.16 Influent and effluent quality from Rouge RTBs (courtesy of Rouge data set, C. Hufnagel personal files).

11.3 Treatment Shafts

Treatment shafts provide volume for disinfection and screening. Several are under construction in Dearborn, Michigan, although none are operational as of 2010 (Ghalib et al., 2006).

12.0 SOLIDS REMOVAL

Solids removal can be accomplished through a number of different processes. The most common and easiest to accomplish at an unstaffed facility that only operates during wet weather is settling. Other processes have increased effectiveness; however, they are also more complex and more demanding from an operational perspective. Selection of treatment for wet weather flows should consider not only the potential technology effectiveness, but also the effectiveness of consistently performing at that level of treatment under real-world conditions. As a result, some of the more complex treatment approaches are primarily implemented for peak flows at WWTPs.

12.1 Settling

12.1.1 Plain Sedimentation

Sedimentation is removal of suspended particles from combined wastewater by gravitational settling in tanks or similar structures. Wet weather flows have highly varying influent characteristics and true effectiveness is difficult to measure. Studies have shown that well-designed plain sedimentation reduces biochemical oxygen demand (BOD) concentrations by approximately 30% and total suspended solids (TSS) by up to 50%.

12.1.2 Chemically Enhanced Sedimentation

The sedimentation process is adaptable to the addition of chemicals, such as lime, alum, ferric chloride, and polymers, which provide higher suspended solids and BOD removal at the same overflow rates compared to plain settlings. Effective sedimentation basin footprints with chemically enhanced sedimentation may be one-third to one-fifth of the size of sedimentation without chemical addition. However, chemically enhanced sedimentation typically is preceded by injection, mixing, and floc maturation areas (King County Department of Natural Resources and Parks, 2010).

12.2 Vortex Units

Vortex units are physical treatment devices that promote the settling of solids from wet weather flows. They are often referred to as "swirl concentrators" due to the tangential nature of the flow. A centripetal force is applied to solids in the flow, forcing them to the inside of the tank where velocities are slower and settling can occur (Brashear et al., 2002). Vortex units have been used both as standalone technologies and in conjunction with sedimentation basins or RTBs. Use of vortex units has generally declined as emphasis on total volumetric capture has increased. Nevertheless, their potential for use as either a standalone treatment unit or in conjunction with other facilities continues.

Three components of vortex units provide pollutant removal. These include capture of a concentrated underflow, capture of volume with associated pollutant load, and concentration of solids within the unit. An evaluation of available performance data indicated that gross removal of solids varied from 25% to 50% in a variety of storm conditions. Net removal of solids (removal in excess of volume capture) was between 7% and 15% (Brashear et al., 2002).

The following points should be considered with vortex units:

- Underflow is required to maximize unit performance. If interceptor capacity is inadequate, then storage for the underflow is required for best performance.
- Vortex units provide volume that supports CSO capture. Evaluation of total performance should anticipate the capture that will be provided by the storage volume and treatment effectiveness.
- Chemical addition can assist in solids and associated pollutant removal. Vendor performance data suggest that this application may be effective. Performance of in-place units should be evaluated.
- Vortex units coupled with other facilities may provide effective treatment consistent with the 85% total pollutant capture aspect of the CSO policy and reducing solids to enhance disinfection. This concept should be considered for CSO facility planning.
- Combined sewer overflow facility operators report anecdotally that the effluent from the swirls is visually superior to the influent, and that cleaning requirements for downstream facilities is reduced.

12.3 Enhanced Solids Removal

Enhanced solids removal may be required to meet minimum treatment requirements to protect water quality standards. This section describes options for enhanced solids removal.

12.3.1 Chemical Enhancement

Chemical addition of coagulants (typically, ferric chloride and polymer) helps to accelerate the sedimentation of solids. Chemical addition can increase the settling characteristics of solids in settling basins or vortex facilities.

12.3.2 Enhanced Physical Settling

Inclined plate sedimentation is a fundamental component of ballasted flocculation systems, but can also be used independently. Inclined plate sedimentation is a technique that can increase the effective settling area of an existing unit by insertion of plates set at a 50 to 60 deg. incline. Effective surface overflow rates can be increased by a factor of 10. This method has primarily been considered relative to WWTPs and, generally, as a retrofit. However, this technology has also been considered for CSO treatment (King County Department of Natural Resources and Parks, 2010).

12.3.3 High-Rate Clarification

High-rate clarification systems used for wet weather treatment in the United States include the proprietary systems of Actiflo (Krüger, Inc., Cary, North Carolina) and DensaDeg (Infilco Degrémont, Inc., Richmond, Virginia). These systems, typically installed at WWTPs produce high-quality effluent through a process that uses chemical addition.

The Actiflo system is a high-rate clarification process that uses microsand to promote flocculation and create a dense floc with a high settling velocity. In this process, effective solids removal relies on particle destabilization through neutralization of charges on the particle surface. This is achieved through the addition of a primary inorganic coagulant such as aluminum or iron salts and high-molecular-weight polymer. The polymer aids in the attachment of the floc to the microsand via polyelectrolyte bridging. The microsand is then settled in a clarifier equipped with lamella plate-settling modules. Settled microsand is continuously pumped to hydrocyclones. Sludge and microsand particles are separated by centrifugal force in the hydrocyclones. Clarified effluent is discharged and the sludge is forwarded for additional solids thickening or treatment outside the Actiflo system; cleaned microsand is returned and injected back into the Actiflo system.

DensaDeg is a high-rate clarification process that produces a dense sludge through a proprietary blend of energy input and high-density solids recirculation. In addition to reducing the volume of waste sludge, the densification process produces a rapidly settling sludge, which allows the facility to operate at high clarifier loading rates. The process combines mixing, internal and external solids recirculation, sludge thickening, and tube-settler clarification in two conjoined vessels, which reduce the size of the treatment unit. Effective solids removal for the DensaDeg process relies on particle destabilization through neutralization of charges upon the particle surface. This is achieved by the addition of a coagulant, such as aluminum or iron salts, and high-molecular-weight polymer. The DensaDeg process is shown in Figure 7.17. Key design parameter differences are shown in Table 7.11.

The city of Toledo performed side-by-side evaluation of the two processes during winter conditions. Their findings included

- Actiflo and DensaDeg systems are both technically capable of performing primary and wet weather treatment. Both units successfully demonstrated treatment of primary influent and wet weather flows for TSS and carbonaceous biochemical oxygen demand (CBOD) under a variety of operating conditions.

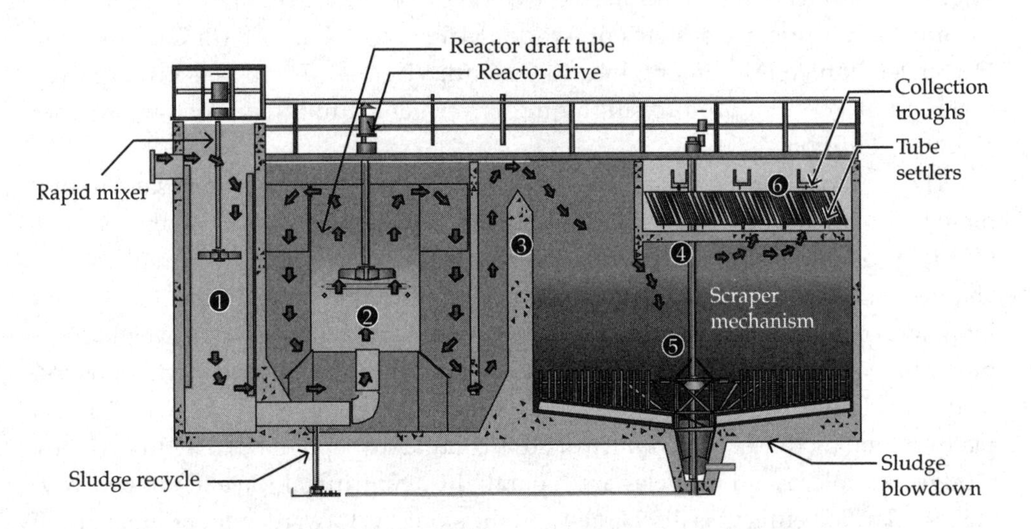

FIGURE 7.17 DensaDeg process (courtesy of Infilco Degremont Inc.).

TABLE 7.11 Actiflo and DensaDeg comparative information. (Solids concentrations from wet weather runs performed at Toledo, Ohio [City of Toledo, 2003]. DensaDeg sludge in Toledo was lower than other pilot sites.)

Design consideration	Actiflo	DensaDeg
Screening upstream of unit	Fine screening	Coarse screening
Ballast	Microsand	Sludge produced
Startup time	10–15 minutes	20–30 minutes
Sludge rate (concentration)	0.1%–0.2% solids	0.9%–1.3% solids

- Chemical addition testing showed good success with ferric chloride and poor performance with ferrous chloride. Ferric chloride combined with aluminum sulfate resulted in some low pH.
- Actiflo and DensaDeg achieved comparable levels of treatment under a variety of conditions. Actiflo operated at higher hydraulic loading rates than DensaDeg. DensaDeg underflow (sludge) was four to six times more concentrated than Actiflo underflow.

12.3.4 *Compressed Media Filtration*

Compressed media filtration uses a bed of synthetic fiber media balls in a chamber compressed through external pressure. Compression reduces the volume of the media bed and, thus, the size of pores to support retention of small particles. To clean (or backwash) the filter, external pressure is removed to expand the chamber. The media bundles are agitated and circulated using compressed air and water to carry away particles retained during filtration. The technology was first introduced in Japan in the 1980s as a high-rate tertiary filter and, by the mid-1990s, it was successfully used for treatment of waters with higher solids loadings such as CSO and industrial wastewater. The first known application for treatment of stormwater was implemented around 2002.

12.3.4.1 *Design Considerations*

Factors considered in sizing a compressed media filter include solids and hydraulic loading rates, available hydraulic head, and the characteristics of the particles targeted for removal. For wastewater flow applications such as CSOs and primary influent wastewaters, the filter is generally designed at 6.8 $L/m^2 \cdot s$ (10 gpm/sq ft), with 1 to 1.5 m (3 to 5 ft) of head plus one-third excess filter cells to accommodate backwash requirements at a continuous peak condition. When treating wastewater or CSOs, the backwash will typically be performed using filter effluent. Other applications can use filter influent water.

Pretreatment requirements for wastewater or CSO flows should include 12-mm (0.5-in.) screening and grit removal. Excess primary or biological flows do not require additional pretreatment before compressed media filtration. Combined sewer overflows with variable solids loadings typically result in less than a 5% wastage rate for average conditions, but may be up to 20% during the peak flush period.

12.3.4.2 Technology Performance

Filtration of CSOs and primary influent typically reduces TSS concentrations by 80% to 90%, with filter effluent concentrations in the range of 15 to 25 mg/L. Filtration of primary effluent and dilute SSOs can be expected to have comparable effluent concentrations (in the range of 15 to 25 mg/L), however, with a lower percent removal because of lower influent concentrations. Compressed media filtration typically removes discrete influent wastewater particles down to the 10 to 20 micron range.

Backwash performance is dependent on both the nature of the waste stream and the backwash method used, although backwash volumes can be as little as 0.5% of the filtered volume for tertiary applications and 3% for CSO applications. Backwash is typically performed in cycles of 20 to 30 minutes in duration, resulting in backwash flow with solids in the range of 2000 to 5000 mg/L.

12.3.4.3 Manufacturers

The WWETCO Filter (WWETCO, LLC, Wet Weather Engineering & Technology, Roswell, Georgia) is a compressed media filtration system that uses the hydraulic head of incoming fluid against a flexible membrane to compress a synthetic fiber media bed. The flexible membrane is compressed from the sides in the lower portion of the media bed and creates a vertical compression gradient, from a loose to a tight media bed, in the down-flow direction. This technology has no mechanical moving parts and the media is completely contained in a chamber with the compressed shape optimized for filtration and the expanded shape optimized for backwashing. The backwashing method uses low head air (24.1 kPa [3.5 psi]) to create a distinct flow path that circulates, agitates, and impinges the media against a perforated plate as it lifts spent water and solids into a backwash trough. Using air as the principal means to release and lift out particles and limiting makeup water minimizes backwash time and waste volumes. The WWETCO Filter is shown in Figure 7.18.

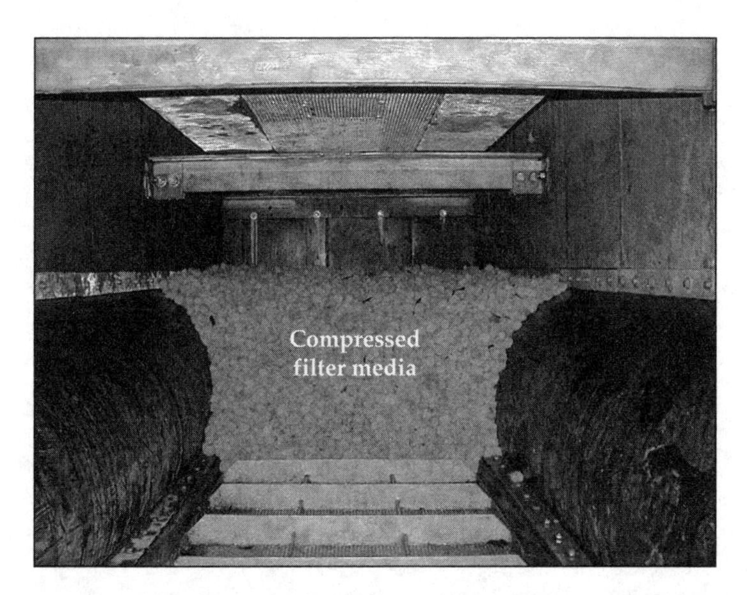

FIGURE 7.18 WWETCO 10-mgd FlexFilter™ installation (courtesy of Columbus Water Works; photo by Mark Boner).

The Schreiber Fuzzy Filter® (Schreiber LLC, Trussville, Alabama) is a compressed media filtration system that uses a vertical moving upper plate to compress flow downward in an upflow filter. The downward compression and upflow hydraulics create a compression gradient from less compressed to more compressed in the direction of the flow. During backwash, the upper plate is extended to provide a completely mixed media bed and makeup water is added to raise the spent water to an outlet pipe. The Schreiber Fuzzy Filter® is shown in Figure 7.19.

13.0 DISINFECTION

Disinfection is used to reduce the concentration of pathogens in wastewater before it is discharged to receiving water. The required level of disinfection varies by waterbody and state, depending on the designated use of the waterbody and which indicator the state uses to measure bacterial contamination. Disinfection of combined flow has been shown to be effective as long as chemical feed systems are functioning properly. Results from the Rouge National Wet Weather Demonstration Program

FIGURE 7.19 Schreiber 85-mgd Fuzzy Filter® installation (courtesy of city of Atlanta; photo by Mark Boner).

(Rouge Watershed CSO Workgroup, 2001) show that fecal coliform levels below 400 cfu/100 mL (the permit standard) could be consistently achieved in RTBs.

13.1 Available Methodologies

To date, the most frequently used chemical method for disinfection is sodium hypochlorite. This chlorine-based liquid is an effective disinfectant and is relatively safe to handle. The chemical is readily available, comparatively low cost, and can be stored for reasonable durations without significant deterioration. The primary disadvantage to the use of sodium hypochlorite is that persistent chlorinated compounds are created when it is used to treat wastewater and that degradation of the chemical in storage makes performance unreliable. Chlorine gas is not used in CSO facilities.

Other disinfectant chemicals have been shown to be effective in laboratory studies and in pilot testing, but have yet to be implemented in routine operation. Peracetic acid is an effective disinfectant, with similar effectiveness at comparable contact times to sodium hypochlorite during bench testing (DWSD, 2000a). Louisville Metropolitan Sewer District (Louisville, Kentucky) evaluated peracetic acid at wastewater treatment facilities, using Proxitane WW-12 (Guthrie et al., 2008). The demonstration was performed on primary effluent. In the demonstration, a 4-log reduction of bacterial levels was achieved in 5 to 10 minutes of contact time. The city of Frankfort, Kentucky, has used commercially available peracetic acid for secondary effluent on an interim basis. The city has reported good results, although the corrosiveness of the chemical is high, requiring special piping and pumps and care in handling by staff. No chlorinated byproducts are generated with this chemical. Drawbacks of peracetic acid use include limited availability in bulk quantities, difficulty of storage and delivery, and cost. Peracetic acid is highly corrosive and requires care in handling and delivery systems. Peracetic acid has been widely used in Europe.

Ultraviolet disinfection has been widely used at WWTPs, but has had limited implementation in standalone CSO control facilities. Ultraviolet disinfection has been applied to partially treated combined wastewater in Columbus, Georgia (Boner, 1997). At this location, UV disinfection is used for lower flows and chemical disinfectant is used for higher flows. Ultraviolet disinfection is also used downstream of ballasted treatment facilities for wet weather flows in Cincinnati, Ohio, and East Bremerton, Washington. Due to the high flowrates associated with wet weather facilities (relative to WWTPs), and the lower transmissivity of the flow, large systems would be required in most locations. Capital and operational costs, space requirements, and energy consumption are all factors that that have tended to limit the use of UV disinfection in the past. Additional interest in UV disinfection to avoid chemical disinfection is being considered at locations that are using more sophisticated wet weather treatment.

13.2 Treatment before Disinfection

In a WWTP application, disinfection is applied downstream of treatment processes. In CSO control, the chemical application is often upstream of the settling. This is a result of facilities being sized for settling to occur coincident with disinfection contact time. Separating these processes would result in larger wet weather facilities, increasing the associated costs. The Grand Rapids, Michigan, Market Avenue CSO basin disinfects in the third of three compartments, allowing for sedimentation to occur before chemical addition. Other facilities (such as the Weiss Street CSO facility in Saginaw, Michigan) feed the chemical following passage through a swirl concentrator. In most instances, both with RTBs or with screening and disinfection facilities, screening of the flow occurs before chemical addition.

Facilities with a first-flush capture component do not require disinfection feed of the initial event flows. These compartments capture small events and the most variable portion of larger storm events. Operational experience with disinfection chemical feed systems at intermittently operating facilities with highly variable flowrates has shown that applying a proper dosage of disinfectant is challenging (Moffa et al., 2007).

Increasingly, more sophisticated systems such as the Actiflo or DensaDeg processes and compressed media filtration are considered for wet weather treatment application, and the removals they are able to achieve would provide a better quality flow stream for disinfection. A separate contact chamber or channel may be required to provide desired contact time for disinfection following such facilities. Chemical addition used in these facilities may affect the effectiveness of disinfection.

13.3 Dosing

Application of disinfectant chemical requires a flexible system capable of variable dosing rates. Chlorine (or oxidant) demand is variable over the course of a wet weather event; for example, chlorine feed rates in the range of 5 to 30 mg/L were considered necessary for the Baby Creek CSO facility (DWSD, 2000a), dependent on quality of the influent. Additionally, the rate of flow to the facility varies over extremely wide ranges. Consideration of flow ranging from 1 to 100% of design flowrate is necessary. Finally, concentration of chemical degrades over time, resulting in less-effective dosage per volume of chemical dosed. As a result, increased delivery following long storage periods may be required.

Methodology for determining dosing is recognized as relative to both concentration and time. Extensive evaluation of disinfection was evaluated for the Conner Creek CSO facility in Detroit, Michigan (Fujita et al., 2006). The disinfection system was designed to maintain a CT value of 125 (where C = concentration of chlorine feed, mg/L, and T = time in minutes) to accomplish a 4-log reduction. Maintaining this dosing scheme requires good flow monitoring to support the dosage. Bench testing used a minimum dosage of 5 mg/L. There is a significant chlorine demand associated with solids in the flow, and dosage rates below a point may not be effective.

De-chlorination typically relies on sodium bisulfite as a chemical addition to counteract the sodium hypochlorite. Effluent concentrations of total residual chlorine (TRC) are significantly reduced, but may not reliably meet strict TRC limits.

13.4 Contact Chamber

Contact chambers for chlorine (or oxidant) contact have been sized for anywhere from 5 to 30 minutes, based on the design flowrate. Typically, most events are much lower intensity than the design event, so effective contact times for most events are longer. Contact chambers have been configured as basins, flow-through shafts, and channels or pipelines (either existing outfalls or newly constructed tunnels or pipelines).

Separate contact chambers provide additional storage. They need to be designed to drain back to the interceptor at the end of the event. Many of the design considerations for a storage basin apply to a chlorine contact chamber.

13.5 Chemical Feed Systems

Chemical feed systems for disinfection are made up of storage tanks, piping systems, pumps, valves, and flow metering equipment. The following considerations relative to operational issues should be provided:

- Tanks should have dual containment so that any leakage will not damage equipment. Tank material has typically consisted of fiberglass-reinforced plastic or fiber-glass-lined tanks. Tanks should be sized based on chemical delivery batches and allow for dilution to a stable concentration (typically 5%). Design criteria for chemical storage should evaluate total volume of discharge through a continuous simulation approach for a critical period. The period should be sufficient for resupply.
- Stratification has been noted in storage tanks and a means to mix the chemical during storage is suggested. Deterioration of chemical concentration over time occurs, and an ability to test the concentration is beneficial in setting feed rates.
- Pumping systems should be provided with variable-speed pumps or valving to support design turndown ratios. Various sizes of pumps may be required.
- Valves on chemical feed systems with intermittent operation have been problematic relative to leakage. As such, investigation of valve performance in similar facilities is recommended.
- Mixers or diffusers to distribute the chemical within the combined wastewater flow should be provided. Vertical orientation of mixers should be appropriate for filling conditions and a facility that has already filled.

14.0 ANCILLARY COMPONENTS

14.1 Grit Removal

Grit removal is specifically called for as pretreatment to Actiflo and DensaDeg processes. In other applications, it can be combined with other solids removal. Grit generation rates in combined systems are high, as wet weather events provide a scouring action in the sewer system. As such, any grit removal systems provided should consider rates of generation significantly higher than typical WWTP design.

Grit handling at wet weather facilities that are remote to the treatment plant typically provides for sending grit back to centralized treatment facilities along with dewatered flow through the interceptor.

14.2 Cleaning (Flushing) Systems

After dewatering CSO and SSO storage and treatment facilities, settled solids and debris sometimes remain for disposal. Various systems are used to move retained materials to the interceptor for transport and treatment at the WWTP. The type of flushing system depends on several factors, including the type of retained material, structure configuration, source of flushing water, and available personnel.

Retained materials can range from large debris and heavy grit typical of large metropolitan sewer systems to lighter, easily moved solids typical of smaller sewer systems, with rags, strings, and so on being common to all. Some facilities are equipped with coarse bar screens to remove large debris from the influent flow stream. Compartmentalization of the facility will assist in limiting the amount of debris throughout the structure as a whole, confining the bulk of the material to the first compartment to fill. Mechanical mixers can be used to help keep solids in suspension, although this does not replace the need for cleaning systems.

In the absence of screens, large items such as branches, timber, and so on are typically removed by hand. Heavy deposited solids typically can be moved with sufficient repetitive flushing, but occasionally require the use of mechanical equipment such as end loaders or blade-mounted vehicles. Pumps and piping should be designed in anticipation of heavy settled solids and debris.

Most automated flushing systems are intended to clean the floor of the storage and treatment facility. Manual flushing capability or pipe header and nozzle systems for wash down of walls may also be provided.

Types of flushing systems commonly used include

- Manually directed flushing from fire hoses or wall-mounted water cannon,
- Pipe header and nozzle system for facility wash down,
- Tipping buckets,
- Flushing gates, and
- Vacuum-assisted flushing systems.

14.2.1 *Manually Operated Flushing*

Manually operated flushing requires adequately spaced flushing water cannons that can deliver the necessary force to retained solids to dislodge them and direct them to a collection point. These systems require an adequate flushing water source to power the water cannon. Flushing systems of this type may be provided for parts of facilities that also have other flushing apparatuses. For example, wet wells, junction chambers, and static screen locations may be provided with manual flushing capability.

14.2.2 *Pipe Header and Nozzle Systems*

Pipe header and nozzle systems use high-pressure water delivered sequentially through a system of pipe headers to different locations within the structure. This system offers considerable flexibility in that it can be adapted to any tank configuration and can use either city water, strained final effluent, or strained receiving water. To avoid heavy instantaneous demands on the flushing water source, storage is often provided. Such systems are typically automated to provide sequential flushing. Nozzles must be placed where they provide adequate scouring to each section of floor. Nozzles may be mounted close to the floor to deliver better velocity. Various types of nozzles are available to optimize flushing action. This type of system is relatively capital-intensive, with a significant amount of piping, valves, nozzles, and booster pumps necessary to operate the system. Floor-positioned systems may result in hang up of debris.

14.2.3 *Tipping Buckets*

Tipping buckets are mounted in tanks above the maximum water surface. The device is a cylindrical container slung on an off-center swivel mounting at the end of the tank. After the tank is drained, the tipping bucket is filled at a slow rate, typically using potable water, although other water sources could be used. At a fill point just before overtopping, the center of gravity shifts and causes the tipping bucket to rotate about its central axis, thereby spilling its contents along the back wall of the tank. A fillet at the intersection of the back wall and tank floor redirects the flush water across the floor of the tank. The flushing force removes the sediments from the tank floor and transports it to a sump located at the far end of the tank. The amount of water required to clean a tank depends on the amount of accumulated debris, distance to be flushed, the height of the tipping bucket above the tank floor, and the slope of tank floor in the direction of flushing. The maximum flushing length is approximately 50 m (170 ft), as the wave cannot be sustained and tends to break up into rivulets at greater flushing lengths. Two flushes are often necessary when operating at the maximum flushing length.

14.2.4 Flushing Gates

Flushing gates consist of a hydraulically operated flap gate, a flush water storage area created by the erection of a concrete wall section, a float or pump to supply hydraulic pressure, and valves controlled by either a float system or an electronic control panel. The water level in the sewer or tank is used to activate the release and/or closure of the gate using a permanently sealed, float-controlled hydraulic system. The flushing system is designed to operate automatically whenever the in-system water level reaches a predetermined level, thereby releasing the gate and causing a "dam break" flushing wave to occur. Activation by remote control is also possible. Depending on physical constraints, the flushing system may include one, few, or many gates. The units may operate either in a fully automated mode, where the receding water level triggers the start of the preprogrammed flushing sequence, or a manual mode, where the tanks are flushed at the discretion of an operator. By design, flushing gates use stored water from the CSO or SSO storage to clean the debris from the floor of the tank. The gates are located in the tank or storage pipeline, so maintenance of the gates would require potential confined space entry. An approximate effective flushing length for flushing gates is 75 m (250 ft). Flushing gates are shown schematically in Figure 7.20.

Circular tank cleaning systems are available commercially. The concept of the Hydroself system (COPA) uses a centrally located water storage container that fills, lifts, and generates a radially oriented cleaning wave that is similar in flush characteristics to that produced by flushing systems used in rectangular chambers.

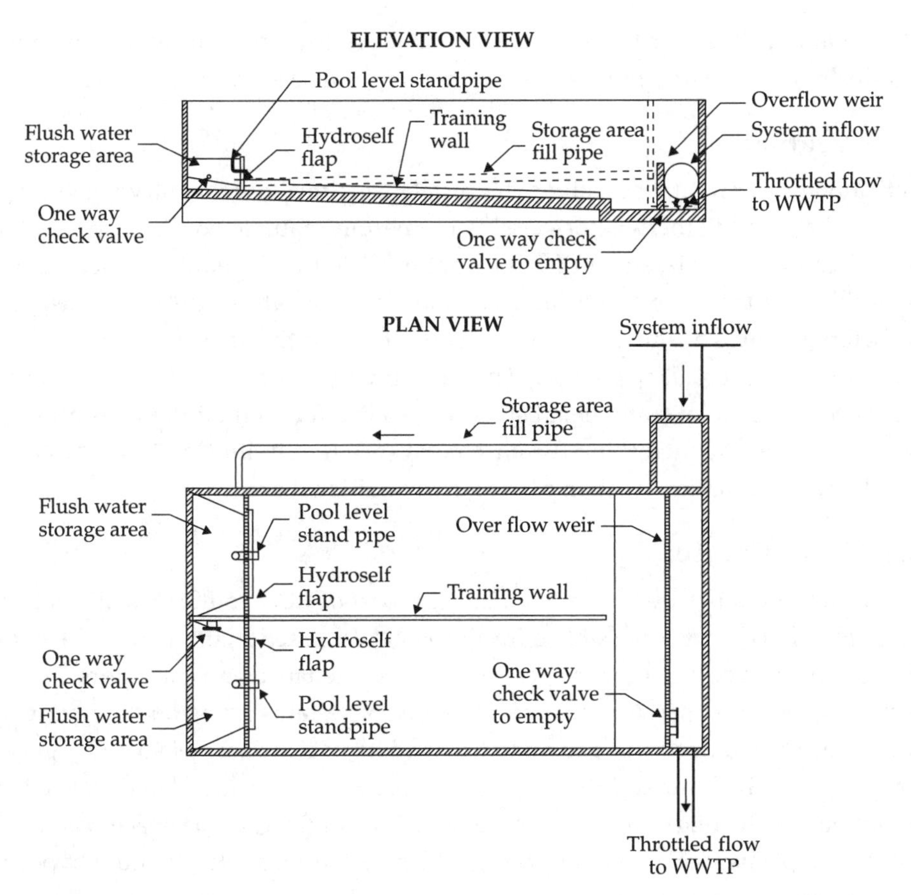

FIGURE 7.20 Flushing gates.

14.2.5 *Vacuum-Assisted Flushing System*

Vacuum-assisted flushing systems use retained wastewater to flush the basin floor. Vacuum-assisted flushing systems use a vacuum pump to elevate a column of water above the floor of the basin. This column of water is suitable for flushing longer flush lengths than tipping buckets or tipping gates. Estimated flushing length capability is approximately 100 m (325 ft) or more. The technology can be used for rectangular channels and circular tanks. Equipment (e.g., vacuum pump) is located at the top of the column and is more convenient for maintenance. The concept requires ensuring that the system is air-tight to sustain the water column. There is limited ability to resupply the vacuum column once it is emptied. A relatively small amount of water in the column will remain behind the weir that forms the lower air seal on the column after flushing. Efforts to drain this area may compromise the air seal.

14.3 Ventilation

Ventilation of wet weather facilities is typically used to allow for entry of operations staff in what would otherwise be classified as permit-required confined spaces. Most subsurface storage or treatment facilities are provided with ventilation according to applicable standards for wastewater facilities (such as NFPA, 2008). This aspect of the facility requires air-intake and exhaust locations. If the area above the facility is to be used for a secondary purpose (such as parking or a soccer field), the ability to locate needed air-intake and exhaust structures is often limited to the location of a service building. By increasing the air-conveyance requirements, the cost of a centrally located ventilation location can result in additional cost.

14.4 Odor Control

Odor control systems are typically an integral part of CSO and SSO facilities to satisfy public concern over possible odor problems. In practice, odor control facilities often go unused because proper operation and maintenance prevents odors. Even so, most designers and operators believe odor-control systems are a good investment in up-front public relations and good practice if odors become a problem. Odor generation on-site can be limited if the facilities are promptly dewatered and cleaned and any screenings or solids are promptly removed or confined to the space where odor control is implemented. Retention treatment basins that dose with sodium hypochlorite help limit the generation of odors as well.

Design air volumes for odor control systems can be large due to the size of facilities. A reasonable basis for sizing odor control could be determined based on the rate at which air is typically displaced when the facility fills during a wet weather event.

A multitude of odor control systems are available that can effectively treat foul air. These may require the use of media, chemicals, or both. Technologies used for odor control include the conventional wet air scrubber system, perhaps the most common odor-control method, wherein the air stream is passed through a packed tower through counter-flowing droplets of hypochlorite solution. Activated carbon has been used because of its capacity for adsorbing hydrogen sulfide and organics and its relative ease of regeneration or reactivation. Many facilities effectively control odor by directing air through a granular activated carbon adsorber system. Compost biofilters are typically not used at CSO treatment facilities because they are not suitable for intermittent use applications. Water Environment Federation's *Control of Odors and Emissions from Wastewater Treatment Plants* (WEF, 2004) contains additional information on odor control in wastewater facilities

14.5 Pumping Systems

Pumping systems in CSO facilities are predominately used for dewatering to the interceptor, although pumping may be required at the influent of the facility. With influent pumping, large-capacity pumps may be required.

In most instances, screens or bar racks should be provided to protect the pumps from damage that could be caused by oversized debris. Pumps designed for solids handling should be used. Generally, dewatering pumping stations are sized to be able to dewater the facility in 24 to 72 hours. Consideration should be given to providing pumping capacity in excess of the "normally" anticipated operation to maximize flow to treatment when interceptor capacity is available. Generally, dewatering facilities are supplied with fixed-speed pumps with equal capacity. If interceptor capacities are limited, the ability to pump a smaller dewatering flow may also be beneficial.

Influent pumping stations would need to lift the total flow generated in the combined sewer area. As such, these stations may be quite large. For relatively small lifts, axial-flow or mixed-flow pumps have been used in many applications to lift flow into an influent channel. Influent pumping stations may have complex hydraulics and, as such, physical hydraulic modeling should be considered.

14.6 Weir Design and Underflow Baffles

Weir design in CSO facilities needs to be consistent with the weir function. Weirs may be used to split flow, serve as a break in tank function, or support settling. Weirs may need to pass significant flow, resulting in relatively large head loss over the weir. In instances where weirs control the discharge from sedimentation basins, enough weir length should be provided to limit approach velocities. Weirs with extensive length may be interrupted by structural support columns, limiting the total effective length.

Weir submergence may be acceptable depending on the purpose for the weir. In some instances, the primary function of weirs is to compartmentalize the facility to limit the necessary extent of cleaning. When the facility is full, submergence of the weir installed for this purpose would not be problematic, although short-circuiting may limit settlement that occurs.

In locations with limited availability of hydraulic gradient, consideration should be given to the application of bending weirs to prevent upstream surcharging.

14.7 Valves and Gates

Generally, valves and gates used in wet weather facilities should be used for isolation purposes in the process flow path rather than flow modulation. Wet weather flows are highly variable and there is minimal opportunity for feedback and valve or gate repositioning before a change in flow conditions. Consideration should be given to gate and valve operational speeds (e.g., time to open or close) to ensure proper operation. Gates that are provided for backwater isolation or bypass should be designed to provide fail-safe operation

Valves may be used in dewatering applications to control the rate of flow return to the collection system. In some applications, both the rate of dewatering and head conditions may be highly variable. Evaluation of the valve under consideration should be evaluated over the entire design range of conditions. Cavitation or vacuum pressure may be present under some circumstances, negatively affecting valve performance.

Valves used with chemical feed systems that operate intermittently have a history of problems, even in applications consistent with manufacturers' suggested uses. Leakage of valves can result from a variety of factors. Before valve selection, the application in a similar facility should be identified and interviews of the operators should be conducted to identify any performance concerns.

14.8 Flow Splits and Hydraulics

Flowrates in CSO or SSO facilities are highly variable. Hydraulic elements of facilities should be designed to accommodate these conditions. In most instances, passive flow-splitting devices in conjunction with open and shut gates should be considered. Use of modulating valves or gates must be used with care. Modulating weirs are typically safer and easier to control than sluice gates because they do not close the flow opening completely.

14.9 Flow Monitoring and Sampling

Flow monitoring in wet weather facilities must encompass a wide range of flow conditions in locations where hydraulic turbulence and backwater conditions may be significant. Flow monitoring should use various techniques and include points of redundancy to accomplish desired objectives.

Accurate flow metering may require dual techniques to accomplish measurements over the entire flow range. Hydraulic design typically results in pipes that are sized for peak design flows, which may be ineffective at measuring during lower flow conditions. Many flow metering installations that rely on velocity measurements have been limited in effectiveness due to a variety of challenging situations, including entrained air (even when not anticipated), influent channel upstream of the weir, and backwater condition through facility and solids accumulation. In conditions where facility hydraulics can accommodate the associated head loss, primary flow devices (such as flumes) should be considered for flow measurement. Other techniques that have been used include the following:

- Volume displacement (rate of fill in compartment based on depth measurement);
- Pump discharge curves tied to operating time to document total volume of pumped flow;
- Primary devices such as Parshall, cut throat, or other types of flumes;
- Depth over weirs (limited accuracy if the depth of flow is small or the weir is interrupted by structural columns);
- Area and velocity flow meters; and
- Transit time or velocity "bin" meters.

Flow metering is particularly critical when data are used for dosage of chemical feed systems. Redundancy of metering (e.g., influent and effluent) helps to provide a cross check of the equipment accuracy.

Most facilities have requirements for sampling of the quality of influent and effluent flow. Installed automatic samplers can collect samples on timed intervals without an attendant present at the facility. Sample collection location should be representative of the desired flow stream. Effluent samples, for example, should be taken in the proximity of an overflow weir (high in the depth orientation of the tank) rather than low in the tank. Typically, sample pumps are used at the point of collection with sample tubing or piping extending to a refrigerated sampler location. Refrigerated samplers should consider flexibility desired for discrete or composite samples and necessary sample volumes anticipated based on projected parameter lists.

Bacteria sampling can be problematic. Technically, bacteria samples should be grab samples direct from the flow stream. In practice, however, they are often collected from the automatic samplers. If direct grab samples are important, provision of access to a collection point should be provided for the operator.

14.10 Dewatering Systems

Generally, dewatering systems are designed to allow for the facility to be drained within a set period of time (such as 24, 48, or 72 hours). Dewatering may be accomplished by gravity or pumping or a combination of the two. Dewatering systems should provide flexibility to accelerate or slow dewatering rates based on interceptor conditions. Dewatering gravity systems need to have throttling capability to match rates to capacity. Pumped systems need to have the ability to deliver varying rates. Typically, this will be accomplished with multiple pumps and/or throttling valves. Dewatering systems should be able to operate during periods of influent flow as interceptor capacity may be available for a portion of this duration.

Small pumps to complete the dewatering and cleaning of certain system components (such as wet wells) will help to fully dewater and clean the facility and avoid odors.

14.11 Instrumentation and Control

Control systems for wet weather facilities should be fully capable of monitoring, data collection, and operation from either on-site or remote location. Control system requirements depend on the algorithms and means specified for operating the controllable elements of the system (see Section 7.0). Control systems for wet weather facilities need to be designed to be fail-safe in the event of communications or equipment failure. Control systems rely on sensor data received. As such, the means of monitoring should be capable of providing the necessary precision over the range of flows for the control of the facility. Beyond the facility itself, control systems will likely require information about conditions in the interceptor (for dewatering), the WWTP (to ensure there is capacity), and, potentially, influent sewers.

14.12 Data Retention

Data storage for regulatory reporting or analysis of operations should be considered as the control system is designed. Frequently, discharge permits require reporting of overflow volume and duration.

14.13 Siting

Siting of wet weather facilities can be extremely challenging, although when well-executed the projects can be used as a catalyst for local improvements and structures can be designed so that they are aesthetically pleasing. Most facilities require parcels of 3 ac or more that are in proximity to existing watercourses or interceptors. Sites that are level, accessible (construction traffic), free from contamination, and contain good geotechnical conditions are ideal, although it is typical that at least one of these parameters is not obtainable. Sites should allow for construction staging, access roads, and proper setbacks per zoning ordinances.

Following completion of the project, most facilities have service buildings that house equipment, pumps, screens, electrical and chemical feed equipment, and control room and staff facilities. Some of the equipment or the operating water surface in the basin may result in high-profile buildings. Building requirements are lessened for storage facilities (as opposed to treatment facilities).

14.14 Access

Access to facility components should be provided such that operations and maintenance can be executed safely. Typical access to basin components is by stairway to a catwalk, with ladders to storage areas. Equipment that requires maintenance should be accessible.

14.15 Staff Facilities

Basic restroom and staging facilities should be provided at facilities for the comfort of staff providing maintenance. Some facilities also provide meeting rooms or laboratory areas for staff.

15.0 REFERENCES

Akridge, A.; Bingham, B.; Carty, D.; Colas, H. (2006) An Operational Perspective to Real Time Control for Consent Decree Compliance. *Proceedings of the Water Environment Federation Collection Systems Specialty Conference—Infrastructure Stewardship: Partnering for a Sustainable Future;* Detroit, Michigan, Aug 6–9; Water Environment Federation: Alexandria, Virginia.

Bhattarai, R.; Guthikonda, G. (2008) Everything Is Big in Texas—Even a Private Lateral Program. *Proceedings of the Water Environment Federation Collection Systems Specialty Conference;* Pittsburgh, Pennsylvania, May 18–21; Water Environment Federation: Alexandria, Virginia.

Brashear, R.; Vitasovic, C.; Johnson, C.; Clinger, R.; Rife, J.; Lafitte, D. (2002) *Best Practices for the Treatment of Wet Weather Wastewater Flows;* 00-CTS-6; Water Environment Research Foundation: Alexandria, Virginia.

Boner, M. (1997) *Initial Test Results from the Columbus Advanced Demonstration Project, Wet Weather Rx;* Water Environment Research Foundation: Alexandria, Virginia.

City of Toledo (2005) Long Term Control Plan, CSO Control Technologies, Chapter 7; City of Toledo: Toledo, Ohio.

Construction Innovation Forum (2008) Construction Innovation Forum, Green Roofs for Storm Water, 2008 Nova Award Nomination 29, Seattle Green Roof Evaluation Project. http://www.cif.org/noms/2008/29_-_Green_Roofs_for_Storm_Water.pdf (accessed Oct 2010).

Courter, C.; Bradley, S.; Vatter, B. (2008) Floatables Control is Surfacing Again, What Can You Do? *Proceedings of the Water Environment Federation 81st Annual Technical Exposition and Conference* [CD-ROM]; Chicago, Illinois, Oct 18–2; Water Environment Federation: Alexandria, Virginia.

Detroit Water and Sewerage Department (2000a) *DWSD Contract No. CS-1286, Baby Creek CSO Facility Basis of Design Report;* Prepared by Tetra Tech: Detroit, Michigan.

Detroit Water and Sewerage Department (2000b) *DWSD Contract No. CS-1329, Installation of In-System Storage Devices;* Prepared by Applied Science, Inc.: Detroit, Michigan.

Domenick, N. (2007) Creating Storage Capacity via CSO Regulator Modifications. *Proceedings of the Five Cities Conference;* Louisville, Kentucky, July 18–20.

Enfinger, K. L.; Cook, S. L. (2008) The Road Less Traveled—Inspection of Trunk Sewers and Stream Crossings Yields Positive Results. *Proceedings of the Water Environment Federation 81st Annual Technical Exposition and Conference* [CD-ROM]; Chicago, Illinois, Oct18–22; Water Environment Federation: Alexandria, Virginia.

Fujita, G.; Rabbaig, M.; Moffa, P.; Davis, D.; Courter, C.; Hocking, C. (2006) The Challenges of Abating a Very Large Combined Sewer Overflow (CSO)—Conner Creek, Detroit, Michigan. *Proceedings of the Water Environment Federation Collection Systems Specialty Conference—Infrastructure Stewardship: Partnering for a Sustainable Future;* Detroit, Michigan, Aug 6–9; Water Environment Federation: Alexandria, Virginia.

Ghalib, S.; McCormack, C.; Eloubaidy, A.; Kalinowski, S. (2006) What Results After a CSO Storage Tunnel is Flipped. *Proceedings of the Water Environment Federation Collection Systems Specialty Conference—Infrastructure Stewardship: Partnering for a Sustainable Future;* Detroit, Michigan, Aug 6–9; Water Environment Federation: Alexandria, Virginia.

Gonwa, W.; Ellis, D. (2006) Preventing Increased Infiltration/Inflow from Residential Storm Water Best Management Practices. *Proceedings of the Water Environment Federation Collection Systems Specialty Conference—Infrastructure Stewardship: Partnering for a Sustainable Future;* Detroit, Michigan, Aug 6–9; Water Environment Federation: Alexandria, Virginia.

Gray, T.; Hedin, R. (2008) Storage of Wet Weather Overflows in Abandoned Coal Mines. *Proceedings of the Water Environment Federation Collection Systems Specialty Conference;* Pittsburgh, Pennsylvania, May 18–2; Water Environment Federation: Alexandria, Virginia.

Gray, W. S.; Barnes, G.; Pope, R. L.; Sawhney, B. (2006) City of Atlanta: East Area CSO–Custer Avenue CSO Storage and Dechlorination Facility. *Proceedings of the Water Environment Federation Collection Systems Specialty Conference—Infrastructure Stewardship: Partnering for a Sustainable Future;* Detroit, Michigan, Aug 6–9; Water Environment Federation: Alexandria, Virginia.

Great Lakes-Upper Mississippi River Board of State and Provincial Public Health and Environmental Managers (2004) *Recommended Standards for Wastewater Facilities;* Health Research, Inc.: Albany, New York.

Grumbles, B. (2007) Using Green Infrastructure to Protect Water Quality in Stormwater, CSO, Nonpoint Source and other Water Programs. U.S. Environmental Protection Agency Memorandum, Office of Water, Assistant Administrator, Washington, D.C., Mar 5 .

Guthrie, D.; Melisizwe, L.; Sydnor, W.; Kraus, T. (2008) Solids and Floatables Control—A Coordinated Effort: 78 Installations in 36 Days. *Proceedings of the Water Environment Federation 81st Annual Technical Exposition and Conference* [CD-ROM]; Chicago, Illinois, Oct 18–22; Water Environment Federation: Alexandria, Virginia.

Guthrie, D.; Schardein, B.; Pavoni, J. (2008) Disinfection of Wet Weather Bypasses and Overflows Using Peracetic Acid. *Proceedings of the Five Cities Plus Conference;* Columbus, Ohio, June 4–6.

King County Department of Natural Resources and Parks, Wastewater Treatment Division (2010) Combined Sewer Overflow Treatment Systems Evaluation and Testing—Phase 2 Pilot Test Report; Prepared by CDM: Seattle, Washington.

Kurtz, T. (2007) Monitoring Results: Green Solution Benefits to the Combined Sewer System. *Proceedings of the Water Environment Federation Collection Systems Specialty Conference: Pioneering Trails to Collection Systems Excellence;* Portland, Oregon, May 13–16; Water Environment Federation: Alexandria, Virginia.

Landers, J. (2006) Test Results Permit Side-by-Side Comparisons of BMPs. *Civ. Eng.*, **76** (4), 34–35. http://www.unh.edu/erg/civil_eng_4_06.pdf (accessed June 2010).

Liebe, M.; Collins, D. (2007) Modeling of Stormwater Removal and Peak/Volume Reduction Effects of Green Solutions (Inflow Controls) Using an Explicit Combined/Sanitary Sewer Model. *Proceedings of the Water Environment Federation Collection Systems Specialty Conference: Pioneering Trails to Collection Systems Excellence;* Portland, Oregon, May 13–16; Water Environment Federation: Alexandria, Virginia.

Massachusetts Water Resources Authority (2006) *Combined Sewer Overflow Control Plan Annual Progress Report 2005;* Massachusetts Water Resources Authority: Boston, Massachusetts.

Moffa, P.; McCormack, C.; Moore, T. (2007) Operations and Maintenance Requirements of Vortex with Disinfection and Screening with Disinfection CSO Facilities. *Proceedings of the Water Environment Federation 80th Annual Technical Exposition and Conference* [CD-ROM]; San Diego, California, Oct 13–17; Water Environment Federation: Alexandria, Virginia.

Nordstrom, O.; Zawacki, J.; Warrow, A.; Sherman, B. (2009) What Site Characteristics Explain Variability of Peak Footing Drain Flows? *Proceedings of the Water Environment Federation Collection Systems Specialty Conference;* Louisville, Kentucky, April 19–22; Water Environment Federation: Alexandria, Virginia.

Oakland County Drain Commissioner (2000) *Retention Basin Evaluation for the Acacia Park CSO RTB;* Prepared by Hubbell, Roth & Clark, Consulting Engineers. http://www.rougeriver.com/pdfs/cso/tr16.pdf (accessed Oct 2010).

Owen, B. (2007) Weaving Green Solutions Into CIP Projects. *Proceedings of the Water Environment Federation Collection Systems Specialty Conference: Pioneering Trails to Collection Systems Excellence;* Portland, Oregon, May 13–16; Water Environment Federation: Alexandria, Virginia.

Recos, M.; Carrier, A. J.; Duggan, E. W.; Liu, D. (2008) Quantitative Assessment of Sewer Separation for the Reserved Channel in South Boston. *Proceedings of the Water Environment Federation Collections Systems Specialty Conference;* Pittsburgh, Pennsylvania, May 18–21; Water Environment Federation: Alexandria, Virginia.

Rouge Watershed CSO Workgroup (2001) *Goal 2—Protection of Public Health, Elimination of Raw Sewage. Evaluation of Wayne County and Oakland County CSO Facilities;* RPO-SR30; Rouge Program Office: Detroit, Michigan.

Ruggaber, T. (2009) CSOnet: An Innovative Solution for the Real Time Monitoring and Control of Combined Sewer Systems. *Proceedings of the Water Environment Federation Collection Systems Specialty Conference;* Louisville, Kentucky, April 19–22; Water Environment Federation: Alexandria, Virginia.

Shoemaker, L.; Riverson, J.; Alvi, K.; Zhen, J.; Paul, S.; Rafi, T. (2009) *SUSTAIN—A Framework for Placement of Best Management Practices in Urban Watersheds to Protect Water Quality;* EPA-600/R-09-095; U.S Environmental Protection Agency: Washington, D.C.

Stinson, M.; Vitasovic, Z. (2006) Real Time Control of Sewers: U.S. EPA Manual. *Proceedings of the Water Environment Federation Collection Systems Specialty Conference—Infrastructure Stewardship: Partnering for a Sustainable Future;* Detroit, Michigan, Aug 6–9; Water Environment Federation: Alexandria, Virginia.

Strause, A.; Smith, C.; Mahmutoglu, S. (2008) Confirming Post-Construction CSO Infrastructure and Validating Design Goals. *Proceedings of the Water Environment Federation 81st Annual Technical Exposition and Conference* [CD-ROM]; Chicago, Illinois, Oct 18–22; Water Environment Federation: Alexandria, Virginia.

Struck, S. D.; Jacobs, T.; Moore, G.; Pitt, R.; Ports, M. A.; O'Bannon, D.; Schmitz, E.; Field, R. (2009) Green Infrastructure for CSO Control in Kansas City, Missouri. *Proceedings of the Water Environment Federation 82nd Annual Technical Exposition and Conference* [CD-ROM]; Orlando, Florida, Oct 17–21; Water Environment Federation: Alexandria, Virginia.

Trypus, J.; Mausbaum, B.; Oakley, J.; McKelvey, J. (2008) A Practical Guide—Planning Considerations for Mega Dollar CSO Projects. *Proceedings of the Water Environment Federation 81st Annual Technical Exposition and Conference* [CD-ROM]; Chicago, Illinois, Oct 18–22; Water Environment Federation: Alexandria, Virginia.

University of New Hampshire Stormwater Center (2006) *2005 Data Report;* University of New Hampshire: Durham, New Hampshire.

Vander Tuig, K.; Hufnagel, C.; Christian, D.; Carrier, A.; Struck, S. (2009) The Great Sewer Separation Debate—Pros, Cons and Ways to Do It Right. *Proceedings of the Water Environment Federation 82nd Annual Technical Exposition and Conference* [CD-ROM]; Orlando, Florida, Oct 17–21; Water Environment Federation: Alexandria, Virginia.

U.S. Environmental Protection Agency (1994) *Combined Sewer Overflow (CSO) Policy; Fed. Regist.*, 59, 75, 18688–18698; Apr 19.

U.S. Environmental Protection Agency (1995) *Combined Sewer Overflows: Guidance for Nine Minimum Controls;* EPA-832/B-95-003; U.S. Environmental Protection Agency: Washington, D.C.

U.S. Environmental Protection Agency (1999) *Combined Sewer Overflows Technology Fact Sheet: Netting Systems for Floatables Control;* EPA-832/F-99-037; U.S. Environmental Protection Agency: Washington, D.C.

U.S. Environmental Protection Agency (2009a) *National Pollutant Discharge Elimination System BMP Fact Sheet: Green Roof.* http://cfpub.epa.gov/npdes/stormwater/menuofbmps (accessed Oct 2010).

U.S. Environmental Protection Agency (2009b) *National Pollutant Discharge Elimination System BMP Fact Sheet: Porous Pavement.* http://cfpub.epa.gov/npdes/stormwater/menuofbmps (accessed Oct 2010).

U.S. Environmental Protection Agency; National Association of Clean Water Agencies; Natural Resources Defense Council; Low Impact Development Center; Association of State and Interstate Water Pollution Control Administrators (2007) *Green Infrastructure Statement of Intent;* Washington, D.C.

Walker, D.; Heath, G. (2008) Design Considerations for Off-Line Storage Tanks. *Proceedings of the Water Environment Federation 81st Annual Technical Exposition and Conference;* Chicago, Illinois, Oct 18–22; Water Environment Federation: Alexandria, Virginia.

Washington State Legislature (2000) Washington Administrative Code, *Submission of Plans and Reports for Construction and Operation of Combined Sewer Overflow Reduction Facilities,* Chapter 173, Section 245. http://apps.leg.wa.gov/WAC/default.aspx?cite=173-245-020 (accessed Sept 2010).

Water Environment Federation (2004) *Control of Odors and Emissions from Wastewater Treatment Plants;* WEF Manual of Practice No. 25; McGraw-Hill: New York.

Water Environment Federation (2006) *Guide to Managing Peak Wet Weather Flows in Municipal Wastewater Collection and Treatment Systems;* Water Environment Federation: Alexandria, Virginia.

Water Environment Federation (2010) Water Environment Federation, *Private Property Virtual Library* Web Site. http://www.wef.org/PrivateProperty/ (accessed June 2010).

Water Environment Federation; American Society of Civil Engineers; Environmental and Water Resources Institute (2009) *Existing Sewer Evaluation and Rehabilitation,* 3rd ed.; WEF Manual of Practice No. FD-6; ASCE Manual and Report on Engineering Practice No. 62; McGraw-Hill: New York.

Westmoreland, P.; Culpepper, M.; Melchert, R. (2006) Lessons Learned During the Bloomfield Orchards Footing Drain Disconnection Program in the City of Auburn Hills. *Proceedings of the Water Environment Federation Collection Systems Specialty Conference—Infrastructure Stewardship: Partnering for a Sustainable Future;* Detroit, Michigan, Aug 6–9; Water Environment Federation: Alexandria, Virginia.

16.0 SUGGESTED READINGS

Alsaigh, R. (1994) *CSO Demonstration Facilities Design Parameter Report;* RPO-CSO-TR02.00; Rouge Program Office: Detroit, Michigan.

Brink, P.; Bryson, D.; Kluitenberg, E.; Johnson, C.; Hufnagel, C. (2005) *CSO Basins: Getting the Most Performance from Your Pollution Control Dollar;* RPO-TR52; Rouge Program Office: Detroit, Michigan.

Duffy, S.; Hyland, R.; Lienhard, E. (2008) Application of Modeling Shows that Downspout Disconnect Program Can Eliminate Sensitive CSO. *Proceedings of the Water Environment Federation 81st Annual Technical Exposition and Conference* [CD-ROM]; Chicago, Illinois, Oct 8–22; Water Environment Federation: Alexandria, Virginia.

Farnan, J.; Sobanski, J.; Venusa, N.; Gronski, A. (2004) Chicago's Deep Tunnel—History and Evolution. *Proceedings of the Water Environment Federation Collection Systems Specialty Conference—Innovative Approaches to Collection Systems Management;* Milwaukee, Wisconsin, Aug 8–11; Water Environment Federation: Alexandria, Virginia.

Federation of Canadian Municipalities and National Research Council (2003) *Infiltration/ Inflow Control/Reduction for Wastewater Collection Systems.* http://sustainablecommunities.fcm.ca/files/infraguide/storm_and_wastewater/inflow_infiltr_control_reduct_wastewtr_coll_syst.pdf (accessed Sept 2010).

Guthrie, D.; Melisizwe, L.; Sydnor, W.; Kraus, T. (2007) Solids and Floatables Control—A Coordinated Effort: 78 Installations in 36 Days. *Proceedings of the Five Cities Plus Conference;* Louisville, Kentucky, July 18–20.

Heath, G.; Roberts, A.; Schimmel, J.; Stoops, R.; Minard, D.; Walker, D. (2007) Detailed System Understanding Supports Project Refinement During CSO LTCP Implementation Phase. *Proceedings of the Water Environment Federation Collection Systems Specialty Conference: Pioneering Trails to Collection Systems Excellence;* Portland, Oregon, May 13–16; Water Environment Federation: Alexandria, Virginia.

Heath, G.; Walker, J.; Adams, R.; Hedetniemi, M. (2009) Construction and Start-Up of a High-Rate Wet Weather Treatment Facility in Nashua, New Hampshire. *Proceedings of the Water Environment Federation Collection Systems Specialty Conference;* Louisville, Kentucky, April 19–22; Water Environment Federation: Alexandria, Virginia.

Hufnagel, C. (2001) *Rouge Watershed CSO Workgroup Operators' Forum Report;* RPO-TR39; Rouge Program Office: Detroit, Michigan.

Hufnagel, C.; Catalfio, C. (1999) *CSO Basin Evaluation Plans–Data Collection and Transfer Guide;* RPO-NPS-TM33.00; RPO-TR39; Rouge Program Office: Detroit, Michigan. http://rougeriver.com/pdfs/cso/tm33.pdf (accessed June 2010).

Hyland, R.; Zuravnsky, L. (2007) Green Infrastructure Approaches for CSO Control in Urban Areas. *Proceedings of the Water Environment Federation 80th Annual Technical Exposition and Conference* [CD-ROM]; San Diego, California; Oct 13–17; Water Environment Federation: Alexandria, Virginia.

Munsey, F.; Roddy, M.; Jankowski, J. (2004) Deep Tunnel O&M—Milwaukee's 10 Years of Lessons Learned. *Proceedings of the Water Environment Federation Collection Systems Specialty Conference—Innovative Approaches to Collection Systems Management;* Milwaukee, Wisconsin, Aug 8–11; Water Environment Federation: Alexandria, Virginia.

National Fire Protection Association (2008) NFPA 820: Standard for Fire Protection in Wastewater Treatment and Collection Facilities; National Fire Protection Association: Quincy, Massachusetts.

Pennington, T.; Cook, R.; Gray, B. (2007) Design and Construction of an Underground Cavern for CSO Control. *Water Environment Federation Collection Systems Specialty Conference: Pioneering Trails to Collection Systems Excellence;* Portland, Oregon, May 13–16; Water Environment Federation: Alexandria, Virginia.

Pisano, B.; Matthews, S.; Gray, S.; Stoney, E. (2009) Capital Cost Savings Using Passive Controls in Atlanta's Utoy Creek Basin. *Proceedings of the Water Environment Federation Collection Systems Specialty Conference;* Louisville, Kentucky, April 19–22; Water Environment Federation: Alexandria, Virginia.

Ridgway, K.; Rabbaig, M. (2006) Use of Inflatable Dams for CSO Control. *Proceedings of the Water Environment Federation 79th Annual Technical Exposition and Conference* [CD-ROM]; Dallas, Texas, Oct 21–25; Water Environment Federation: Alexandria, Virginia.

Sherrill, J.; Fujita, G. (2006) Budget Development for Operations/Maintenance Requirements for CSO/SSO Control Facilities. *Proceedings of the Water Environment Federation Collection Systems Specialty Conference—Infrastructure Stewardship: Partnering for a Sustainable Future;* Detroit, Michigan, Aug 6–9; Water Environment Federation: Alexandria, Virginia.

Southeastern Michigan Council of Governments (2008) *Low Impact Development Manual for Michigan: A Design Guide for Implementors and Reviewers;* Southeastern Michigan Council of Governments: Detroit, Michigan.

Stoner, N.; Kloss, C.; Calarusse, C. (2006) *Rooftops to Rivers—Green Strategies for Controlling Stormwater and Combined Sewer Overflows*. Natural Resources Defense Council: New York.

United Kingdom Industry Research Limited (1999) *CSO Research Group WW-08 Screen Efficiency (Proprietary Designs)*; Report Reference No. 99/WW/08/5; Prepared by Professor Adrian J. Saul, University of Sheffield: Sheffield, U.K.

U.S. Environmental Protection Agency (1999) *Combined Sewer Overflows Technology Fact Sheet: Floatables Control*; EPA-832/F-99-008; U.S. Environmental Protection Agency: Washington, D.C.

U.S. Environmental Protection Agency (1999) *Combined Sewer Overflows Technology Fact Sheet: Screens*; EPA-832/F-99-040; U.S. Environmental Protection Agency: Washington, D.C.

U.S. Environmental Protection Agency (1999) *Combined Sewer Overflows Technology Fact Sheet: Retention Basins*; EPA-832/F-99-042; U.S. Environmental Protection Agency: Washington, D.C.

U.S. Environmental Protection Agency (2009) *National Pollutant Discharge Elimination System BMP Fact Sheet: Dry Detention Pond*. http://cfpub.epa.gov/npdes/stormwater/menuofbmps (accessed Oct 2010).

U.S. Environmental Protection Agency (2009) *National Pollutant Discharge Elimination System BMP Fact Sheet: Grassed Swale*. http://cfpub.epa.gov/npdes/stormwater/menuofbmps (accessed Oct 2010).

U.S. Environmental Protection Agency (2009) *National Pollutant Discharge Elimination System BMP Fact Sheet: Infiltration Basin*. http://cfpub.epa.gov/npdes/stormwater/menuofbmps (accessed Oct 2010).

U.S. Environmental Protection Agency (2009) *National Pollutant Discharge Elimination System BMP Fact Sheet: Infiltration Trench*. http://cfpub.epa.gov/npdes/stormwater/menuofbmps (accessed Oct 2010).

U.S. Environmental Protection Agency (2009) *National Pollutant Discharge Elimination System BMP Fact Sheet: Rain Barrels*. http://cfpub.epa.gov/npdes/stormwater/menuofbmps (accessed Oct 2010).

U.S. Environmental Protection Agency (2009) *National Pollutant Discharge Elimination System BMP Fact Sheet: Rain Gardens*. http://cfpub.epa.gov/npdes/stormwater/menuofbmps (accessed Oct 2010).

U.S. Environmental Protection Agency (2009) *National Pollutant Discharge Elimination System BMP Fact Sheet: Stormwater Wetland.* http://cfpub.epa.gov/npdes/stormwater/menuofbmps (accessed Oct 2010).

U.S. Environmental Protection Agency (2009) *National Pollutant Discharge Elimination System BMP Fact Sheet: Wet Pond.* http://cfpub.epa.gov/npdes/stormwater/menuofbmps (accessed Oct 2010).

Weinstein, N.; Glass, C.; Heaney, J.; Lee, J.; Huber, W.; Jones, P.; Kloss, C.; Quigley, M.; Strecker, E.; Stephens, K. (2005) *Decentralized Stormwater Controls for Urban Retrofit and Combined Sewer Overflow Reduction;* 03-SW-3; Water Environment Research Foundation: Alexandria, Virginia.

Wilber, C.; Hyland, R. (2007) Green Infrastructure in CSO Cities: National Overview and Strategic Considerations. *Proceedings of the Wet Weather Partnership & NACWA Conference;* Chicago, Illinois, April 26–27.

Chapter 8

Overflow Mitigation Plan Development and Implementation

(continued)

1.0 INTRODUCTION

This chapter describes the process of developing and implementing a plan to manage overflows and provide mitigation measures to help protect the environment and human health. The information discussed in this chapter should provide timely guidance for those just getting started in the planning process and those that have already developed plans or are in the process of implementation.

Plans differ from engineering projects because they are dynamic and typically need to be updated on a regular schedule. The schedule for updating a plan may be driven by a regulatory agency or may be needed because of changes in plan assumptions (e.g., changes in service area growth) or changes in technology (e.g., modeling software that provides more comprehensive analysis of a system or new abatement technologies). Therefore, a plan is never finished; it is a cyclical process of development, implementation, and refinement.

The topic of overflow mitigation is extensive; however, the intent of this chapter is to discuss the history of overflow issues and provide elements that could be included in a municipality's individual plan. Not all elements discussed in this chapter may be applicable to every system, but should give individuals that are developing, reviewing, and implementing overflow mitigation plans consolidated information that can help in further development of planning documents and provide references to material that include more in-depth discussion on key plan elements.

1.1 History and Background

Plans for managing overflows were completed long before regulations required the documents. Historically, master plans have been completed by municipalities to plan for future sewer services as the service areas expand because of growth. These types of plans typically concentrated on estimating future sanitary sewer flows, or dry weather flows, and applying wet weather peaking factors to account for some infiltration and inflow, both for new and existing sewers. These plans are typically referred to as *master plans*.

As systems age and experience increased infiltration and inflow, planning for wet weather flows becomes more of an issue in master plans. With the advent of sophisticated tools such as geographic information systems and hydraulic models, more detail could be provided on system flows and facilities and that could be a driver for additional analysis into existing and potential future overflows.

Newer collection systems are typically designated as separate sewer systems because they were not designed to convey stormwater runoff. Collection systems in older cities tend to be combined sewer systems (CSSs). Many cities have a combination of CSSs and separate sewer systems, where older parts of the city are still combined whereas newer (or rehabilitated) portions are separated.

The Clean Water Act (CWA) of 1972 established the need to make all U.S. waters fishable and swimmable. One of the intents of CWA was to eliminate untreated combined sewer overflows (CSOs) and sanitary sewer overflows (SSOs) throughout all systems in the United States because this would directly contribute to the main goal of CWA. However, urban stormwater point discharges and nonpoint pollution were two other pollution sources that were not initially regulated or still have not been directly regulated.

Combined sewer overflows were first targeted by the U.S. Environmental Protection Agency (U.S. EPA) due to the significant impacts caused by the volume of runoff discharged to receiving waters in comparison to SSOs. U.S. EPA'S 1994 *combined sewer overflow (CSO) control policy* (CSO policy) established the need for each CSO community to implement nine minimum controls (NMCs) and develop a long-term control plans (LTCPs) for their CSS.

However, separate sewer systems were still only regulated under the general conditions of CWA that explicitly forbid SSOs. Since the implementation of the CSO policy, municipalities and regulators alike have been struggling with how best to regulate SSOs in light of the fact that all SSOs cannot be eliminated in perpetuity. Detailed background and regulatory information on CSS and separate sewer systems can be found in Chapters 1 and 3.

1.2 Similarities and Differences between Combined Sewer Overflows and Sanitary Sewer Overflows

Similarities and differences in the management of SSOs and CSOs are also highlighted throughout this chapter because mitigation measures can be similar, with differences found mainly in regulations.

The following are some physical and regulatory similarities and differences between CSOs and SSOs:

- Physical similarities—both CSOs and SSOs are caused by excessive infiltration and inflow due to rain events. Inflow is typically the largest contributor to CSOs, while either inflow or infiltration can be the driving factor in wet weather SSOs. Both types of overflows can occur during dry weather if capacity is significantly restricted by transient blockages (e.g., roots, grease, sediment, etc.) or equipment failure (e.g., pumping station shutdown, etc.).
- Physical differences—combined sewer overflows are based on historic construction of the system and occur at predetermined structures. Sanitary sewer overflows are typically not designed into the system, but are relief points (manholes) that overflow due to excessive infiltration and inflow and elevated hydraulic grade lines. Sanitary sewer overflows can contain higher levels of bacteria due to a lower amount of dilution from inflow sources. Combined sewer overflows can contain higher levels of other pollutants related to higher flows from inflow sources (e.g., urban runoff from impervious areas).

- Regulatory similarities—proposed capacity, management, operation, and maintenance (CMOM) is similar to NMCs, but the direction for comprehensively managing SSOs is lacking compared to CSO regulations (e.g., development of a CSO LTCP).
- Regulatory differences—combined sewer overflows regulated under a specific structure since 1994, which included NMCs and a LTCP; Sanitary sewer overflows are still considered to be unlawful, even though they can be caused by the same processes as CSOs, because no specific structure for management is included in CWA and recent regulatory modifications have been delayed by the Office of Management and Budget (OMB is part of the Executive Branch, which supervises federal budget and oversees the Administration's procurement, financial management, information, and regulatory policies).

1.3 Management Versus Elimination

It is no coincidence that the terms *manage* and *control* are used when referencing SSOs and CSOs throughout this chapter. This is because overflows will rarely be eliminated from a system even though this has historically been U.S. EPA's intent. There are too many variables that contribute to overflows and simply having the goal of elimination is unrealistic and can hinder the process of creating a cost-effective control plan that meets the needs of the public and the environment. Once this limitation is accepted, a realistic plan can be developed to minimize and mitigate the effects of overflows when they do occur.

2.0 PLANNING METHODS

Three types of planning methods will be discussed in this section. These include traditional planning, collaborative planning, and adaptive management.

2.1 Traditional Facilities Planning

Traditional facilities planning is considered a type of "command and control" planning technique in that a problem is identified, alternatives are evaluated, and a solution is recommended. If this solution is implemented, the problem is solved. However, this planning technique is typically too simplistic to encompass the many challenges that affect management and regulation of overflows.

This type of linear approach is much more applicable to a specific design project where many of the variables have already been constrained to best direct the outcome to a specific product (e.g., a primary clarifier or storage basin). Application of this technique to an entire system-wide plan probably stemmed from the design approach. However, planning, by definition, is much different than design and requires other approaches and methods to provide for uncertainties that are inherent in any sewer system planning effort.

2.2 Collaborative Planning

Planning is the rational, deliberate activity and process that can help in making decisions to solve known problems and assist in developing methods to address and adapt to potential future problems. It is a formal process of evaluating future consequences of our decisions (USACE, 2009). Planning is a step-by-step repetitive process; it is not a one-time effort or project. A planning process is a structured approach to problem solving and can be distilled down to a six-step process commonly used in the water industry. Figure 8.1 illustrates the iterative process of planning (USACE,

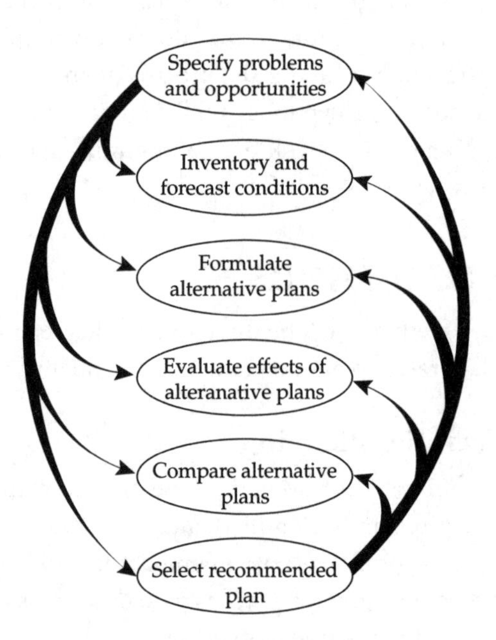

FIGURE 8.1 Iterative planning process (USACE, 1997).

1997). Other useful documentation on planning is provided in U.S. Army Corps of Engineers' (1996) *Planning Manual.*

2.2.1 *Definition*

Collaborative planning differs from traditional planning in that it involves a collaboration of different groups of individuals. While master plans can be developed by a municipal agency or by a consultant and municipal agency, collaborative planning involves a larger audience. Collaborative planning involves a group, team, or partnership of people working and learning together to find ways to address shared problems and opportunities through a joint decision-making process. Decisions made through collaborative planning result in a decision that reflects broader interests than those of the lead agency alone, and typically includes a public involvement program to both inform the public and provide opportunities for the broader public to influence decisions made during the planning study (USACE, 2009).

Collaborative planning builds on the typical planning steps outlined in Figure 8.1 by identifying critical stakeholders up front and providing a process to engage these stakeholders throughout the planning process. Involving those parties that are responsible for developing and implementing the plan (i.e., municipalities and consultants), those that will have to pay for and live with the solutions (i.e., public and environmentalists), and those that have to regulate the plan (i.e., federal, state, and/or local regulators) is a necessary approach for today's complex master plans. The collaborative planning process helps direct when a decision is made so that all critical parties feel a commitment to the proposed action, which can dramatically increase chances of successful implementation.

2.2.2 *Technical Decisions Versus Values*

Many decisions that municipalities make regarding overflow management and mitigation are thought of as technical decisions. However, due to the atmosphere in which these decisions must be made, the resulting decision is actually a choice between competing values. When an agency chooses a plan it is not just deciding which alternative or alternatives are technically feasible (indeed, all of the alternatives must be technically feasible), rather, it is deciding about what values are important in a particular situation at a particular time.

For example, purely technical decisions typically involve just one value dimension (e.g., what size pipe needs to be designed to convey a necessary flow). Technical decisions are governed by professional standards and procedures that are typical for engineering design; in theory, as in the aforementioned pipe-sizing example, multiple engineers should come up with comparable results. These types of decisions are linear in nature and do not involve multiple or compounding uncertainties.

Value choices, on the other hand, involve multiple, non-linear, and often conflicting elements. Valuing may involve choosing between alternative plans that society thinks of as good, such as a low-cost project versus a project that best protects the environment and public health. Both projects may be good, but, for a particular situation, values must be applied to decide which is most important. Therefore, valuing involves deciding how important one good plan is versus another good plan. Valuing is a key concept in applying level-of-service metrics to making decisions. "Level of service" will be discussed in more detail later in this chapter.

When the difference between technical and values choices are confused, stakeholders begin to not only question an agency's values, but its technical abilities as well. When agencies consult with multiple stakeholders on important issues and clarify that decisions are being made on values choices, the technical competence of the agency can actually increase in the public eye. Municipalities still have to make difficult value choices when planning for major capital improvement projects that can be an outcome of a sewer master plan or overflow mitigation plan, but these decisions are prime candidates for collaboration between agencies, and participation by the public, to make the right value decisions for the community.

2.2.3 *Advisory Committees*

The complexity of the plan typically dictates the extent of stakeholder interaction. For example, a construction project (e.g., pipeline replacement) may not require much, if any, collaboration between an agency and the public if it is located in a field where interruptions to commercial establishments and residents are insignificant. However, a master plan typically involves multiple alternatives and multiple stakeholders (including the public, regulatory agencies, and environmentalists), which may have competing values. Therefore, it is in the agencies' best interest to coordinate with these stakeholders early and often throughout the process of constructing and implementing the plan.

The establishment of a citizen's advisory committee (CAC) at the beginning of the planning process can be advantageous to the agency. A CAC typically consists of public volunteers who form a focus group to be educated on the planning process and alternatives that will be developed. The size of the group will depend on the complexity and size of the project, but typically includes between five and

15 members. This group should consist of a cross-section of the public so that the greater public feels it has been represented. The CAC's role becomes an intermediate, unbiased group representing the concerns of the community and offering focused input throughout the planning process. This group provides the public with a voice in the process and provides the agency with an educated group of non-technical and potentially unbiased individuals who understand the value decisions that are being made and disseminate these decisions to the greater public.

A technical advisory committee (TAC) may also be beneficial to certain plans. A TAC can consist of representatives from other municipalities (particularly those that have recently been involved in a master plan of similar size and complexity) and industry experts (including professors and outside consultants). This group offers opinions (hopefully unbiased) to the agency and its consultants regarding the direction of the plan, reviews draft documents, and offers input on technical decisions. The TAC can also be crucial to providing a CAC (and the public at large) with unbiased information on effectiveness and potential costs of alternatives, thus lending additional credence to the final plan.

2.2.4 *Process and Tools*

Figure 8.2 illustrates significant components that make up the collaborative planning process. The collaborative planning process encompasses the traditional planning process and provides support at critical steps. The collaborative planning process never truly ends once a planning document is complete because the plan itself must be implemented and managed to ultimately provide the results that have been decided on. Adaptive management, discussed in more detail in the following section, provides this mechanism.

2.3 Adaptive Management

Adaptive management began as a resource management technique in the 1970s and was applied to many large-scale projects by agencies such as U.S. EPA, National Oceanic and Atmospheric Association, U.S. Bureau of Reclamation, U.S. Department of the Interior, and the British Columbia Forest Service. Freedman et al. (2004) describe a summary of these cases and how adaptive management can be applied to watershed management specifically through the total maximum daily load process. Freedman et al. (2007) and Nemura et al. (2008) applied adaptive management techniques to a large collection system in Northern Kentucky (Sanitation District No. 1).

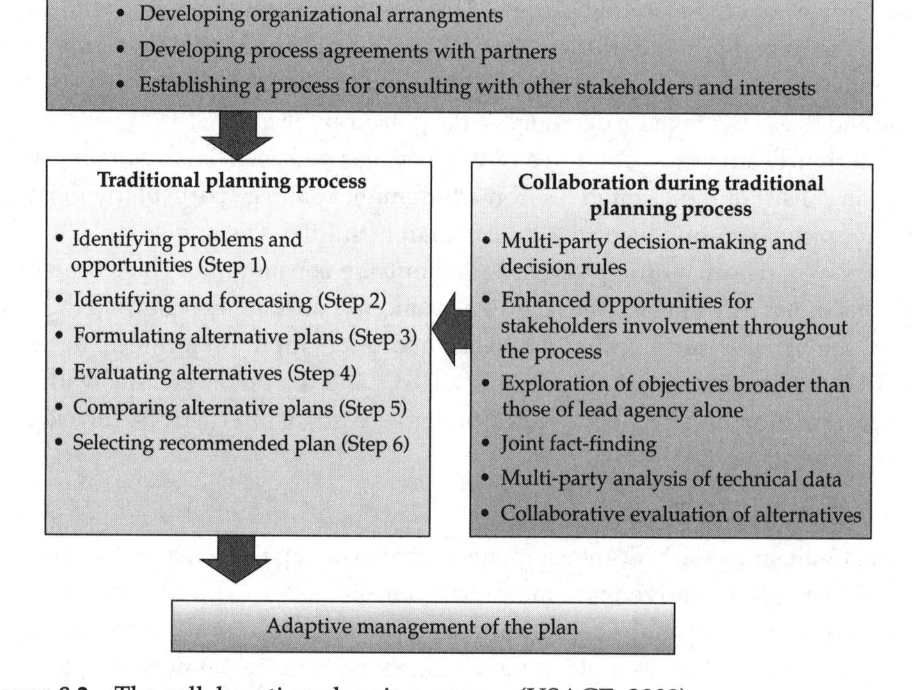

FIGURE 8.2 The collaborative planning process (USACE, 2009).

A form of adaptive management has been integrated into CSO guidelines and will be discussed further in the following section.

2.3.1 *Definition*

Adaptive management can be defined in many ways. However, the definition provided by *Adaptive Management: The U.S. Department of the Interior Technical Guide* (Williams et al., 2007) is most applicable to this discussion. An adaptive approach involves exploring alternative ways to meet management objectives, predicting the outcomes of alternatives based on the current state of knowledge, implementing one or more of these alternatives, monitoring to learn about the impacts of management actions, and then using the results to update knowledge and adjust management actions. Adaptive management focuses on learning and adapting through partnerships of stakeholders who learn together to create and maintain sustainable resource systems (Williams et al., 2007).

Adaptive management also recognizes that solutions targeted at solving one problem may have unanticipated, undesirable effects and, therefore, allows for further testing before proceeding with additional controls (Freedman et al., 2004). Stankey et al. (2005) reference adaptive management as a continuous four-step process of plan, act, monitor, and evaluate. Figure 8.3 illustrates this process, which is further described in literature by Stankey et al. (2005). This process has been updated by Williams et al. (2007), as discussed in the following section.

2.3.2 *Steps in Adaptive Management*

Williams et al. (2007) propose that adaptive management be implemented in a nine-step process. These nine steps include

- *Stakeholder involvement*—ensure stakeholder commitment to adaptively manage the enterprise for its duration;
- *Objectives*—identify clear, measurable, and agreed-upon management objectives to guide decision-making and evaluate management effectiveness over time;

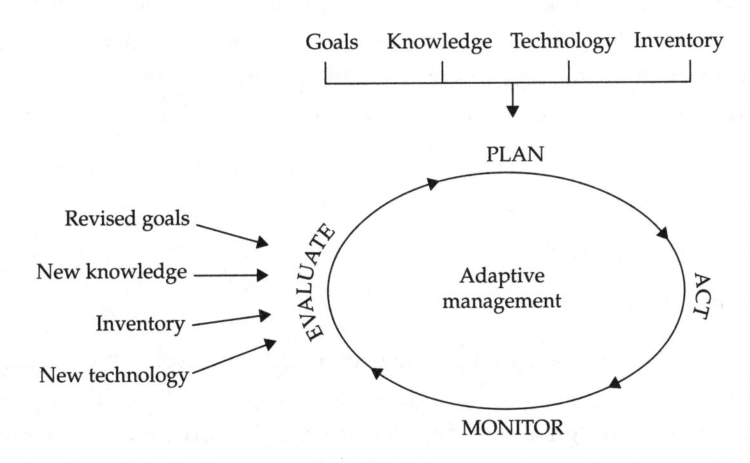

FIGURE 8.3 Adaptive management process (Reynolds, 2005).

- *Management actions*—identify a set of potential management actions (and management responsibilities) for decision-making;
- *Models*—identify models that can characterize different mechanisms about how the system works;
- *Monitoring plans*—design and implement a monitoring plan to track resource status and other key resource attributes;
- *Decision-making*—select management actions based on management objectives, resource conditions, and understanding;
- *Assessment*—improve understanding of resource dynamics by comparing predicted and observed changes in resource status; and
- *Iteration*—cycle back to monitoring plans.

Many of these steps, either purposefully or unknowingly, have been integrated into U.S. EPA's CSO LTCP guidance. More discussion on the steps designated in the CSO LTCP, and how they are similar to the aforementioned steps defined for adaptive management, are covered in detail later in this chapter. The complex process of overflow mitigation, due to the extensive uncertainties and risks involved, requires an adaptive approach.

2.4 Planning Tools

Planning does not require software, but can be greatly enhanced with computer tools that allow for a transparent, repeatable, graphical documentation of the planning process. There are many computer programs available to assist in planning; these tools can be generally classified in the following groups (U.S. Army Corps of Engineers, 2009):

- Problem definition tools,
- Analytical tools, and
- Synthesis tools.

Figure 8.4 illustrates tools that can be used in the collaborative planning process. Problem definition tools allow stakeholders and planners to identify problems and opportunities through a well-defined process that documents the process and defines the reasons for specific decisions. These types of tools may include diagrammatic representations that allow stakeholders to more easily follow the process from

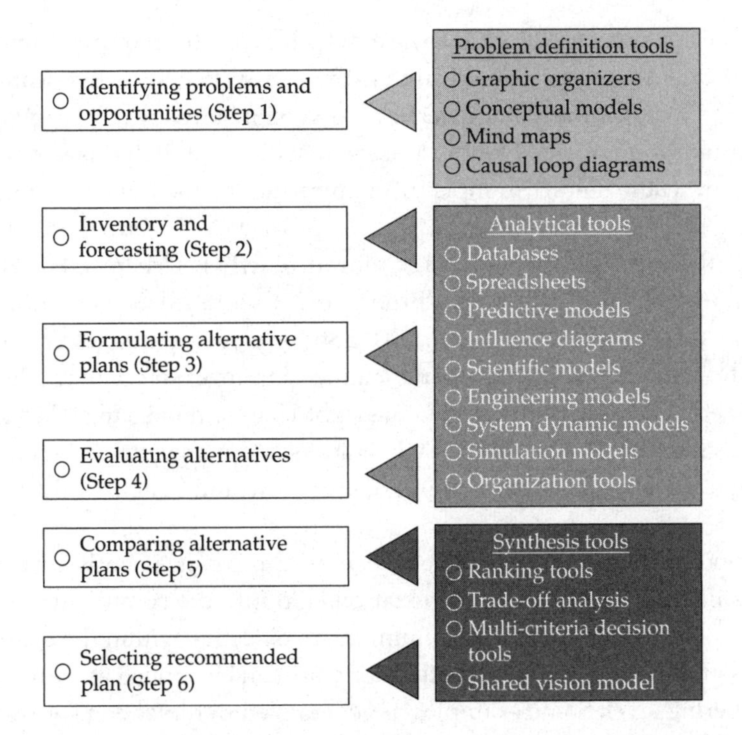

FIGURE 8.4 Tools for collaborative planning (adapted from USACE [2009]).

beginning to end. Common diagrams include mind maps (brainstorming), fishbone, Web, cerebral chart, and causal loop diagrams. Example software packages include Cayra (Management Intalev, http://cayra.net) and MindMapper (SimTech Systems, Inc., Lewisville, Texas). Stella (isee systems, inc., Lebanon, New Hampshire) and Powersim (Powersim Software AS, Norway) are considered both causal loop tools and analytical tools.

Analytic tools typically are thought of as simulation software for evaluating alternatives of a planning study. Simulation software can be applied at a strategic planning level and at a highly detailed engineering level within the planning process. Application at the strategic level typically involves the need to apply this type of software on a large scale that uses significant simplification of many connected processes. Engineering-level tools include detailed, compartmentalized software, such as hydraulic models, that allow detailed analysis of the existing system and exploration of future alternatives.

Examples of strategic-level software, which typically requires some customization (but can also require significant customization), include Extensim (Imagine That Inc., San Jose, California), Powersim (Powersim Software AS, Norway), Matlab and Simulink (MathWorks, Natick, Massachusetts), and Stella (isee systems, inc., Lebanon, New Hampshire). Examples of engineering-level software packages include U.S. EPA's *Storm Water Management Model User's Manual, Version 5.0* (U.S. EPA, 2010), InfoWorks (MWH Soft, Broomfield, Colorado), MIKE Urban (DHI, Horshoom, Denmark), InfoSWMM (MWH Soft, Broomfield, Colorado), Sewer GEMS (Bentley, Exton, Pennsylvania), and XPSWMM (XP Software, Inc., Portland, Oregon). When appropriate, linking strategic-level and engineering-level software can leverage the unique capabilities of each to provide the right level of details to stakeholders during the planning cycle (e.g., outside stakeholders and management would benefit at the strategic level, while engineers and operators would benefit at the engineering level).

Another analytic tool, which could also be applied as a synthesis tool, is optimization software. The goal of optimization is to find the combination of decision variables that either maximizes or minimizes an objective within the confines of the constraints of the problem. Optimization software can be applied at both the strategic and engineering level. Some examples of generalized optimization software include Solver (Frontline Systems Inc., Incline Village, Nevada), RiskOptimizer (Palisade, Ithaca, New York), and Tabu Search (SolveIT Software, Adelaide, Australia). Optimization software also integrates risk analysis into the optimization algorithm so that risk can be explicitly defined within the context of optimization solutions (e.g., RiskOptimizer). Some engineering-level software has started to build in optimization routines (e.g., generic algorithms) to provide search capabilities for optimal sizing of new facilities.

As its name implies, synthesis tools can help to synthesize the multiple alternatives that are typically developed during the planning process. These types of tools are generally applied when goal achievement and muli-criteria decision analysis are warranted. These tools, also known as *interpretive tools*, are used to define specific performance measures for each objective and/or alternative and to quantify the level of achievement. Some examples of this type of software include Criterion Decision Plus (InfoHarvest, Seattle, Washington), and DPL (Syncopation, Concord, Massachusetts). These types of tools provide the benefits of explicitly defining objectives and criteria for each alternative, documentation of qualitative and quantitative data, and the ability to analyze many alternatives in an efficient manner.

3.0 DEVELOPING A PLANNING APPROACH

Whether a municipality is responsible for a separate sewer system, a CSS, or both, the planning approach is similar. As previously discussed, national, state, and local regulations may dictate the steps that must be completed depending on the type of system. However, there are many more similarities between planning for SSOs and CSOs than differences. Therefore, this section discusses the development of a planning approach by first concentrating on the similarities and then discussing differences as they are applicable to separate sewer systems and CSSs.

3.1 Elements of an Overflow Management Plan

There are many steps, or elements, to developing a comprehensive overflow management plan, whether the plan will address SSOs or CSOs. Depending on the needs of an agency, the elements presented in this section may not all be applicable, but are included so that specific elements can be chosen to best fit the needs of the plan.

Current U.S. EPA CSO policy provides specific elements that should be included in an overflow management plan. In contrast, the courts have provided no definitive direction on SSO legal issues and, as a result, U.S. EPA has delayed development of national SSO regulations. As previously mentioned, wet weather SSOs, like CSOs, may not be able to be completely stopped from occurring during significant rainfall events due to the reality of the expansive (and expensive) buried infrastructure and infiltration and inflow. As such, according to the CWA, wet weather SSOs are still unlawful and must be eliminated. Because elimination of all wet weather SSOs in a system (especially very large systems) may not be practical, this chapter provides wet weather overflow mitigation planning elements that apply to both types of systems.

Figure 8.5 illustrates four categories of elements in an overflow management plan, which includes stakeholder elements, regulatory elements, technical elements, and financial elements. This figure is based on the long-term CSO planning approach (Exhibit 1-2) in *Combined Sewer Overflows: Guidance for Long-Term Control Plans* (U.S. EPA, 1995b), with several modifications. Moreover, this figure is broadened to not be specific to CSO planning, but to also include many of the similar elements that are needed for an SSO plan.

The technical elements in Figure 8.5 are discussed in more detail than the other elements because they tend to be the focus of an overflow mitigation plan. However, the other three ancillary elements (i.e., stakeholder, regulatory, and financial elements) are also discussed as they relate to the technical approach. Because these three

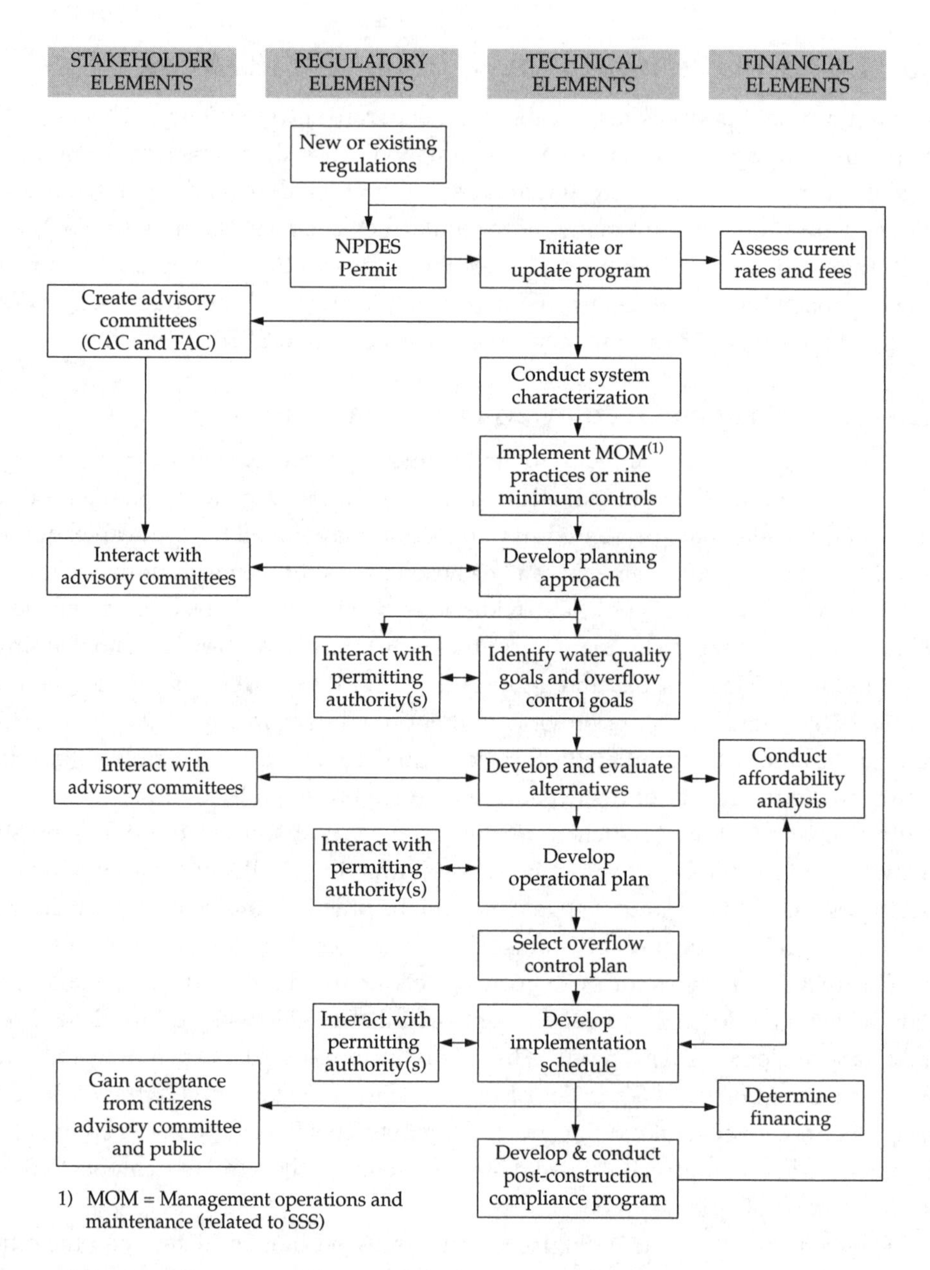

FIGURE 8.5 Overflow management planning approach.

ancillary elements are not the focus of this section, the reader should reference other documents to gain a thorough understanding of the complexities of these elements. References are provided, as available, to direct the reader to other literature on these elements.

3.2 Regulatory Elements

Regulations tend to drive the need for an overflow mitigation plan. Agencies in charge of CSS and separate sewer systems may have had some type of overflow mitigation plan in effect before 1972; however, CWA provided the driving mechanism to start the process of codifying a planning approach. Regulations are discussed in detail in Chapter 3 of this manual; as such, this section will concentrate on how current regulations drive the planning process.

3.2.1 *Regulatory Drivers*

Municipalities are always subject to new and/or existing regulations as they apply to overflow management. These regulations tend to drive the process by which an agency must ultimately comply with CWA and initiate a plan. Generally, an agency is pushed to initiate an overflow control plan due to existing or new regulations. Initiation or updating of a plan may be triggered by either a regulatory agency (e.g., U.S. EPA, state, or local regulators) or a third-party lawsuit. In the case of a third-party lawsuit, the process is simply expedited by a third party (e.g., local water group), which brings regulators into the process on a quicker schedule.

The basis for regulating the agency is the National Pollutant Discharge Elimination System (NPDES) permit. The NPDES permit is the basis by which a plan is developed and executed. Whatever the outcome of the plan, it must be reflected in the agencies' permit because the permit provides the legal basis by which the agency will be judged to be in or out of compliance.

3.2.2 *Compliance Planning*

Two methods that are used to bring a permittee into compliance include an administrative order and a consent decree. An administrative order or consent decree is rarely the last step in the process. Completion of an administrative order can lead to a subsequent order or decree that incorporates findings and plans from the just-completed administrative order or consent decree. A wet weather decree is the program that stipulates sewer overflow reductions, achievement of water quality standards (WQS), infiltration and inflow reduction, and what it means to reach ultimate "compliance" with the terms of decree (AMSA, 2003). Two typical approaches exist for a municipality to take when entering into a consent decree.

One approach is to enter a decree that orders the permittee to conduct studies and planning to determine elements of the compliance program. These elements are then approved by the decree parties and the court, and the permittee is left to implement the program. This approach provides the municipality with flexibility in defining the plan, but can also be too open ended, which will impact overall program costs and schedules.

Another approach incorporates full details of the required improvements in the decree. If the municipality has completed system evaluations and alternative analyses before the degree process is initiated, then specific capital project improvement descriptions can be included in the degree. This gives the permittee the advantage of having realistic costs and schedules to implement the decree.

As described in the *Wet Weather Consent Decrees: Protecting POTWs in Negotiations—An AMSA Handbook* (AMSA, 2003), it likely will be in the permittees' best interest to make compliance standard provisions as specific and narrow as possible to minimize the likelihood of diverse interpretations about the compliance goal. Broad compliance standards such as "... achievement of all applicable water quality standards (WQS)" or "... full compliance with the CWA" should be avoided because these standards are difficult to measure and depend on the nature of the regulatory authorities at the time of compliance measurement (which may not be the same regulators that helped develop the permit) (AMSA, 2003).

Further information on initial elements that drive the wet weather overflow planning process, such as NPDES permits and consent decrees, can be found in *Combined Sewer Overflows: Guidance for Permit Writers* (U.S. EPA, 1995d) and the aforementioned handbook from the Association of Metropolitan Sewerage Agencies (2003). Further information on measuring plan compliance will be discussed in Section 4.0 of this chapter.

3.3 Technical Elements

The goal of an overflow management plan is to bring a municipality into compliance with their permit, order, and/or decree. To achieve this goal, many technical elements must be addressed. The CSO policy is detailed in previous chapters, but, for a discussion on planning elements, reference to the NMCs and the significant elements of a LTCP helps to structure what should be included as the technical elements in an overflow mitigation plan (whether for CSOs or SSOs).

Figure 8.5 lists nine technical elements after initiation or updating of the overflow control program. These nine elements provide a path to follow to answer the following questions:

- What is the status of the current system?
- What immediate non-capital changes can be made to improve the performance of the system (e.g., management, operation, and maintenance [MOM])?
- What capital elements must be completed to bring the system into compliance after non-capital programs are initiated (e.g., capital improvement plan [CIP])?
- How much does the CIP cost and how will it be paid for (e.g., financing)?
- How quickly does the CIP need to be implemented (e.g., schedule), which also affects annual sewer rates and affordability? and
- How and when will compliance be measured?

These questions and the nine elements listed in Figure 8.5 will be discussed by addressing non-capital, capital, and compliance planning elements.

3.3.1 *Non-Capital Planning*

Non-capital elements typically consist of identifying management, operation, and maintenance techniques that can be immediately implemented to maximize existing facility capacity and available manpower to minimize existing overflows and to better meet regulations. The NMCs are an example of non-capital planning for CSO communities, and are required under CSO policy. Similar programs for separate sewer systems have been advanced such as MOM activities. Management, operation, and maintenance activities were initially pioneered by U.S. EPA Region 4, and were later expanded on by other states such as California's current Sanitary Sewer Management Plan.

Capacity, management, operation, and maintenance may be typically thought of as MOM, but the capacity "part" is really a capital-planning element and, as the CSO policy did, leaves capital improvements to a longer-term plan (i.e., an LTCP) and separates out the MOM elements into the NMCs. U.S. EPA should consider this same approach for SSOs because capital projects must be planned for on a longer timeline and typically incur much larger costs than MOM activities. Therefore, these two types of planning should be separated accordingly.

Major elements of non-capital planning include

- Proper operation and regular maintenance programs for sewer infrastructure (including pipelines, pumping stations, and appurtenances);
- Maximum use of existing facilities and maximizing flow to the treatment plant (full conveyance capacity and in-system storage, if possible);
- Monitoring of overflow activity (e.g., location, volume, duration, cause [e.g., wet weather]); and
- Public notification of overflows.

Further information on these types of activities can be found in *Combined Sewer Overflows: Guidance for Nine Minimum Controls* (U.S. EPA, 1995c), NPDES SSO and peak flows documents (visit http://cfpub.epa.gov/npdes/whatsnew.cfm?program_id=4), and California's sewer system management plan (visit http://www.waterboards.ca.gov/water_issues/programs/sso/index.shtml#plan; see also Dent et al. [2007] for a summary). Completion and constant improvement of these elements can make a difference in overflow occurrences, but are almost always not enough to solve systemic overflows over a variety of wet weather events. Therefore, capital improvements become the next necessary element of a plan.

3.3.2 *Capital Planning*

Capital planning, or a CIP, is the heart of a wet weather overflow mitigation planning effort and can be a component of a dry weather overflow mitigation effort. The basic purpose of a CIP is to provide an analysis of alternatives and a logical selection of a preferred alternative that includes costs, benefits, and an implementation schedule. The implementation schedule typically includes phasing of projects to make sure the most critical ones are completed first and to provide a mechanism to manage cash flow so that rates are not adversely affected in any one year. Further description of financial elements is provided in later sections of this chapter.

A CIP typically begins with the identification of water quality goals and overflow control goals. Water quality goals are usually already defined by the designated uses of the waterbodies that may be affected by overflows. For CSO planning, the presumptive or demonstrative approach (discussed in previous chapters) can be used to define overflow control goals. For SSO planning (and sometimes for CSO planning), a design event is typically defined as an overflow control goal (e.g., all SSOs up to, and including, the 5-year overflow event will be controlled). Selecting a design event for overflow control is summarized in literature by Dent et al. (2002). The risks of not meeting expected goals are also critical to define at this stage. Water Environment Federation's *Guide to Managing Peak Wet Weather Flows in Municipal Wastewater Collection and Treatment Systems* (WEF, 2006) provides extensive information on applying a risk-based, level-of-service approach to wet weather flow management.

Development and evaluation of alternatives is the heart of a CIP. Alternatives for a comprehensive overflow mitigation plan must consider the systemic reasons for wet weather overflows and examine not only the collection system, but also the treatment plant. This system-wide analysis may be complicated by the fact that one agency may own and operate the collection system, while another owns and operates the treatment plant. This is especially complicated for large systems where multiple satellite collection systems discharge to a regional system and treatment plant. A National Association of Clean Water Agencies' (2009) white paper provides several examples of satellite collection systems and how agencies across the country are dealing with these specific issues.

There are many overflow mitigation technologies that have been described in detail in previous chapters of this manual. For planning purposes, these wet weather mitigation technologies can be grouped into the following four strategies: infiltration and inflow reduction, increased conveyance, storage, or treatment. Most plans typically include a combination of these strategies to develop a variety of alternatives that meet plan objectives while providing maximum benefits for the lowest cost. Determining the right mix of these strategies to best control wet weather overflows can be difficult.

A "knee-of-the-curve" analysis can be used to predict where a cost-effective solution could exist for a mix of alternatives. This type of analysis typically plots the costs of alternatives against different design events (or level of service). The knee-of-the-curve is said to be where the cost curve inflects, typically in the vicinity of a 5-year design event or less, where costs tend to start increasing at a much higher rate after this point. This type of analysis has also historically been used to find a minimum cost point in a cost curve when analyzing rehabilitation costs versus transport and treatment costs. Dent et al. (2003) and Wright et al. (2003) provide discussions on optimization techniques that can be used to help define the appropriate mix of strategies for an SSO program in California. Varghese et al. (2008) provide a similar approach for a separate sewer system in Kansas.

An operation plan is often an overlooked element of a CIP, but is critical to providing staff and regulators a means for judging if the newly required facilities are able to meet plan objectives. Extensive new facilities will incur more labor and costs to properly operate. Wet weather operational plans should include policy decisions that are codified to not only provide clear guidance to operators during wet weather events, but also protect the agency if questions arise after an overflow occurs.

The last technical element of a CIP is the development of an implementation schedule. The purpose of an implementation schedule for a wet weather CIP is to detail the order of project completion to meet regulatory milestones while leveling-out expenditures over the period of implementation. Figure 8.6 illustrates an example schedule for a CSO LTCP. The schedule must be designed to allow critical projects to proceed as quickly as possible without overburdening the agency in any one year. An implantation schedule is also typically written into the NPDES permit, which becomes a critical compliance metric for the regulatory agency. Further information on how schedule and costs affect an overflow mitigation plan is included in the following section.

3.4 Financial Elements

The financial elements of an overflow mitigation plan include program cost (in total and as annual expenditures), affordability issues, and financing options.

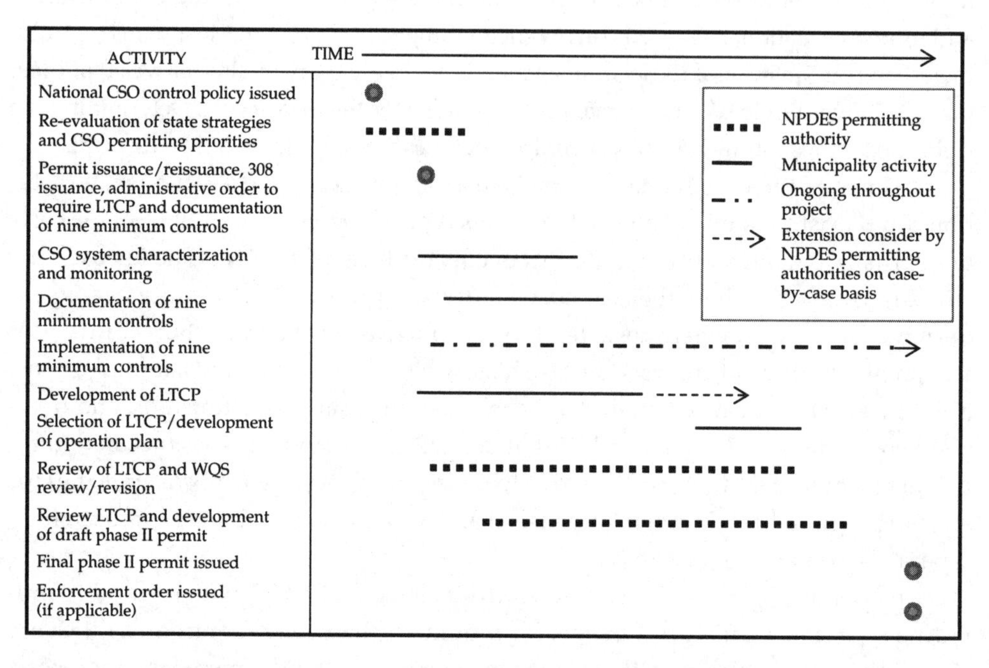

FIGURE 8.6 Example schedule for a CSO LTCP (U.S. EPA, 1995b).

3.4.1 *Program Costs*

Overall program costs include the costs of developing the plan, capital costs for new facilities, and annual operation and maintenance (O&M) costs for the new facilities. Plan development costs may include engineering costs for developing the overflow mitigation plan, field services costs (e.g., flow metering and surveying), and legal costs (e.g., permit negotiations). Capital costs may include engineering costs for design and construction oversight, construction costs, legal and administrative costs, and a contingency.

Typically, a contingency is included to cover the many unknowns that occur during construction. Contingencies can be developed as a percentage of construction costs and combined with construction costs and engineering, legal, and administrative (ELA) costs. Sometimes an additional contingency may be added to ELA costs when the program is very large and significant uncertainties may be involved in completing the plan. Annual O&M costs typically include estimates for labor, utilities, chemicals, spare parts, and other items needed to operate and maintain facilities over the program life. Capital costs plus O&M costs projected over the design life of the project are referred to as *life-cycle costs*. Further discussion of costs can be found in U.S. EPA's *Combined Sewer Overflows: Guidance for Long-Term Control Plans* (1995b).

Cost estimates are typically prepared at several points during project planning and design. The expected level of accuracy is directly proportional to the degree of engineering effort applied and the accuracy of the known details. The American Association of Cost Engineers (AACE) recommends four levels of accuracy in estimating construction costs. The AACE-recommended cost categories are as follows (AACE, 2005):

- Category 1: conceptual estimate,
- Category 2: study estimate,
- Category 3: preliminary estimate, and
- Category 4: detailed estimate.

The accuracy of construction cost estimates will increase as the project moves from planning through the final design estimate prepared prior to bidding for construction. As can be expected, planning estimates will have a wide range of accuracy relative to the construction contract value due to the lack of design features and details available early in a project. Table 8.1 summarizes the anticipated cost-estimating accuracy for each of the four major categories.

Table 8.1 Summary of cost-estimating accuracy (adapted from AACE [2005]).

Category	Project level	Expected accuracy based on estimated cost
1. Conceptual	Prestudy estimates and screening of alternatives	+50% to –30%
2. Study estimate	Pipe-routing studies and master plans	+30% to –20%
3. Preliminary estimate	Predesign report	+20% to –10%
4. Detailed estimate	Completed plans/specifications	+15% to –5%

3.4.2 *Affordability Issues*

Affordability, or the financial capability of a municipality to pay for overflow mitigation expenditures, is detailed in the U.S. EPA's *Combined Sewer Overflows: Guidance for Financial Capability Assessment and Schedule Development* (U.S. EPA, 1997). The initial focus of this guidance document is on the annual percentage of median household income (MHI) dedicated to sewer bills, referred to as the *residential indicator.* This measure serves as an overall indicator of community purchasing capacity and the burden imposed on households as a result of increased sewer costs. Figure 8.7 illustrates two example graphs showing how affordability by MHI and annual sewer bills can be reported.

The guidance document classifies the impact of the household bill on MHI into three categories: less than 1%, between 1 and 2%, and above 2%. Anything below 1% is considered to be affordable. The range between 1 and 2% is considered to be moderately impactful and subject to judgment. Anything above 2% is considered to be significantly impactful.

An affordability assessment is typically performed in two phases. Phase one establishes the severity of the impact. Phase two evaluates how widespread the effect of the CIP will be by using a variety of debt, socioeconomic, and financial management indicators. Phase one impacts are referred to as "substantial", and phase two impacts are referred to as "widespread". Further detail can be found in *Combined Sewer Overflows: Guidance for Financial Capability Assessment and Schedule Development* (U.S. EPA, 1997).

3.4.3 *Financing Options*

Options available to a municipality for funding necessary capital improvements directly affect the affordability of the program. There are a variety of funding options

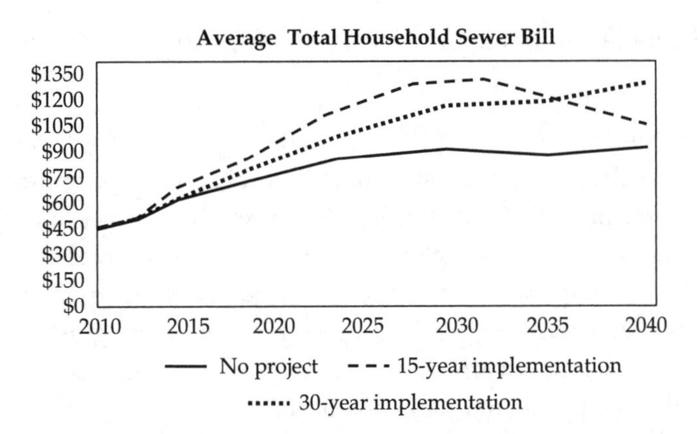

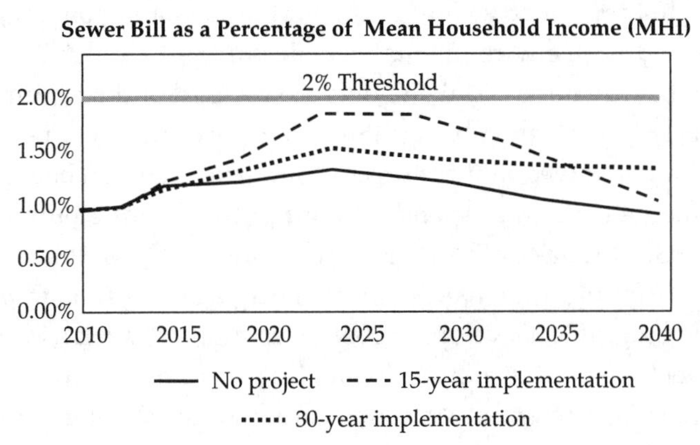

FIGURE 8.7 Example affordability graphs.

available, although many state programs such as a state revolving fund seem to have decreased in resources over the last few years. Further details on financing options will not be detailed here, but can be found in the U.S. EPA's *Combined Sewer Overflows: Guidance for Funding Options* (1995a).

3.5 Stakeholder Elements

An important element that should not be overlooked during development of an overflow mitigation plan is the inclusion of stakeholders at key points within the planning process. Stakeholders are those individuals or groups that have a direct stake in the outcome of the program. Stakeholders can include the general public, environmental groups, regulators (federal, state, and local), municipal personnel, satellite agencies, and so on.

3.5.1 *Stakeholders*

The public is typically the primary stakeholder in this type of program because they are the largest contributor of funds (through taxes, service charges, etc.) that will ultimately pay for the program. As such, the sooner they can be involved in the process the better understanding they will have of the need and complexities of the project, which should translate into a higher level of support for the program. The public will be most concerned that the overflow mitigation program is not too costly while still providing public health and environmental benefits.

Other stakeholders, such as environmental groups and some regulators, may be more concerned with environmental benefits and put a much lower priority on costs. This may be true if these stakeholders are not part of the community that will be directly affected by potentially substantial increases in sewer rates. When large sums of money are involved in these types of projects, there can be many conflicting opinions on what is enough to spend without putting undue burden on the public. Although U.S. EPA has defined a general guide of 2% MHI as previously discussed, this does not preclude regulators or third-party lawsuits to push for much more expensive programs. Developing specific committees such as CACs and TACs can be extremely beneficial to a municipality by providing outside (and potentially unbiased) support for a program. As stated earlier in this chapter, the choice of plan comes down to value-based decisions, not technical decisions.

3.5.2 *Citizens Advisory Committee*

A CAC is typically made up of members of the community that have an interest in the program and have time to devote to gain a more in-depth understanding of the issues than the typical rate payer. The purpose of a CAC is to provide a focus group of individuals who can be more deeply educated about the complexities of the project, which would not be possible to provide to the public at large. The CAC should be made up of individuals that represent a cross-section of the public and should be limited to approximately 10 individuals.

The CAC's responsibility should include interaction with the municipality to help develop the values of the community toward the program and provide an intermediary between the municipality and the public, especially at public meetings. Because the CAC will be more deeply involved with the program, it can provide a buffer between the municipality and the public when contentious issues arise. If the CAC is truly brought in as a stakeholder in the process, and agrees with the direction of the program the municipality is developing, then the CAC can present at public meetings and lend more confidence to the general public that their concerns have been addressed. This can take some of the burden off the municipality's management when pressed by board members or the public.

3.5.3 *Technical Advisory Committee*

The purpose of a TAC is to provide unbiased input on the technical validity of the plan as it is being developed. A TAC can consist of outside consultants and academics that have extensive experience in these types of projects. A TAC provides quality control and quality assurance and, if made up of individuals with national or international experience, can bring new ideas to the overall program. Technical advisory committee members should be invited to initial strategy workshops to review important technical documents. The use of a TAC, like the use of a CAC, lends more confidence to the other stakeholders that the municipality is providing high-quality analysis to the process.

As stated earlier in this chapter, when the difference between technical and values choices are confused, stakeholders begin to not only question an agency's values, but its technical abilities as well. When agencies consult with multiple stakeholders on important issues and clarify that decisions are being made on values choices, the technical competence of the agency can actually increase in the public eye.

3.6 Integration with Other Programs

A wet weather overflow mitigation program should also be integrated with other programs that the municipality has, or will be, implementing. Two examples of programs that have direct ties to the overflow mitigation program include asset management and watershed management programs. The scope of these programs represent a much bigger issue than can be adequately covered in this chapter, but should be investigated as part of an overflow mitigation program. Nemura et al. (2008) and Dent et al. (2001) describe SSO management as part of a watershed program.

3.7 Example Plan Table of Contents

Table 8.2 summarizes an example of the elements that may be included in overflow mitigation program as a table of contents for a planning report. Not all

TABLE 8.2 Example plan table of contents.

Chapter	Subsection
1 – Introduction	
	1.1 Purpose
	1.2 Scope and Authorization
	1.3 Plan Organization
	1.4 History of Service Area
	1.5 Authority and Management
	1.6 Related Planning Studies
2 – Basis for Planning	
	2.1 Introduction
	2.2 Service Area Characteristics
	2.3 Neighboring Jurisdictions (Satellite Communities)
	2.4 Physical Features of Service Area
	2.5 Service Area Goals and Policies
	2.6 Rules and Regulations
3 – Demographic Analysis	
	3.1 Introduction
	3.2 Population and Employment (existing and future)
	3.3 Land use (existing and future)
4 – Flow and Load Projections	
	4.1 Introduction
	4.2 Planning Basis
	4.3 Historic and Existing Flows
	4.4 Historic and Existing Wastewater Loads
	4.5 Flow and Load Projections
5 – Collection System Inventory	
	5.1 Introduction
	5.2 System Description
	5.3 Historic and Existing Deficiencies
	5.4 Operations and Maintenance

Table 8.2 (Continued)

Chapter	Subsection
6 – Collection System Analysis	
	6.1 Introduction
	6.2 Modeling Approach
	6.3 Performance Criteria
	6.4 Peak Flow Modeling
	6.5 Long Term Simulation Modeling
	6.6 Peak Flow Statistics
	6.7 Historic Peak Flow Management Approach
	6.8 Future Peak Flow Management Alternatives
7 – Recommended Collection System Improvements	
	7.1 Introduction
	7.2 Recommended Collection System Improvements
	7.3 Peak Flow Management Alternatives
8 – Existing Treatment Facility	
	8.1 Introduction
	8.2 Design Criteria
	8.3 Plant Components
	8.4 Solids Production and Management
	8.5 Pretreatment Program
	8.6 Operations and Maintenance
9 – Treatment Plant Analysis	
	9.1 Introduction
	9.2 National Pollutant Discharge Elimination System Permit Requirements
	9.3 Existing Plant Performance
	9.4 Reliability Requirements
	9.5 Capacity Evaluation and Rating

(*continued*)

TABLE 8.2 (Continued)

Chapter	Subsection
10 – Recommended Treatment Plant Improvements	
	10.1 Introduction
	10.2 Basis of Planning
	10.3 Upgrades Required for Maximum Month Conditions
	10.4 Upgrades Required for Peak Flows
11 – Capital Improvements Plan	
	11.1 Introduction
	11.2 Overall CIP Project Summary
	11.3 Basis of Estimated Costs
	11.4 Project Phasing Plan
	11.5 Capital Improvement Plan
12 – Financial Plan	
	12.1 Introduction
	12.2 Past Financial Performance
	12.3 Financial Plan
	12.4 Available Funding Assistance and Financing Resources
	12.5 Financial Forecast
	12.6 Current and Projected Rates
	12.7 Affordability Analysis
	12.8 System Development Charges
	12.9 Conclusions and Recommendations

elements may be necessary depending on the size and complexity of the plan. However, the information in this table should provide an overall guide for developing a plan write-up. Another document that provides guidance on developing a LTCP report for small communities can be found in *The Long-Term Control Plan-EZ (LTCP-EZ) Template: A Planning Tool for CSO Control in Small Communities* (U.S. EPA, 2007).

4.0 PLAN IMPLEMENTATION

Once an overflow mitigation plan is developed and accepted by stakeholders the next step is to begin implementation to meet the goals of the program. The schedule developed as part of the plan becomes a prime tool for management of all capital projects. Metrics are also applied as the plan is implemented to judge whether the selected alternative is performing as expected. A post-construction compliance program should, or may be required to, be completed as the program is implemented.

4.1 Schedule

The implementation schedule typically includes phasing of projects to make sure the most critical ones are completed first and providing a mechanism to manage cash flow so that rates are not adversely affected in any one year. Key milestones within the implementation schedule are typically included in the NPDES permit. A typical NPDES permitting cycle is every 5 years, which provides the regulating agency with a means of checking the progress of the program. Many times, implementation schedules must contain some flexibility because pilot-scale projects may be necessary.

4.1.1 Schedule Impacts of Pilot-Scale Projects

Pilot-scale projects may be necessary within the overflow mitigation program to reduce uncertainties and provide more system-specific information for greater application of a technology throughout a system. A pilot-scale project might be needed to verify infiltration and inflow reduction in a separate sewer system. Similarly, a sewer separation pilot might be needed to prove that water quality of the receiving water is not further degraded if the stormwater portion in a CSS is routed to a separate storm sewer and, therefore, does not receive treatment. Other relatively new treatment technologies may also need to be piloted to prove they work as assumed during wet weather conditions (e.g., ballasted flocculation).

4.1.2 Flexibility and Adaptation

A complex overflow mitigation program schedule must contain flexibility when projects are completed, especially pilot-scale projects, so adjustment can be made if specific assumptions made during the planning process are not realized. For example, an infiltration and inflow reduction-effectiveness pilot program may be initiated to identify how effective infiltration and inflow removal techniques will be within a select basin. If significant infiltration and inflow reduction can be achieved in one basin, and the techniques are repeatable throughout other basins, then an assumption such as 30% reduction in infiltration and inflow can be applied across the system at a given cost. If this reduction factor corresponds with the assumptions made during the plan development (or exceeds the assumed reduction), then no further action is required. However, if it does not meet an assumed reduction, then further rehabilitation projects may need to be shifted to other projects to achieve a designated level of wet weather reduction (e.g., storage). A discussion of this type of pilot program is included in literature by Dent et al. (2004) and Lukas (2007).

4.2 Metrics for Measuring Success

Measures of success as defined in the CSO LTCP guidance document (U.S. EPA, 1995b) are objective, measurable, and quantifiable indicators (or metrics) that illustrate trends over time. These metrics can be grouped in four categories: administrative, point of overflow, receiving water, and human health measures. The following measures present examples of these metrics:

- Administrative measures:
  - Capital improvement plan program cost normalized on a per-person or household basis,
  - Sewer rates per person or normalized to as a percentage of MHI, and
  - Controlling overflows up to a certain design-level event.
- Point-of-overflow measures:
  - Reduction peak flow and volume of wet weather overflows;
  - Reduction in number and frequency of wet weather overflows (e.g., no more than four overflows per year);
  - Reduction in pollutant loadings (conventional and toxics); and
  - Reduction of wet weather overflows in sensitive areas (e.g., areas of human contact or sensitive ecologic areas).
- Receiving water measures:
  - Reduced in-stream concentration of pollutants after rainfall events;
  - Attainment of narrative or numeric water quality criteria;
  - Additional river miles (or estuary area) in compliance;

- Return of once-present species, restored habitat, and improved biodiversity indices; and
- Improved or expanded shellfish harvesting.
- Human health measures:
  - Reduced flooding and drainage problems;
  - Reduced beach closures;
  - Reduced contact recreation illnesses; and
  - Reduced fish consumption advisories.

These measures also provide a comparison to other communities of similar size and characteristics. Figure 8.8 illustrates graphical examples for measuring the effectiveness of an overflow mitigation program. These metrics should be used to first produce a baseline of the existing system before improvements so that future improvements can be reasonably compared.

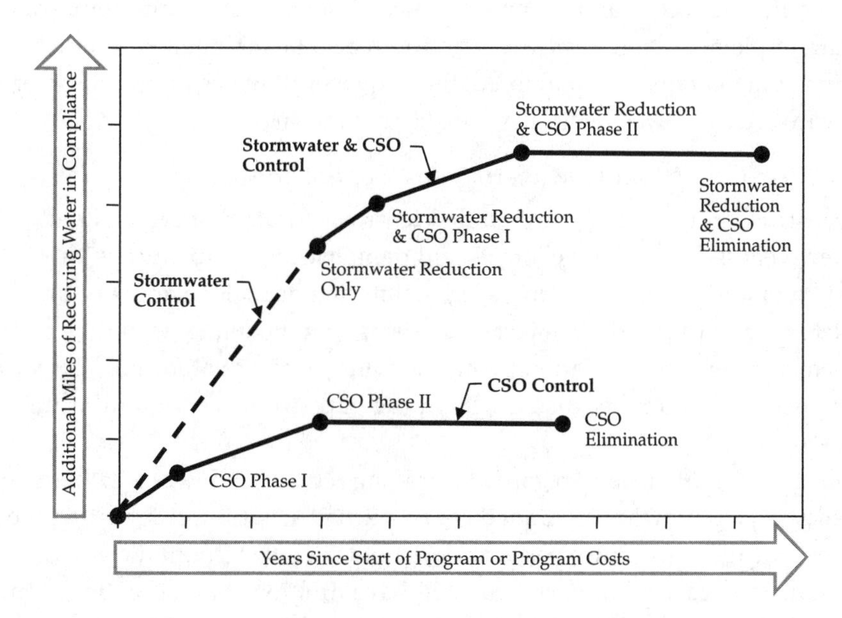

FIGURE 8.8 Example metrics for measuring the effectiveness of a mitigation program to control wet weather pollution (adapted from USACE [2006]).

4.3 Post-Construction Compliance Program

Combined Sewer Overflows: Guidance for Long-Term Control Plans (U.S. EPA, 1995b) discusses post-construction compliance monitoring programs as a necessity to judge how well an overflow mitigation plan has been implemented. This nomenclature was used in the original guidance document, but may be too narrow. Although monitoring is a primary component of judging the effectiveness of a program, it is not the only tool to evaluate the program. Post-construction monitoring is necessary, but application of models built during the planning phase should also be used and updated as necessary to help prove what benefits have been realized in overflow reductions and effects.

4.3.1 Post-Construction Monitoring

The structure of a post-construction monitoring program may be dependent on how monitoring was first done during the planning phase. To properly compare pre- and post-construction conditions, locations and sampling techniques should be consistent. For example, every effort should be taken to monitor flows and pollutants at the same overflow locations and in the same receiving water segments before and after program implementation. Consistency should also be maintained as much as possible, including parameters monitored, the frequency at which monitoring was done, and the frequency of wet versus dry weather monitoring.

4.3.2 Post-Construction Modeling and Evaluation

Post-construction monitoring is intended to provide data that can be used to verify the effectiveness of overflow controls and demonstrate compliance with WQS and protection of designated uses. However, monitoring data alone does not provide the complete picture (if it did, then just monitoring, not modeling, would be needed to complete an overflow mitigation plan). Modeling is essential for post-construction compliance evaluation because it fills in the gaps from what monitoring cannot provide.

This is especially true in regards to proving reduction in wet weather overflow frequency, especially when examined over a yearly basis. Simply reporting the number of wet weather overflows per year does not take into account the amount of rain that occurred in each year. Each year will have different rainfall patterns and volumes and must be normalized to overflow frequency to provide useful statistics. This normalization can be done in a simple way by expressing overflow volumes in a year divided by rainfall volume. This simplistic analysis will show overall trends, but will not show the complexities of how different rainfall patterns affect overflows. Therefore, the model used to project existing overflows should be used and run with measured rainfall after the overflow mitigation projects are implanted to provide a fair comparison of peak flows, volumes, and frequency of overflow events.

As of the writing of this chapter, U.S. EPA was drafting a CSO Post Construction Compliance Monitoring Guidance report. This report should be referenced for specific requirements of CSO programs and can also be used as a guide for SSO programs.

5.0 ADDITIONAL CONSIDERATIONS

Two additional considerations are included in this chapter in relation to overflow mitigation planning and implementation. These include possible changes in system infiltration and inflow issues over a planning period and the potential effects that climate change may have on overflow mitigation plans.

5.1 Future Infiltration and Inflow Issues

There are many methods for predicting existing infiltration and inflow, but very little guidance is available to help predict how infiltration and inflow may change in the future. As an industry, are we underestimating the amount of future infiltration and inlfow in our master plans that lead to underestimating the extent of rehabilitation, replacements, and improvements needed to manage infiltration and inflow? Future budgets may be at risk, as well as water quality violations, if these key assumptions are not applied to size and cost out improvements.

Little information is available from industry literature to guide municipalities in selecting and applying valid methods for predicting future infiltration and inflow. Recently, several methods for predicting how infiltration and inflow may change in the future for a sanitary sewer system were included in literature by Dent and Wright (2008). Although methods in their paper were applied to three collection systems, they can be applied to any collection system. Other sources that discuss future infiltration and inflow estimation include literature by Morgan et al. (2004) and Perry et al. (2007).

Past infiltration and inflow was examined to try to predict how future infiltration and inflow might change the amount of equalization that may be needed to manage the increase in flow over a 35-year period. The results indicate that infiltration and inflow can significantly increase in the future depending on which assumptions are made for system growth and net system deterioration (e.g., overall deterioration less rehabilitation and replacement). If no increase in future infiltration and inflow were assumed for the combination of the three sewer systems analyzed, no additional equalization would be needed in 2040. If a more realistic assumption of service-area growth plus deterioration were applied, 15 mil. gal of equalization should be planned.

The results of this study indicate that the risk of not preparing for the future in a realistic way could result in significant shortcomings in future capital improvement budgets. The results could be built on by other studies to eventually provide comprehensive guidelines for sufficiently addressing future infiltration and inflow for collection system master planning efforts.

5.2 Climate Change

Much has been publicized in recent years regarding the potential for future climate change. Although global warming and climate change have been debated, the vast majority of the scientific community and, more recently, the military (e.g., U.S. Army Corps of Engineers) agree that it is real and must be planned for accordingly.

U.S. EPA released a report that assessed, at a screening-level, the potential implications of future climate change on CSOs in the New England and Great Lakes regions. The purpose of this report is to determine whether the potential implications of climate change on CSOs in these regions warrant further consideration and study and, secondly, to evaluate the need for decision support tools and information enabling CSS managers to better incorporate consideration of climate change into their decision-making processes (U.S. EPA, 2008).

Because investments in infrastructure are long-term, capital-intensive, and, in many cases, irreversible in the short-to-medium term, it is necessary to understand that today's decisions could influence the ability of wastewater infrastructure to accommodate changes in climate for decades into the future and, therefore, should be planned for accordingly. Although this U.S. EPA report (2008) is only a first step toward understanding a complex issue, the implications of which will vary significantly in different locations and for different systems, the results indicate that there will be an effect of climate change on CSOs.

The report concludes that in the Great Lakes region, projected long-term (2060 to 2099) changes in precipitation suggest that if CSO mitigation efforts are designed based on historical precipitation, many systems could experience increases in the frequency of CSO events beyond their design capacity, resulting in increases in overflow volume discharged to receiving waters. In the New England region, projected near-term (2025 to 2050) changes in precipitation are inconclusive and thus require further study.

The real question that faces the industry is that, faced with the prospect of future climate change, do opportunities exist where current CSO mitigation efforts can be upgraded at little additional cost to provide an added margin of safety to account for both near-term extreme events and the potential future effects of climate change? Faced with an uncertain and non-stationary environment, it is best to apply adaptations that provide more resilience, which can be an integrating principle for a comprehensive overflow mitigation plan.

6.0 REFERENCES

American Association of Cost Engineers (2005) *Cost Estimate Classification System—As Applied in Engineering, Procurement, and Construction for the Process Industries*; International Recommended Practice No. 18R-97, TCM Framework: 7.3—Cost Estimating and Budgeting. http://www.aacei.org/technical/rps/18r-97.pdf (accessed June 2010).

Association of Metropolitan Sewerage Agencies (2003) *Wet Weather Consent Decrees: Protecting POTWs in Negotiations—An AMSA Handbook;* Association of Metropolitan Sewerage Agencies: Washington, D.C.

Dent, S. A.; Berlin, J.; McDonald, S.; Kalkman, T.; Hoffman, R. (2007) California's New Sanitary Sewer Collection System Requirements—An SSMP Opportunity. *Proceedings of the Water Environment Federation Collection Systems Specialty Conference: Pioneering Trails to Collection Systems Excellence;* Portland, Oregon, May 13–16; Water Environment Federation: Alexandria, Virginia.

Dent, S. A.; Sathyanarayan, P.; Ohlemutz, R. (2004) How Effective Is Collection System Rehabilitation? *Proceedings of the Water Environment Federation Collection System Specialty Conference;* Milwaukee, Wisconsin, Aug 8–11; Water Environment Federation: Alexandria, Virginia.

Dent, S. A.; Sathyanarayan, P.; Wright, L. T.; Mosely, C.; Smith, D. (2001) A Watershed Approach to SSO Management. *Proceedings of the Water Environment Federation Specialty Conference: A Collection System Odyssey: Integrating Operations & Maintenance Manual and Wet Weather Solutions;* Milwaukee, Wisconsin, July 8–11; Water Environment Federation: Alexandria, Virginia.

Dent, S. A.; Wright, L. T. (2008) What Is the Future of I/I in Your Collection System? *New Eng. Water Environ. Assoc. J.*, **42** (3).

Dent, S. A.; Wright, L. T.; Sathyanarayan, P.; Matheson, R.; Ohlemutz, R. (2003) Strategic Elimination of SSOs: Optimization Model Supports District-Wide Capital Improvement Program. *Proceedings of the Water Environment Federation 76th Annual Technical Exposition and Conference* [CD-ROM]; Los Angeles, California, Oct 11–15; Water Environment Federation: Alexandria, Virginia.

Dent, S. A.; Wright, L. T.; Sathyanarayan, P.; Ohlemutz, R. (2002) Developing a Wet Weather Design Event for a Collection System and WWTP. *Proceedings of the Water Environment Federation 75th Annual Technical Exposition and Conference* [CD-ROM]; Chicago, Illinois, Sept 28–Oct 2; Water Environment Federation: Alexandria, Virginia.

Freedman, P. L.; Eger, J. A.; Gibson, J. P.; Clements, N.; Nemura, A. D. (2007) The Role of Adaptive Watershed Management Concepts in Wet Weather Consent Decrees. *Proceedings of the Water Environment Federation 80th Annual Technical Exposition and Conference* [CD-ROM]; San Diego, California, Oct 13–17; Water Environment Federation: Alexandria, Virginia.

Freedman, P. L.; Nemura, A. D.; Dilks, D. W. (2004) Viewing Total Maximum Daily Loads as a Process, Not a Singular Value: Adaptive Watershed Management. *J. Environ. Eng.*, **130** (6).

Lukas, A. (2007) Update On a Nationwide I/I Reduction Database. *Proceedings of the Water Environment Federation Collection Systems Specialty Conference: Pioneering Trails to Collection Systems Excellence;* Portland, Oregon, May 13–16; Water Environment Federation: Alexandria, Virginia.

Morgan, M.; Swarner, B.; Lampard, M.; Ji, Z.; Crawford, B.; Pang, J. (2004) Locating the Problem: A Regional Assessment of I/I in King County, Washington. *Proceedings of the Water Environment Federation Collection Systems Specialty Conference;* Milwaukee, Wisconsin, Aug 8–11; Water Environment Federation: Alexandria, Virginia.

National Association of Clean Water Agencies (2009) Working With Satellite Communities on Regional Wet Weather Issues: A NACWA White Paper; National Association of Clean Water Agencies: Washington, D.C.

Nemura, A. D.; Turner, C. L.; Gibson, J. P.; Turner, J.; Vatter, B.; Zettler, D.; Grant, G. M.; Fitzgerald, S.; Lyons, J. (2008) Implementing a Sewer Overflow Consent Decree through Watershed Management. *Proceedings of the Water Environment Federation 81st Annual Technical Exposition and Conference* [CD-ROM]; Chicago, Illinois, Oct 18–22; Water Environment Federation: Alexandria, Virginia.

Perry, D.; Lukas, A.; Bennett, D.; Bate, T. (2007) Sustainable I/I Reduction: Planning Level Cost Estimation Using a Performance-Based Cost Function. *Proceedings of the Water Environment Federation Collection Systems Specialty Conference: Pioneering Trails to Collection Systems Excellence;* Portland, Oregon, May 13–16; Water Environment Federation: Alexandria, Virginia.

Reynolds, K. M. (2005) Integrated Decision Support for Sustainable Forest Management in the United States: Fact or Fiction? *Comput. Electr. Agric.*, **49**, 6–23.

Stankey, G. H; Clark, R. N.; Bormann, B. T. (2005) *Adaptive Management of Natural Resources: Theory, Concepts, and Management Institutions;* General Technical Report PNW-GTR-654; U.S. Department of Agriculture, Forest Service, Pacific Northwest Research Station: Portland, Oregon.

U.S. Army Corps of Engineers (1996) *Planning Manual;* IWR Report 96-R-21; U.S. Army Corps of Engineers: Washington, D.C.

U.S. Army Corps of Engineers (1997) *Planning Primer;* IWR Report 97-R-15; U.S. Army Corps of Engineers: Washington, D.C.

U.S. Army Corps of Engineers (2006) *Final Phase I Report, Merrimack River Watershed Assessment Study*; Prepared by CDM For the USACE, New England District and the Merrimack River Basin Community Coalition; http://www.nae.usace.army.mil/projects/ma/merrimack/LMRBmerrimackfinalreport.pdf (accessed March 2011).

U.S. Army Corps of Engineers (2009) *Collaborative Planning Toolkit.* http://www.svp.iwr.usace.army.mil/CPToolkit/PrintAllContent.asp (accessed June 2010).

U.S. Environmental Protection Agency (1994) *Combined Sewer Overflow (CSO) Control Policy; Fed. Regist.*, **59,** No. 75, Apr 19, Notices (FRL-4732-7).

U.S. Environmental Protection Agency (1995a) *Combined Sewer Overflows: Guidance for Funding Options;* EPA-832/B-95-007; U.S. Environmental Protection Agency, Office of Water: Washington, D.C.

U.S. Environmental Protection Agency (1995b) *Combined Sewer Overflows: Guidance for Long-Term Control Plans;* EPA-832/B-95-002; U.S. Environmental Protection Agency, Office of Water: Washington, D.C.

U.S. Environmental Protection Agency (1995c) *Combined Sewer Overflows: Guidance for Nine Minimum Controls;* EPA-832/B-95-003; U.S. Environmental Protection Agency, Office of Water: Washington, D.C.

U.S. Environmental Protection Agency (1995d) *Combined Sewer Overflows: Guidance for Permit Writers;* EPA-832/B-95-008; U.S. Environmental Protection Agency, Office of Water: Washington, D.C.

U.S. Environmental Protection Agency (1995e) *Combined Sewer Overflows: Guidance for Screening and Ranking;* EPA-832/B-95-004; U.S. Environmental Protection Agency, Office of Water: Washington, D.C.

U.S. Environmental Protection Agency (1997) *Combined Sewer Overflows: Guidance for Financial Capability Assessment and Schedule Development;* EPA-832/B-97-004; U.S. Environmental Protection Agency, Office of Water: Washington, D.C.

U.S. Environmental Protection Agency (2006) Sanitary Sewer Overflows and Peak FlowsRecent Additions Web Site. http://cfpub.epa.gov/npdes/whatsnew.cfm?program_id=4 (accessed Sept 2010).

U.S. Environmental Protection Agency (2007) *The Long-Term Control Plan-EZ (LTCP-EZ) Template: A Planning Tool for CSO Control in Small Communities;* EPA-833/R-07-005; U.S. Environmental Protection Agency: Washington, D.C.

U.S. Environmental Protection Agency (2008) *A Screening Assessment of the Potential Impacts of Climate Change on Combined Sewer Overflow (CSO) Mitigation in the Great Lakes and New England Regions;* EPA-600/R-07-033F; U.S. Environmental Protection Agency: Washington, D.C.

U.S. Environmental Protection Agency (2010) *Storm Water Management Model User's Manual, Version 5.0;* EPA-600/R-05-040; U.S. Environmental Protection Agency, Water Supply and Water Resources Division, National Risk Management Research Laboratory: Cincinnati, Ohio. http://www.epa.gov/ednnrmrl/models/swmm/epaswmm5_user_manual.pdf (accessed Oct 2010).

Varghese, V.; Coleman, T.; Witt, A.; Nelson, R. (2008) Application of Genetic Algorithm Optimization for Collection System to Johnson County's Turkey Creek Sewer-Shed. *Proceedings of the Water Environment Federation Collection*

Systems Specialty Conference; Pittsburgh, Pennsylvania, May 18–21; Water Environment Federation: Alexandria, Virginia.

Water Environment Federation (2006) *Guide to Managing Peak Wet Weather Flows in Municipal Wastewater Collection and Treatment Systems;* Water Environment Federation: Alexandria, Virginia.

Williams, B. K.; Szaro, R. C.; Shapiro, C. D. (2007) *Adaptive Management: The U.S. Department of the Interior Technical Guide;* Adaptive Management Working Group, U.S. Department of the Interior: Washington, D.C.

Wright, L. T.; Heaney, J. P. (2009) Risk Optimization of a Sanitary Sewer Overflow Control Plan; Submitted to the *ASCE J. Water Resour. Plan. Manage.*

7.0 SUGGESTED READINGS

California Environmental Protection Agency State Water Resources Control Board (2009) Sanitary Sewer Overflow (SSO) Reduction Program Web Site. http://www.waterboards.ca.gov/water_issues/programs/sso/index.shtml#plan (accessed Sept 2010).

Maser, C.; Bormann, B. T.; Brookes, M. H.; Kiester, A. R.; Weigand, J. F. (1994) Sustainable Forestry through Adaptive Ecosystem Management is an Open-Ended Experiment. In *Sustainable Forestry: Philosophy, Science, and Economics;* Maser, C., Ed.; St. Lucie Press: Delray Beach, Florida; pp 303–340.

U.S. Environmental Protection Agency (1999) *Combined Sewer Overflows: Guidance for Monitoring and Modeling;* EPA-832/B-99-02; U.S. Environmental Protection Agency, Office of Water: Washington, D.C.

U.S. Environmental Protection Agency (2001) *Guidance: Coordinating CSO Long-Term Planning with Water Quality Standards Reviews;* EPA-833/R-01-002; U.S. Environmental Protection Agency: Washington, D.C.

Wright, L. T.; Heaney, J. P.; Dent, S. A. (2003) Risk-Based Design of a Sanitary Sewer Overflow Control Plan. *Proceedings of the American Society of Civil Engineers World Water and Environmental Resource Congress;* Philadelphia, Pennsylvania, June 22–26.

Wright, L. T.; Heaney, J. P.; Dent, S. A. (2006) Prioritizing Sanitary Sewers for Rehabilitation Using Least-Cost Classifiers. *ASCE J. Infrastruct. Syst.,* **12** (3), 174–183.

Index